Passive Radars on Moving Platforms

Other volumes in this series:

Passive Radars on Moving Platforms

Edited by
Diego Cristallini and Daniel W. O'Hagan

The Institution of Engineering and Technology

Published by SciTech Publishing, an imprint of The Institution of Engineering and Technology, London, United Kingdom

The Institution of Engineering and Technology is registered as a Charity in England & Wales (no. 211014) and Scotland (no. SC038698).

The Institution of Engineering and Technology
Futures Place
Six Hills Way, Stevenage
Herts SG1 2AU, United Kingdom

www.theiet.org

British Library Cataloguing in Publication Data
A catalogue record for this product is available from the British Library

ISBN 978-1-83953-118-7 (hardback)
ISBN 978-1-83953-119-4 (PDF)

Typeset in India by MPS Ltd

Cover Image: Light trails at Night – Zhe Yong/EyeEm via Getty Images

Contents

About the editors

Diego Cristallini is the head of the Passive Radar Group at Fraunhofer FHR, Germany. His research interests are in multi-channel radar signal processing, in STAP, GMTI, SAR, and passive radar. Dr Cristallini serves as a reviewer for several international technical journals. He is also a regular lecturer at the Fraunhofer International Summer School on Radar and SAR. Dr Cristallini is the co-chair of the NATO group SET-242 on "Passive radars on moving platforms."

Daniel W. O'Hagan is the head of the Passive Radar and Anti-Jamming Techniques Department (PSR) at Fraunhofer FHR, Germany. He is associate professor of radar at the University of Cape Town (UCT) and professor at the University of Birmingham, UK. He has been lecturing full postgraduate (Masters and PhD) courses on radar systems and antenna array synthesis since 2014. Professor O'Hagan is the chair of two full-status NATO groups, SET-268 and SET-296.

Preface

I find the topic of passive radar fascinating. The idea of simply exploiting signals that already exist, and of trying to imitate what the "luxury colleagues" do with active radars is indeed as challenging as it seems! I say "luxury colleagues" because the freedom of designing your own transmitted waveform is a real luxury that one only appreciates when you don't have one!

The special case of passive radar mounted on moving platforms is, under most conditions, even more challenging. First, because the typical radar applications performed from moving platforms are *per se* quite challenging. Second, because the signals of opportunity typically exploited for passive radar purposes are not designed for reception while in motion.

These few considerations should provide the reader with some insights about the research topics presented in this book. As this book is intended for both passive radar experts and readers less familiar with the general topic of passive radar, we aimed to provide background information in the first three chapters. Specifically, Chapter 1 covers the principles of operation of passive radar and it derives the bistatic geometry and bistatic resolutions. Thereafter, some insights on the derivation of the cross-ambiguity function, which constitutes the core of any passive radar signal processing, are presented. Chapter 2 deals with the applications that can be performed when a passive radar is mounted on a moving platform. The applications are clearly platform specific. Some examples are reported with corresponding results to illuminate the capabilities and the potentials offered by this technology. Chapter 3 is devoted to an overview of existing illuminators of opportunity typically exploited for passive radar in general and from moving platforms in particular. The portfolio of available signals is continuously expanding, so the chapter offers a snapshot of the current situation.

The following block of Chapters 4–9 are specifically focused on research activities conducted by different research groups worldwide. Chapter 4 presents a statistical analysis of the clutter characteristics of passive radar data returns. The analysis is focused on airborne data and land clutter, as this is evidently one of the most challenging and relevant clutter scenarios. Chapter 5 deals with the challenges in reconstructing a digitally modulated reference signal such as DVB-T from a fast moving platform. Chapters 6 and 7 refer to the two most typical airborne radar applications, that is synthetic aperture radar (SAR) and ground moving target indication (GMTI), and specifically with their adaptations to the case of passive radar.

Chapters 8 and 9 relate mostly to operational systems. In particular, they provide the reader with some hints and considerations on multiple receiving channel calibration (Chapter 8), and on hardware realization of radar systems based on the software-defined radio (SDR) principle (Chapter 9).

The concluding Chapter 10 offers some outlook on what passive radar could be in the near future, namely a component of a bigger architecture usually referred to as system of systems (SoS). Additionally, results of on-going activities related to new potential illuminators of opportunity for passive radar are covered.

I hope the reader will find in this volume a better understanding of the "silent" world of passive radar on moving platforms, and find inspiration for new ideas. The battle against clutter is far from won, and new challenges lie ahead!

I would like to make some acknowledgements and thanks to many people that helped and supported me in editing this book and in my professional career. The first big thank you goes to my wife Eva, to my two children Elisa and Federico, and to my parents. I know how difficult I can be at times, and I love you all deeply. A special mention goes to my cousin Simone, who will always be an example and a model of how engineers should be. I acknowledge a few people whom I have met in my career that believed in me and made me fall in love with radar systems: Pierfrancesco Lombardo, Debora Pastina, Fabiola Colone, Alfonso Farina, Andreas Brenner, Richard Klemm and Heiner Kuschel. Without you, I would have probably ended up doing something else! A special thank goes to Daniel O'Hagan for giving me the time to work on this volume. Thanks to all the contributing authors for taking part in this work – that there now exists a book on this topic is thanks to your ideas and great results. Thank you to the entire PSV research group at Fraunhofer FHR for your support and for the outstanding work you are providing every day. Before leaving the reader to the book, I would like to thank the IET for giving me the opportunity of editing this volume, and a special thank goes to Nicki Dennis and Olivia Wilkins for supporting me in this adventure.

And, as always, Forza Fere!

Chapter 1

Introduction to passive radar

Marco Martorella[1] and Diego Cristallini[2]

This chapter provides a first insight into the principle of operation of a passive radar. The bistatic geometry will be introduced as well as the concepts of bistatic range and bistatic Doppler resolution. The well-known radar equation is also derived for the specific case of a bistatic passive radar. In addition, basic schemes for signal processing are presented together with a glimpse of the main issues related to the direct signal suppression.

1.1 Principle of operation

The expression *passive radar system* indicates a class of bistatic radar system that does not transmit a dedicated electromagnetic signal, but instead it exploits electromagnetic signals emitted by other sources for other purposes. Such sources are usually referred to as illuminator of opportunity (IO) and examples include other radars, communication systems, and broadcast systems for public utility. It is therefore clear that as a passive radar it is constituted of only the receiver part of a radar system. This peculiarity opens the way to several unique advantages. Among them, it is clear that it reduced hardware complexity, which often comes along with reduced development costs. From the regulatory point of view, the lack of a dedicated transmitting part enables passive radar operation without any license for the usage of electromagnetic spectrum. For this reason, passive radar is sometimes also referred to as *green radar*, since it does not contribute to increase the electromagnetic pollution. Hence, a passive radar could be safely operated in urban and highly populated areas, which can be a strategic benefit in several civilian operational scenarios. Also in the military context, the lack of a transmitter unit offers strategic advantages. That is, a passive radar is *silent* and therefore difficult to detect and eventually to jam. On the other hand, the lack of a transmitting part also offers several challenges that will be deeply analysed throughout this chapter and the following ones. Here, we simply mention the most technical ones.

[1]University of Pisa, Italy
[2]Fraunhofer FHR, Germany

The transmitted signal is not designed for radar purposes and it is not under the control of the passive radar operator, which requires *ad hoc* signal processing stages to be implemented. In general, the passive radar needs a mean to retrieve information about the transmitted signal (also called *reference signal*) in order to cross-correlate it with the signal in the *surveillance channel* (as defined below), in which echoes from potential targets might be present. Usually (but not necessarily), the reference signal is received through a dedicated reference channel. This is directly connected to an antenna that points towards the transmitter, with the aim of collecting a copy of the transmitted signal as clean as possible from external noise and interfering sources. Typically directive antennas are exploited for the reference channel. The signal received by the reference channel will be a superposition of the direct signal [i.e. the transmitted signal received by the passive radar receiver by direct line of sight (LOS) propagation], plus inevitable thermal noise contribution. Multipath contributions, that is reflections of the direct signal by stationary objects, are usually attenuated at the reference channel by the directivity of the reference antenna. A second dedicated antenna is connected to a so-called surveillance channel. The antenna of the surveillance channel (also referred to as surveillance antenna) is pointed towards the area to be monitored. In this case, there is often the need to deploy a less directive antenna, in order to broaden the surveillance area. The signal received by the surveillance antenna will be a superposition of at least the following components: (i) the direct signal; (ii) the *reflected signal* due to target echo reflection; (iii) multipath contributions (also known as the clutter); and (iv) thermal noise. A typical geometry is sketched in Figure 1.1.

By looking at the signal in the surveillance channel, the reflected signal is expected to be masked by the direct signal. In fact, the direct signal is attenuated by a simple one-way propagation from transmitter to receiver, whereas the reflected signal undergoes a two-way propagation attenuation from transmitter to target and

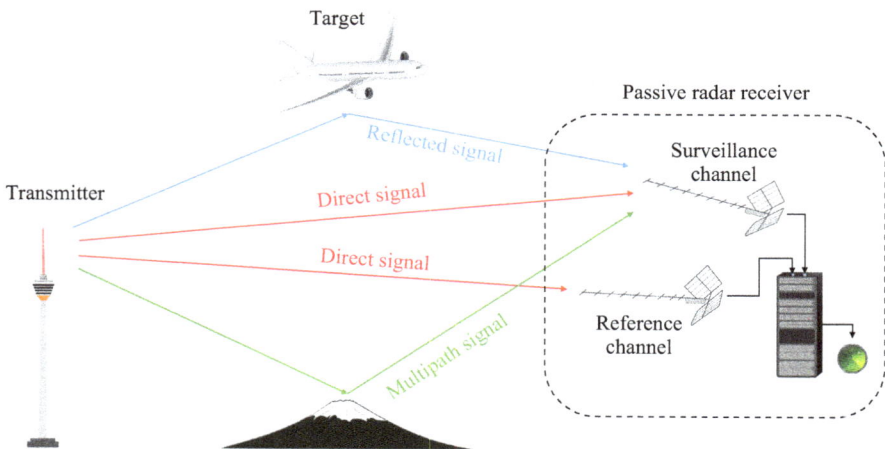

Figure 1.1   Sketch of the passive coherent location (PCL) geometry

then from target to receiver. The power difference between the direct signal and the reflected signal might be of several tens of decibels, posing stringent requirements on the dynamic range of the surveillance channel receiver. For this reason, a dedicated stage of the passive radar signal processing will deal with adequate cancellation of the direct signal component in the surveillance channel.

Figure 1.2 shows the general processing chain of a passive radar system. The signal at the reference channel is first exploited to clean the surveillance channel from direct and multi-path signal contributions. This suppression is usually done in the time domain, by coherently subtracting time-delayed replicas of the direct signal from the surveillance channel. Once this step is performed, the cleaned signal at the surveillance channel can be cross-correlated with the signal at the reference channel. This cross-correlation will perform a range of compressions, as it will be better discussed later on. Doppler processing is usually then performed, to coherently integrate over a coherent processing interval (CPI). After Doppler processing is performed, the signal is conveniently displayed in range-Doppler domain. Target detection [for instance by resorting to constant false alarm rate (CFAR) approaches] is then performed over each range-Doppler map independently, that is for each and every CPI. Detections collected from multiple CPIs can be further processed to produce tracks of detected targets. Target tracking can be performed in the range-Doppler domain, or in Cartesian coordinates if target geo-localisation is available. When multiple PCL receivers are deployed (i.e. multistatic configuration), a two-stage tracking approach is usually applied to reduce the resulting number of false alarms. This two-stage

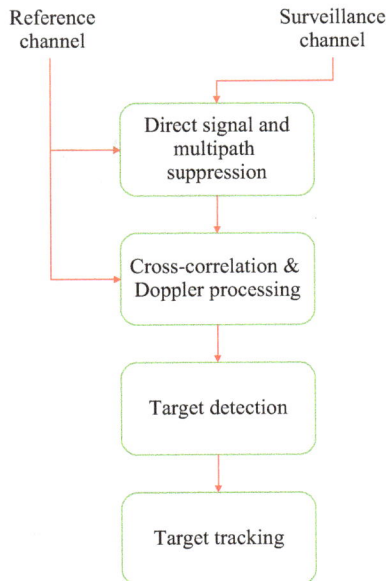

Reference channel

Surveillance channel

Direct signal and multipath suppression

Cross-correlation & Doppler processing

Target detection

Target tracking

Figure 1.2   Basic block diagram of PCL signal processing

tracking is based on tracking first in the range-Doppler domain for each receiver and then tracking in Cartesian coordinates fusing the tracks from different PCL receivers.

1.2   Bistatic geometry

Cross-correlation between the reference signal and the signal received through the surveillance channel allows to measure the bistatic range of a given target echo. In other words, the PCL receiver measures the time difference of arrival (TDOA) between the direct signal and the reflected target echo. The bistatic range measurement is retrieved from a TDOA measurement considering the electromagnetic wave propagation speed. It is well known that a bistatic range measurement locates the target onto an ellipsoid that has the transmitter and the receiver located at its foci. A two-dimensional sketch of the bistatic geometry of a PCL radar system in Cartesian coordinates is depicted in Figure 1.3. The plane containing the transmitter, the receiver, and the target is called bistatic plane. Note that in the usual case of ground based transmitter/receiver and airborne target, the bistatic plane is not parallel to the Earth surface. The line between the transmitter and the receiver is denoted as the baseline, L. The range from the transmitter to the target is R_T, while the range from the target to the receiver is R_R. The angle between the transmitter and the receiver with vertex at the target is denoted as the bistatic angle β. The bistatic range R_{bist} is calculated through TDOA measurements corresponding to $R_{bist} = R_T + R_R - L$. The sum $R_T + R_R$ equals twice the semi-major axis of the corresponding ellipse. An important property is that the bisector of the bistatic angle is always orthogonal to the tangent of the iso-range contour in any point of the ellipse.

1.2.1   Bistatic range resolution

The range resolution in the monostatic case is defined as the minimum distance between two targets that still allows the radar to distinguish the corresponding returns. In meters, this corresponds to the well-known formula $\delta R_{mono} = c\tau/2$, where c is the

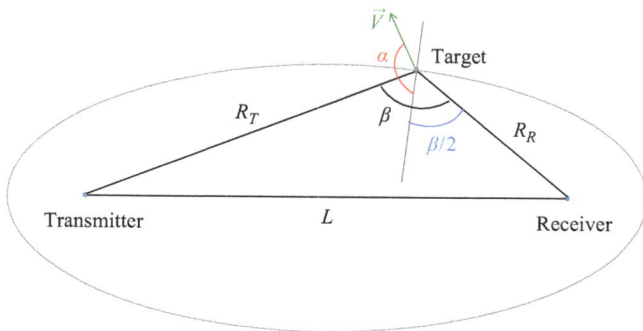

Figure 1.3   *Sketch of the bistatic geometry in two-dimensional Cartesian coordinates*

speed of light in vacuum, and τ is the temporal duration of the pulse (if range compression is applied, $\tau = 1/B$ where B is the processed transmitted signal bandwidth). In other words, in the monostatic case, δR_{mono} is defined as the difference of two concentric circles having a difference in radius of $c/2B$. An extension of this concept in the bistatic case is possible. To do that, let us define a pseudo-monostatic range resolution δR_{pseudo} as the separation between two concentric iso-range ellipses having the semi major axis a and a', where $\delta R_{pseudo} = a - a' = c/2B$. The two aforementioned concentric ellipses are shown in Figure 1.4. From Figure 1.4, it is clear that the bistatic range resolution δR_{bist} not only depends on δR_{pseudo} but also on the specific position within the ellipse. By explicating this latter dependence in terms of the bistatic angle β, one can approximate the bistatic range resolution as [1]

$$\delta R_{bist} \approx \frac{c}{2B \cos(\beta/2)}. \qquad (1.1)$$

1.2.2 Bistatic Doppler

The bistatic Doppler frequency $f_{D,bist}$ is defined as the bistatic range rate R_{bist} over time normalised by the carrier wavelength λ [1]. That is,

$$f_{D,bist} = \frac{1}{\lambda}\left[\frac{\partial}{\partial t}R_{bist}(t)\right] = \frac{1}{\lambda}\left[\frac{\partial}{\partial t}R_T(t) + \frac{\partial}{\partial t}R_R(t)\right]. \qquad (1.2)$$

As is apparent, the bistatic Doppler comprises of two contributions, the first due to the relative radial motion between transmitter and target, and the second due to the relative motion between target and receiver. In the usual PCL radar case, the transmitter and receiver are stationary; therefore, the measured bistatic Doppler only depends

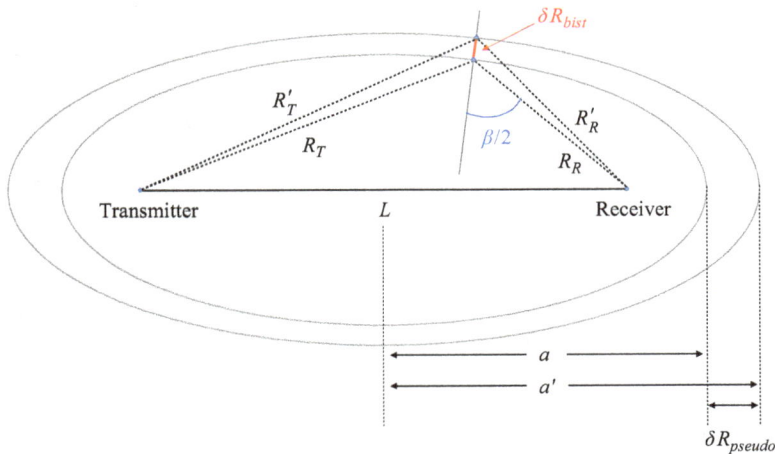

Figure 1.4   Sketch of the bistatic range resolution

on the target motion. In this particular case, the bistatic Doppler frequency can be expressed as

$$f_{D,bist} = \frac{V}{\lambda} \left[\cos(\alpha - \beta/2) + \cos(\alpha + \beta/2) \right] = \frac{2V}{\lambda} \cos\alpha \cos(\beta/2) \qquad (1.3)$$

where V is the modulus of the target velocity vector and α is the orientation of V with respect to the bisector of the bistatic angle (see Figure 1.3). It is important to note that any target moving along the baseline between the transmitter and the receiver will have a zero bistatic Doppler frequency, since the two terms in (1.2) cancel out each other. On the other hand, in the extended baseline case (i.e. when $\beta = 0$), the bistatic Doppler reduces to the monostatic Doppler. By observing (1.2), it is also important to notice that the bistatic Doppler will be zero for all targets moving along the bistatic iso-range, that is the bistatic range $R_{bist}(t)$ does not change over time. In this case, the exploitation of a multi-static configuration would provide geometry diversity gain, thus allowing for detection of all possible motions.

The Doppler resolution of a PCL does not differ in principle from that of a conventional coherent radar. By assuming a CPI duration of T_{CPI} seconds, the Doppler resolution would be $\delta f_{D,bist} = 1/T_{CPI}$ in Hz. What is peculiar of a passive radar, is the possibility to extend the CPI up to durations in the order of a second. This is due to the continuous wave and the broadcast nature of typical IOs.

1.2.3   Bistatic radar equation

The bistatic radar equation defines the basis for a detection performance analysis of a PCL radar system. Specifically, it allows the calculation of the expected signal-to-interference ratio (SIR) as a function of target position and radar cross-section (RCS), as well as transmitter and receiver characteristics (like transmitted power and antenna diagrams). It also allows the estimation of the maximum detection range of the radar system once the desired detection and false alarm probabilities are defined together with target fluctuation models. In PCL systems, the target echo signal always competes with a strong direct signal (or with its residual contribution after direct signal cancellation). With respect to thermal noise, the target echo signal is increased by the processing gain resulting from the correlation process G_{corr}, both due to range compression and Doppler processing. Thus in many situations, specifically when transmitter is close to the receiver, the limiting factor is not thermal noise but rather the correlation side lobes of the (residual) direct signal from the transmitter. For this reason, we consider here the SIR instead of the more conventional signal-to-noise ratio (SNR). A quantitative comparison between direct signal interference and thermal noise level can be found in [2, Chapter 7.2.3]. In order to determine the detection range capability and thus the bistatic coverage of a PCL radar receiver, the power level of the direct signal at the location of the receiver has to be known.

The received power in the passive radar receiver with an antenna gain of $G_R(tx)$ in the direction of the transmitter is then given by

$$P_d = \frac{P_T G_T(rx) F_T \ \lambda^2 G_R(tx)}{4\pi L^2 \qquad 4\pi} G_{corr} \qquad (1.4)$$

where P_T is the transmitted power, $G_T(rx)$ is the transmitter antenna gain in the receiver direction, F_T is the propagation factor on the path from transmitter to the receiver, and L is the baseline. The correlation side lobes of the direct signal, denoted as correlation noise N_{corr}, are reduced with respect to the correlation peak by the factor γ, which may also take into account direct signal cancellation stage. That is,

$$N_{corr} = \frac{P_T G_T(rx) F \; \lambda^2 G_R(tx)}{4\pi L^2 \quad 4\pi} \frac{G_{corr}}{\gamma}.$$ (1.5)

The SIR of a target echo, competing with the correlation noise N_{corr}, depends on the target's RCS σ:

$$SIR = \frac{P_T G_T(tgt) F_T \; \lambda^2 G_R(tgt) \; \sigma F_R \; G_{corr}}{4\pi R_T^2 \quad 4\pi \quad 4\pi R_R^2 \; N_{corr}}$$ (1.6)

with F_T and F_R being the respective propagation factors on the path from the transmitter to the target then on to the receiver, and R_T and R_R being the corresponding distances. $G_T(tgt)$ and $G_R(tgt)$ indicate the transmitter and the receiver antenna gains in the direction of the target, respectively. Inserting (1.5) into (1.6) results in

$$SIR = \frac{G_T(tgt) F_T \; G_R(tgt) \; \sigma F_R \quad L^2 \gamma}{R_T^2 \quad 4\pi \quad R_R^2 \; G_T(rx) F_T G_R(tx)}$$ (1.7)

For free-space conditions, which lead to maximum detection ranges, F, F_T, and F_R are 1. Also assuming the transmitter antenna gain to be omni-directional, that is $G_T(tgt) = G_T(rx) = 1$, we obtain:

$$SIR = \frac{G_R(tgt) \sigma L^2 \gamma}{4\pi R_T^2 R_R^2 G_R(tx)}$$ (1.8)

1.3 Illuminators of opportunity

The passive radar relies on transmitters of opportunity which have primary functions other than to serve as radar illuminators. This section will explain what types of illuminators are available and how their specific properties impact on passive radar processing requirements. After generally classifying the transmitters of opportunity into space borne and terrestrial, a focus will be placed on waveforms, antenna design, and network concepts for terrestrial broadcast services.

1.3.1 Classes of illuminators

There are multiple possible classification criteria for illuminators that one might choose, like signal modulation or purpose of transmission. However, dividing them into terrestrial and space borne classes indicates what is primarily used today and what probably will be desired in the future. Illuminators of opportunity belonging to the terrestrial class are:

- other radars: for example radars used for air traffic control (ATC) or for maritime coastal monitoring;

- mobile communication systems: base stations for Groupe Spécial Mobile (GSM), Universal Mobile Telecommunications System (UMTS), long-term evolution (LTE), fifth generation (5G); access points for wireless fidelity (WiFi), world-wide interoperability for microwave access (WiMAX), high-performance radio LAN (HiperLAN);
- broadcast systems for public utility: frequency modulated (FM) and digital audio broadcast (DAB) radio (analog and digital, respectively); analog TV and digital video broadcast-terrestrial (DVB-T) (also known as digital TV).

Illuminators of opportunity belonging to the space borne class are:

- other radars: for example, radars used for Earth monitoring and remote sensing applications;
- broadcast systems for public utility: such as digital video broadcast-satellite (DVB-S) (also known as digital satellite TV), and its handheld variation digital video broadcast-satellite handheld (DVB-SH);
- mobile communication systems: like Globalstar, Iridium, and Orbcomm;
- geolocalisation system: transmitters like global positioning system (GPS), Globalnaja nawigazionnaja sputnikowaja sistema (GLONASS), Galileo, and Beidou.

1.4   Signal processing for passive radar

PCL is gradually making progress, both in terms of signal processing and system development (demonstrators and prototypes). Several electromagnetic sources, i.e., IOs, can be used to implement such a radar system. Ground-based broadcast transmitters, such as FM and DAB radio, as well as analogue and digital television have been largely exploited to demonstrate their capability to perform air and coastal surveillance [2–8]. Other IOs can also be found in space, where broadcast, navigation and communication systems continuously transmit towards the Earth. Such IOs can potentially enable PCL at a global scale. A question that arises naturally is related to the effectiveness of passive radar when it comes to detecting targets and how this compares to active radar. Fundamental radar detection theory was laid in the early years of radar development and the same concepts can be applied to the PCL case. Therefore, optimality criteria and fundamental mathematical tools can still be exploited to address the detection problem for PCL. Rather than repeating well-known concepts, in this chapter, we will concentrate on those differences and issues that arise when dealing with passive rather than active radar. More in detail, we will recast the matched filter implementation, which is typical of active radar, into the problem of obtaining an equivalent result through the calculation of a cross-correlation function. The problem will be tackled in a two-dimensional space to account for both range and Doppler target's shifts. Efficient algorithms for the calculation of the two-dimensional cross-correlation function (2D-CCF) will be considered since this step is computationally very expensive. The problems of direct signal and clutter cancellation that requires specific solutions typical for PCL will also be addressed. More specifically, we will look at a well-known solution, namely the extensive cancellation algorithm (ECA) [9]. Finally, we will study the problem of reference signal cleaning, which is a necessary step to enable effective detection with PCL systems.

1.4.1 Matched filtering for passive radar and 2D-CCF

This section focuses on the implementation of two-dimensional matched filtering in the PCL case. First, the equivalence between cross-correlation and matched filter will be addressed followed by the concept of 2D-CCF. Efficient methods for the calculation of the 2D-CCF will be then introduced and discussed.

1.4.1.1 Equivalence between matched filter and cross-correlation

It can be easily shown that a matched filter (MF) can be implemented through the calculation of a cross-correlation. A matched filter is represented in Figure 1.5 where $s_R(t)$ represents the signal received by the passive radar through the surveillance antenna, $s_o(t)$ represents the signal at the output of the MF, and $h_{MF}(t)$ represents the MF impulse response.

Equation (1.9) shows the convolution operation that produces the output signal $s_o(t)$

$$s_o(t) = s_R(t) \otimes h_{MF}(t) = \int s_R(\tau) h_{MF}(t - \tau) d\tau \tag{1.9}$$

In order to define the MF impulse response, we need to know the transmitted signal waveform. As this is not generated by the radar itself, a copy of the transmitted signal is obtained through the signal received by the PCL reference antenna, namely $s_{ref}^*(t)$. The MF impulse response is therefore obtained as follows:

$$h_{MF}(t) = s_{ref}^*(-t) \tag{1.10}$$

Therefore, by substituting (1.10) in (1.9), we obtain

$$s_o(t) = s_R(t) \otimes h_{MF}(t) = \int s_R(\tau) s_{ref}^*(\tau - t) d\tau \tag{1.11}$$

Which shows that the same output signal can be obtained by cross-correlating the input signal with the reference signal. This equivalence directly suggests how to implement a MF in the case of a PCL system. A pictorial view is given in Figure 1.6.

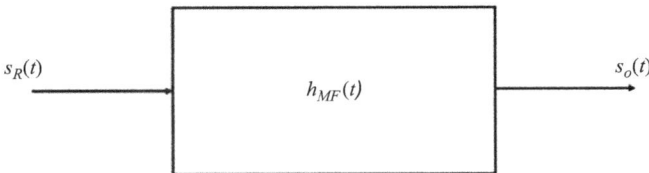

Figure 1.5   Matched filter

Illuminator of opportunity

Target

$s_{ref}(t)$

$s_R(t)$

Cross-correlator

Reference signal

Received signal
(target's echo)

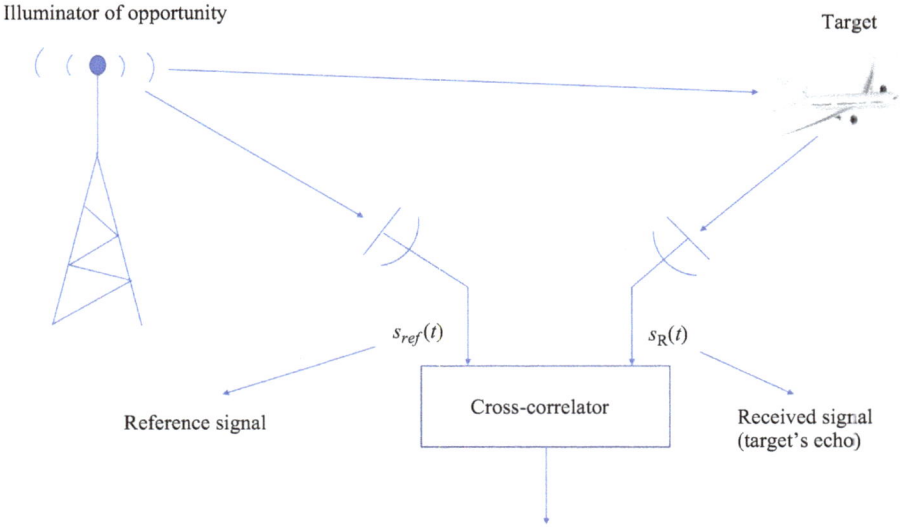

Figure 1.6   Matched filter implementation for a PCL system via a cross-correlation

1.4.1.2   2D-CCF

The MF defined in Section 1.4.1.1 does not take into account Doppler shifts. If a target moves, it can be shown that a Doppler shift is generated. Therefore, a MF must take this into account in order to match the signal at its input. Equivalently, the cross-correlation function must do the same by introducing another variable that accounts for the Doppler shift. This can be done by extending the 1D-CCF to the 2D-CCF as follows:

$$CCF\,(\tau, \nu) = \int s_R\,(\tau)\, s_{ref}^*\,(\tau - t) \exp\left(-j2\pi \nu t\right) d\tau \tag{1.12}$$

where ν represents the Doppler shift. Equation (1.12) can be read as a cross-correlation between a Doppler shift compensated version of the received signal and the reference signal. As it will be clear in the next subsections, different interpretations of the 2D-CCF lead to different efficient implementations of its calculation. It is also important to note that the 2D-CCF is a two-dimensional function with a supporting domain represented by delay-time and Doppler. As the delay-time can directly be transformed into bistatic range, we can conclude that the 2D-CCF can be interpreted as a bistatic range-Doppler map. As signals are sampled and digitised (1.12) should be represented by using a numerical representation as follows:

$$CCF\,(l, m) = \sum_{n=0}^{N=1} s_R\,(n)\, s_{ref}^*\,(n - l) \exp\left(-j2\pi \frac{mn}{N}\right) \tag{1.13}$$

where n represents the time, l represents the delay-time, and m represents the Doppler shift. The total number of time samples is set equal to N, which depends on the coherent processing interval (CPI) considered and the system sampling rate (T_S) as follows:

$$N = \frac{CPI}{T_S} = CPI \cdot F_S \tag{1.14}$$

where $F_S = \frac{1}{T_S}$. An example of 2D-CCF is shown in Figure 1.7. This visualisation result is often addressed as bistatic range-Doppler map. The axes refer to bistatic range and bistatic Doppler.

1.4.2 Efficient calculation of the 2D cross-correlation

Both surveillance and reference signals are sampled based on the signal bandwidth (*B*). Therefore, in order to satisfy Shannon's sampling theorem, complex baseband signals should be sampled at a frequency higher than *B*. As typical IOs transmit continuously and everywhere, CPIs can be made long in order to obtain a high integration gain and therefore improve the SIR. As an example, we may consider a DVB-T signal channel, which has a bandwidth roughly equal to 8 MHz and a CPI of 1 s. In this case, $N = 8$ million complex samples would be acquired during the CPI. The calculation of a cross-correlation function for a number of range bins equal to $M = 1,000$ and a number of Doppler bins of $D = 1,000$, would lead to a number of complex multiplication equal to $NMD = 8$ trillion multiplications! It is clear that the computational load is not to be neglected and that some ways of reducing this computational burden must be found. In the following subsections, we will be looking at two exact methods, namely the correlation fast Fourier transform (FFT) and the Direct FFT, and an approximated method, namely the batched algorithm.

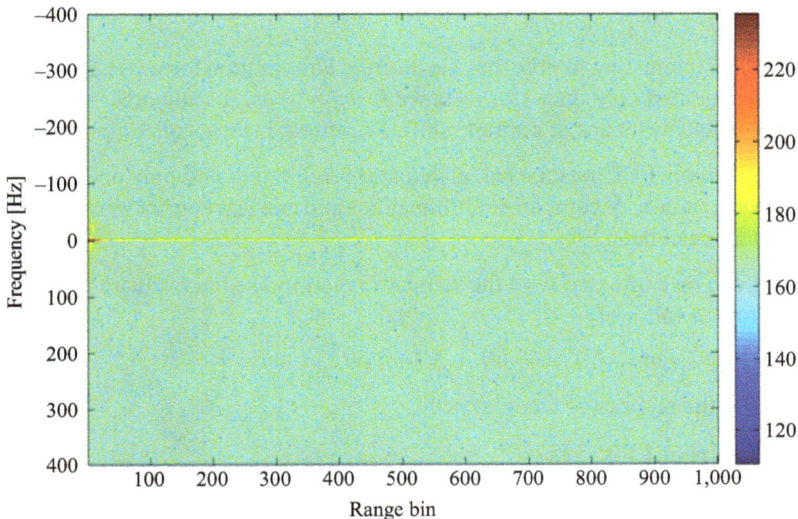

Figure 1.7 Bistatic range-Doppler map in dB scale

1.4.2.1   Correlation FFT

The correlation FFT method derives from an interpretation of (1.14). In fact, by simply swapping the position of the exponential function and the reference signal we obtain

$$CCF\,(l,m) = \sum_{n=0}^{N=1} s_R\,(n)\exp\left(-j2\pi\frac{mn}{N}\right)s^*_{ref}\,(n-l) \qquad (1.15)$$

By substituting (1.16)

$$s_R\,(n,m) = s_R\,(n)\exp\left(-j2\pi\frac{mn}{N}\right) \qquad (1.16)$$

into (1.15), we obtain a new expression that shows that the 2D-CCF can be calculated as the 1D cross-correlation of a Doppler shift compensated signal, namely $s_R\,(n,m)$, and the reference signal. It should be noted that the latter signal depends also on m, which represents the Doppler shift:

$$CCF\,(l,m) = \sum_{n=0}^{N=1} s_R\,(n,m)\,s^*_{ref}\,(n-l) \qquad (1.17)$$

Cross-correlations can be calculated efficiently in the Fourier domain. In particular, we note that

$$CCF\,(l,m) = IDFT\left[s_R\,(k,m)\,s^*_{ref}\,(k)\right] \qquad (1.18)$$

where

$$s_R\,(k,m) = DFT\left[s_R\,(n,m)\right] \qquad (1.19)$$

and

$$s_{ref}\,(k) = DFT\left[s_{ref}\,(n)\right] \qquad (1.20)$$

It is also interesting to note that the discrete Fourier transform (DFT) of $s_R\,(n,m)$ must be calculated only once (for instance for $m=0$) as, for any other value of m, it can be obtained by means of a simple shift. Therefore, in terms of computational load

- two N-points FFT must be calculated at the beginning and only once
- at each iteration, N complex multiplications and one fast Fourier transform (IFFT) must be calculated.

This can be expressed with the standard notation used to evaluate the computational effort as follows:

$$O_{CF} = 2N\log_2\,(N) + N_f\left[N + N\log_2\,(N)\right] \qquad (1.21)$$

where N_f is the number of Doppler cells.

1.4.2.2   Direct FFT

The direct FFT method can be derived by interpreting (1.15) as the DFT of the product of two signals as follows:

$$CCF\,(l,m) = DFT\left[s_R\,(n)\,s^*_{ref}\,(n-l)\right] \qquad (1.22)$$

With this method, no initial FFT must be calculated and, at each iteration, N complex multiplications have to be calculated along with an FFT. In terms of standard notation, the computational effort can be expressed as follows:

$$O_{DF} = N_\tau \left[N + N \log_2 (N) \right] \tag{1.23}$$

The difference in terms of computational load between the two approaches lies on the number of iterations to be completed as, for each iteration, they have the same number of operations. The number of iterations in the direct FFT method is related to the number of range cells (N_τ) of the range-Doppler map that must be obtained, whereas the number of iterations that are required for the correlation FFT method is related to the number of Doppler cells (N_f) that must be displayed in the range-Doppler map. Therefore, in those cases where the number of bistatic range bins is greater than the number of Doppler bins, the correlation FFT method is to be preferred and viceversa. Figure 1.8 shows a pictorial view that explain this concept.

1.4.2.3 Batches algorithm

Both direct and correlation FFT methods may be still computationally too expensive. This is due to the fact that both techniques optimise the calculation only along one dimension, either range or Doppler. In fact the correlation FFT method allows for the number of Doppler cells to be selected but it calculates the output for all range bins even though only a limited amount of range bins are of interest (mainly due to the radar range coverage). On the other hand, the direct FFT method allows for the range bins to be selected but it calculates the output for all Doppler bins, including those that correspond to very high Doppler frequencies and therefore to those targets velocities that are physically impossible to reach. Unfortunately, there is no method that produces an exact solution and at the same time allow for both range and Doppler bins to be selected. Nevertheless, a method, namely the batches algorithm, has been proposed that allows for range and Doppler bins to be selected at the expense of a small SNR loss but with a very large computational load reduction. It should be mentioned that other techniques have been proposed that aim at achieving the same result, such as

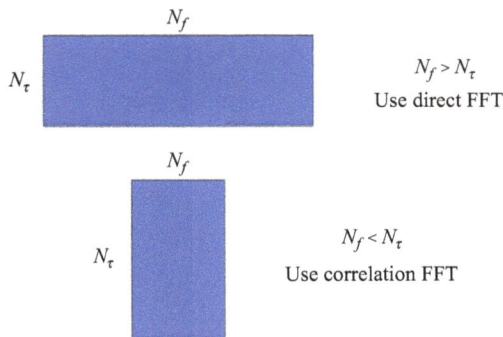

Figure 1.8 *Correlation FFT vs direct FFT. N_τ indicates the number of range cells; N_f indicates the number of Doppler cells.*

the channelisation technique [9]. The batches algorithm is implemented by means of a subdivision of both the reference and surveillance signals into segments (batches). A pictorial view of this segmentation is given in Figure 1.9.

In mathematical terms, this can be represented by rewriting (1.14) as follows:

$$CCF\,(l,m) = \sum_{r=0}^{n_B=1} \exp\left(-j2\pi\,\frac{mrN_B}{N}\right) \cdot \tag{1.24}$$

$$\sum_{p=0}^{N_B=1} s_R\,(rN_B+p)\,s_{ref}^*\,(rN_B+p-l)\exp\left(-j2\pi\,\frac{mp}{N}\right)$$

where n_B is the number of batches, N_B is the batch length and where $n_B \cdot N_B = N$. The assumption that is made for the batches algorithms to be effectively applied is that the Doppler effect within one batch is neglectable, which means that the phase rotation induced by the complex exponential function within the second sum can be ignored. Therefore, (1.24) can be approximated as follows:

$$CCF(l,m) = \sum_{r=0}^{n_B=1} \exp\left(-j2\pi\,\frac{mrN_B}{N}\right)\sum_{p=0}^{N_B=1} s_R\,(rN_B+p)\,s_{ref}^*\,(rN_B+p-l) \tag{1.25}$$

The approximation made in (1.25) can be represented as a step-wise phase function rather than a linear function, as shown in Figure 1.10.

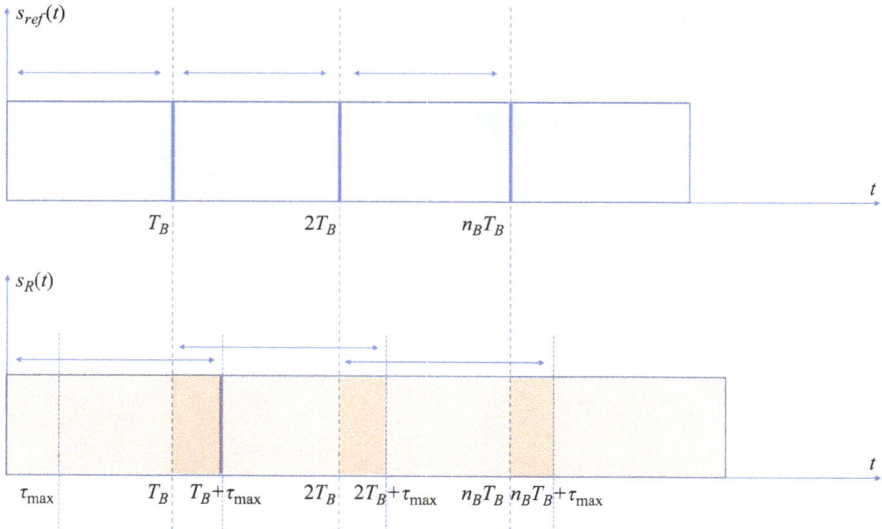

Figure 1.9   Batches algorithm – signal segmentation. T_B indicates the batch duration.

For the sake of compactness, we may rewrite (1.25) as follows:

$$CCF\,(l,m) = \sum_{r=0}^{n_B=1} CCF\,(l,r)\exp\left(-j2\pi\frac{mrN_B}{N}\right) = DFT\,[CCF\,(l,r)] \quad (1.26)$$

where $CCF\,(l,r)$ is the $1D - CCF$ calculated for the rth batch and the $CCF\,(l,m)$ is obtained as a DFT of along the r, which represents a sort of slow-time domain. This processing scheme is similar to that of a frequency modulated continuous wave (FMCW) radar, where each batch can be associated with a FMCW ramp. Figure 1.11 represents a flow chart of the batches algorithms processing scheme.

Some considerations can be made in order to evaluate the SNR losses that are introduced by the approximation in (1.25). Longer batches produce a smaller number of batches. Consequently, the DFT is calculated over a smaller number of points, therefore reducing the computational load. On the other hand, longer bathes introduce a larger error in the phase approximation and therefore heavier losses in terms of SNR, which are due to Doppler phase rotation that would not be compensated. Shorter batches, instead, would produce a larger number of batches and, therefore, the DFT would have to be calculated over a larger number of points, which would increase the computational load.

In this case, the phase errors would be smaller and therefore the SNR losses would be smaller. An extensive analysis of the losses introduced by the batches algorithm has been shown in [10]. Results obtained from real data collected with a passive radar demonstrator owned by the Italian National Consortium for Telecommunications (CNIT) are shown in Figure 1.12. It is quite clear from the plots that losses increase when the target-induced Doppler is higher and when the batch is longer. The plots in

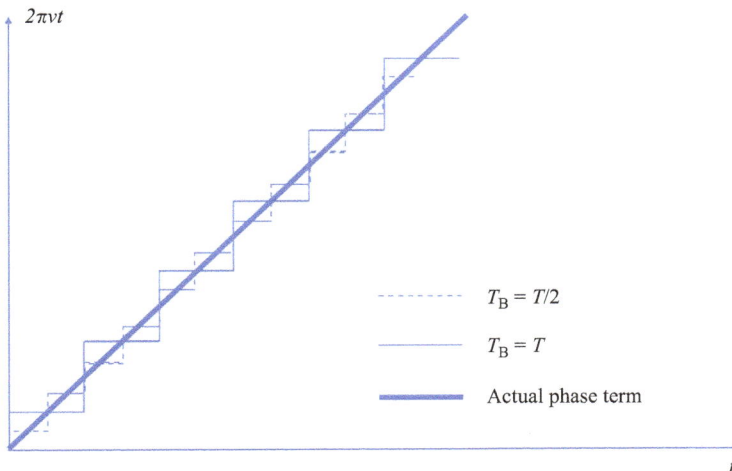

Figure 1.10   Batches algorithm – phase approximation

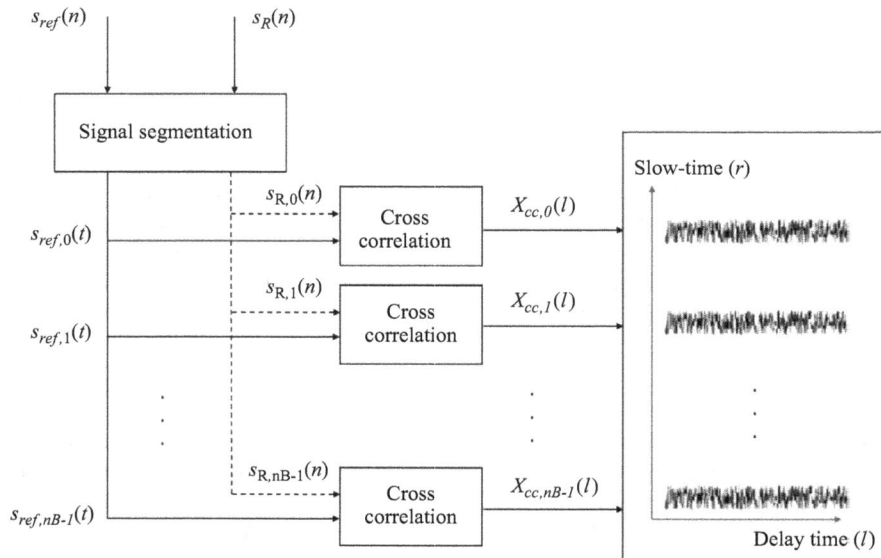

Figure 1.11   Batches algorithm – signal processing scheme

Figure 1.12 also show the theoretical losses (solid line), which are matched by real data (triangle markers). The computational load for the four cases is shown in Table 1.1.

1.4.3   Direct signal and clutter cancellation

The signal received from the surveillance antenna contains the echoes from the targets that are illuminated by the source of opportunity. Unfortunately, this is not the only component that is present in the received signal. In fact the same signal that is emitted by the IO may propagate directly to the surveillance antenna and be captured by an antenna sidelobe. Although sidelobes attenuate the signal significantly with respect to the main lobe, the direct signal may be tens of dBs stronger than any targets echo. Therefore, the direct signal component is by far the strongest signal that is received by the surveillance antenna. Such component is constrained at zero bistatic range and at zero Doppler when both the IO and the surveillance antenna are stationary. Despite the stationarity and the zero-range position of the direct signal response in the 2D-CCF (range-Doppler map), the strength of this signal ripples out to reach other range-Doppler cells and therefore masks targets that may be positioned at other range and Doppler cells. Clutter is also present in PCL as it is in active radar.

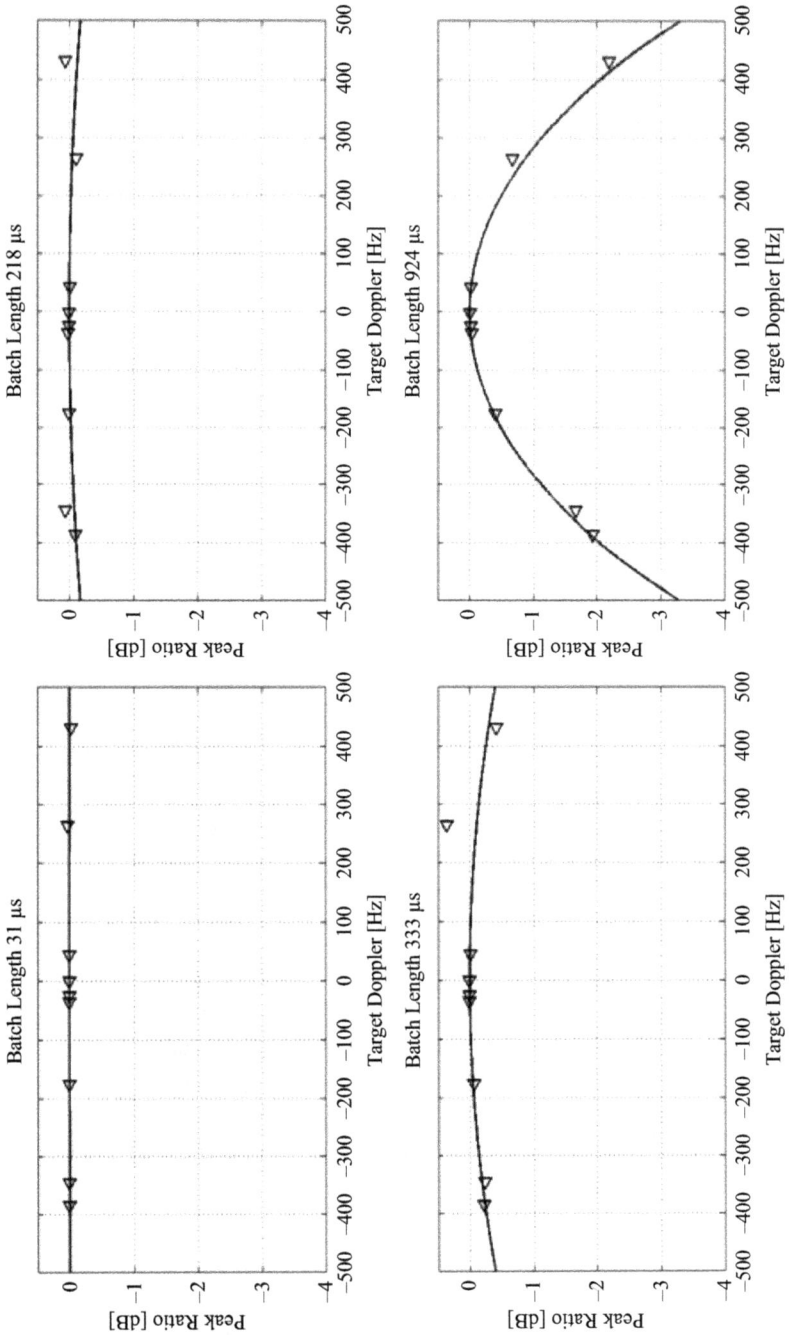

Figure 1.12   Batches algorithm – losses

Table 1.1   Computational load

Batch length (μs)	Processing time (s)
31	4.93
218	0.92
333	0.71
924	0.59

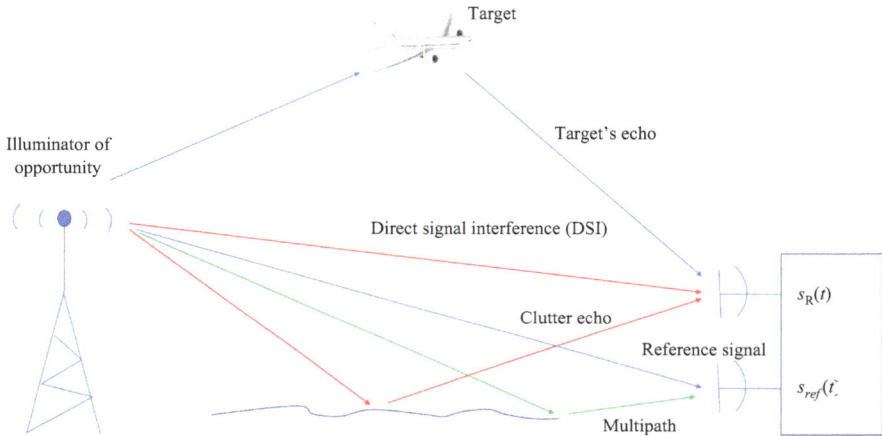

Figure 1.13   PCL scenario: DSI, clutter and multipath components

Nevertheless, it should be noted that due to the bistatic geometry and in particular in the forward or quasi-forward scatter region, clutter can be very strong and appear at several bistatic range cells. This effect, combined with the direct signal effect, may mask several targets and prevent their detection. A pictorial view of this scenario is provided in Figure 1.13.

A complete model for the received signal at the reference channel is shown in (1.27), where the first term represents the direct signal component (also referred to as DSI), the second term (first sum) represents the target's echoes, the third term (second sum) represents the clutter and the last term is noise the component:

$$s_R(t) = A_R s_T(t) + \sum_{m=1}^{N_T} a_m s_T(t - \tau_{T_m}) \exp(j2\pi f_{D_m} t) + \sum_{i=1}^{N_s} b_i s_T(t - \tau_{c_i}) + n(t)$$

(1.27)

where A_R is the complex amplitude of the direct signal at the surveillance channel, $s_T(t)$ is the transmitted signal, N_T is the number of targets, a_m, τ_{T_m}, and f_{D_m} indicate the complex amplitude, the bistatic delay, and the bistatic Doppler of the mth target, respectively. Similarly, N_S indicates the number of multipaths, whereas b_i and τ_{c_i} the complex amplitude and the bistatic delay of the ith multipath, respectively. In order to maximise detection performances, both the direct signal and the clutter components must be removed or at least strongly attenuated. In the next sub-section, we will look at a well-known technique that goes in that direction.

1.4.3.1   Extensive cancellation algorithm

The ECA consists of a projection of the received signal onto a sub-space that is orthogonal to the disturbance sub-space [9]. As the disturbance signals (direct signal and clutter) are much stronger than the target's echoes, it can be argued that the subspace occupied by them can be easily estimated by considering such signal as replicas of $s_T(t)$. In its most simple form, the ECA, considers only stationary replicas, which correspond to zero-Doppler contributions. More complex versions of the ECA also include the case of non-zero-Doppler clutter, such as the ECA batches and ECA batches and stages. A reasonable assumption can be made in order to reduce the computational load that is required by the ECA technique, which consists of limiting the number of range cells where significant clutter contribution is concentrated. Closer bistatic range cells are in the forward and near forward scatter region and they are also at a closer distance from the transmitter and the receiver. Both conditions have the effect of producing a high level of clutter compared to cell at further bistatic ranges. If this assumption can be made, then we can limit the signal analysis to the first K bistatic range cells.

To derive the ECA algorithm, N samples of the surveillance signal are considered

$$\mathbf{s}_R = [s_R[0], s_R[1], ..., s_R[N-1]] \tag{1.28}$$

along with $N + R - 1$ samples from the reference channel

$$\mathbf{s}_{ref} = [s_{ref}[0], s_{ref}[1], ..., s_{ref}[N+R-1]] \tag{1.29}$$

where $R - 1$ is the number of additional reference samples to be considered to obtain a desired integration, which in this case would be equal to N over R time bins. In order to obtain a cancellation of the clutter, the following optimisation problem must be solved, where we aim at minimising the residual energy after removing the clutter from the first K range cells.

$$\min_{\alpha} \left[\|\mathbf{s}_R - \mathbf{X}\alpha\|^2 \right] \tag{1.30}$$

where

$$\mathbf{X} = \mathbf{B}\mathbf{S}_{ref} \tag{1.31}$$

and $\mathbf{B}$ is a matrix that selects only the last N rows of

$$\mathbf{S}_{ref} = \left[\mathbf{s}_{ref}, \mathbf{D}\mathbf{s}_{ref}, \mathbf{D}^2\mathbf{s}_{ref}, ..., \mathbf{D}^{K-1}\mathbf{s}_{ref} \right] \tag{1.32}$$

where $\mathbf{D}$ is a delay matrix that simply shifts the reference signal by one time sample. It should be noticed that the columns of matrix in (1.32) define a basis for a K-dimensional disturbance subspace. The solution for the minimisation problem in

(1.30) is the standard solution of a least square error (LSE) problem, which can be found in (1.33)

$$\alpha = \left(\mathbf{X}^{\mathbf{H}}\mathbf{X}\right)^{-1}\mathbf{X}^{\mathbf{H}}\mathbf{s}_R. \tag{1.33}$$

Therefore, the clutter-free signal can be obtained as follows:

$$\mathbf{s}_{ECA} = \mathbf{s}_R - \mathbf{X}\alpha = \left[\mathbf{I}_N - \mathbf{X}^{\mathbf{H}}\mathbf{X}^{-1}\mathbf{X}\right]\mathbf{s}_R = \mathbf{P}\mathbf{s}_R \tag{1.34}$$

where $\mathbf{P}$ is a projection matrix that projects the received signal onto the subspace orthogonal to the disturbance subspace. Figures 1.14 and 1.15 show a real data range-Doppler map before and after the application of the ECA filter.

1.4.4   Reference signal cleaning

The ECA works effectively when the reference signal is perfectly known. As the reference signal is directly received by the passive radar antenna, we have to assume that such a signal is a perfect replica of the transmitted signal. Unfortunately, multipath effects are present also in the reference signal, as represented in Figure 1.13 in green colour. Therefore, the signal received by the reference antenna is not a perfect replica of the transmitted signal. A realistic model of the reference signal is shown in (1.35)

$$s_{ref} = A_{ref}d\left(t\right) + \sum_{m=1}^{N_M} A_m d\left(t - \tau_m\right) + n_{ref}\left(t\right) \tag{1.35}$$

where A_{ref} is the complex amplitude of the direct signal, N_M is the number of multipath components, A_m are the complex weighs of the multipath components, τ_m are the delays of each multipath component, and $n_{ref}\left(t\right)$ is the noise component. Two methods

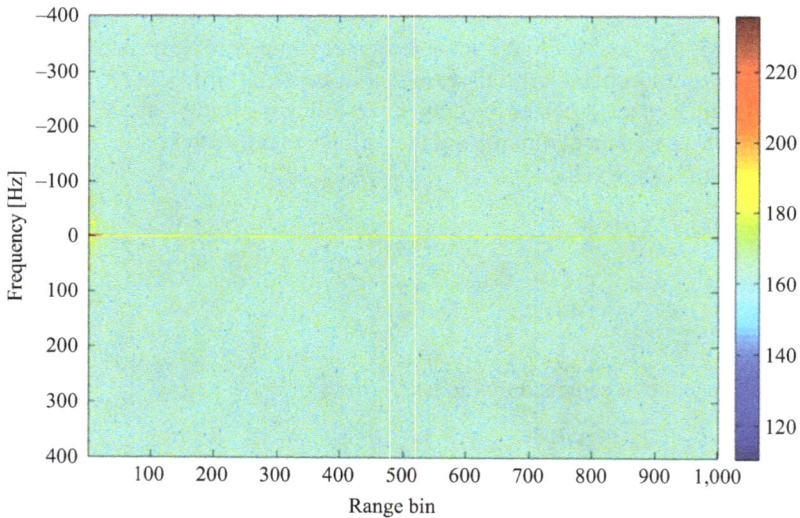

Figure 1.14   Range Doppler map before the application of the ECA in dB scale

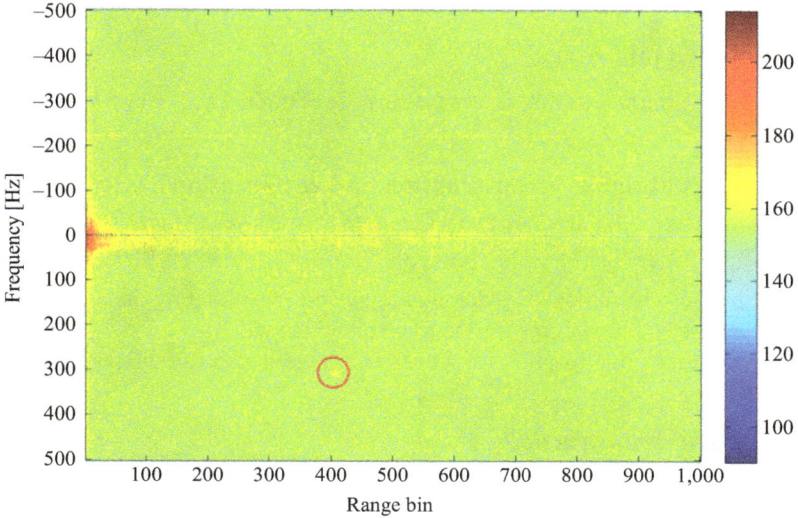

Figure 1.15   Range Doppler map after the application of the ECA in dB scale

will be described in the following subsections. One that is specific for analogue modulations and one that is specific for digital modulations.

1.4.4.1   Constant modulus algorithm

This method, namely the constant modulus algorithm (CMA), is defined for generic analogue-modulated signals [9]. Such an algorithm is effective when several degrees of freedom are available both in the time and space domain. Nevertheless, CMA can be applied to either in the time-only and space-only dimensions. In these cases, the CMA is usually addressed as T-CMA for the time-only version and S-CMA for the space-only version. The goal is to suppress additive interference (multipath) by constraining the output to be constant in modulus. This corresponds to minimising the following function:

$$\hat{y}[n] = \min_{y[n]} \left(E\left[\left(|y[n]|^2 - 1\right)^2 \right]\right) \tag{1.36}$$

which leads to the solution in (1.37)

$$y[n] = \sum_{m=0}^{M-1}\sum_{l=0}^{L-1} w_{m,l}[n]\, s^*_{ref,m}[n-l] \tag{1.37}$$

where m represents the spatial element index, l is the temporal sample index, and $s_{ref,m}$ is the reference signal collected by the mth element and

$$w_{m,l}[n+1] = w_{m,l}[n] - \mu\epsilon[n]\, s^*_{ref}[n-l] \tag{1.38}$$

with

$$\varepsilon[n] = \left(|y[n]|^2 - 1\right) y[n] \tag{1.39}$$

and where μ is a parameter that serves the purpose of trading accuracy for convergence speed.

1.4.4.2 Digital signal reconstruction and equalisation

In this section, we will introduce two techniques to improve the reference signal when the transmitted signal is a digital one. In particular we will show that

- multipath can be removed by demodulating and remodulating the received signal to reproduce a perfect copy of the transmitted signal;
- spurious peaks due to pilot sequences can be removed through an equalisation process.

Digital signal reconstruction

Some digital transmissions, such as DVB-T, offer the possibility to reconstruct the transmitted signal after demodulating and re-modulating the received signal. Part of this operation, and more specifically the demodulation, is typical of any DVB-T receiver, which aims at decoding the quadrature amplitude modulation (QAM) symbols that represent the information content of DVB-T transmissions. Once such symbols are obtained, a perfect copy of the nominal transmitted signal can be obtained. Figure 1.16 shows an example of DVB-T symbols demodulated in the presence of

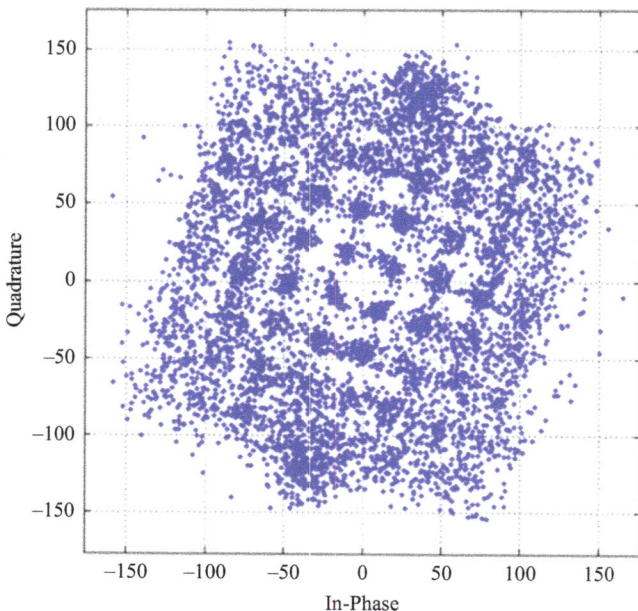

Figure 1.16   Decoded symbols in the presence of multipath

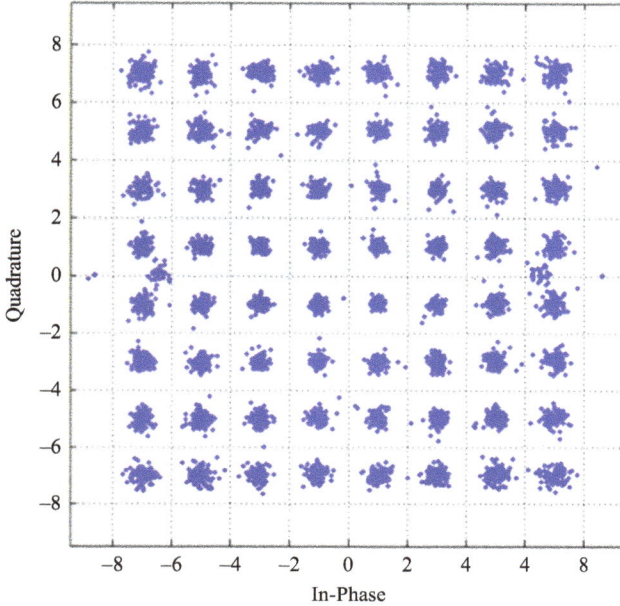

Figure 1.17 Decoded symbols in the absence of multipath

multipath whereas Figure 1.17 shows the same symbols demodulated in the absence of it.

We have to mention here that the re-modulated signal does not bear the distortions that may be introduced by the transmitter power devices. Such distortions, if known, have to be replicated in order to produce a signal that is as close as possible to the transmitted signal. More on this issue will be presented in later chapters of this book.

Digital signal equalisation

In addition to removing multipath, digital signal can be equalised to perform auto-correlation function (ACF) shape control. An undesired effect is provoked by the presence of pilot sequences, which generate spurious peaks in the ambiguity function. An example of such spurious peaks is presented in Figure 1.18, where an ambiguity function is shown relatively to some measured DVB-T signals. An effective algorithm can be implemented that aims at generating locally the power spectrum density (PSD) (PSD-Fourier transform of the ACF) and perform an equalisation by generating a filtering function that aims at transforming the ACF into a spurious-free version of it. Details of the algorithm are shown in Figure 1.19 from [?].

The PSD is estimated for both the reference signal and for the locally generated signal (without pilot sequences). A filtering function **H** is then generated as the ratio of the two PSDs. The corrected PSD is obtained by multiplying the FFT of the

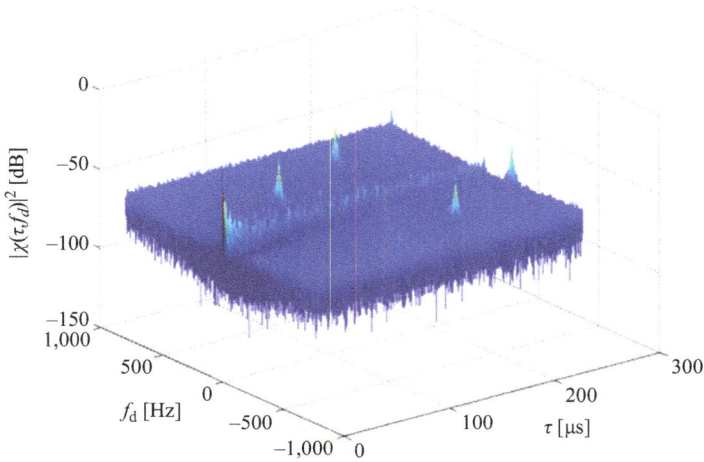

Figure 1.18   Ambiguity function DVB-T signal presence of spurious peaks due to pilot sequences

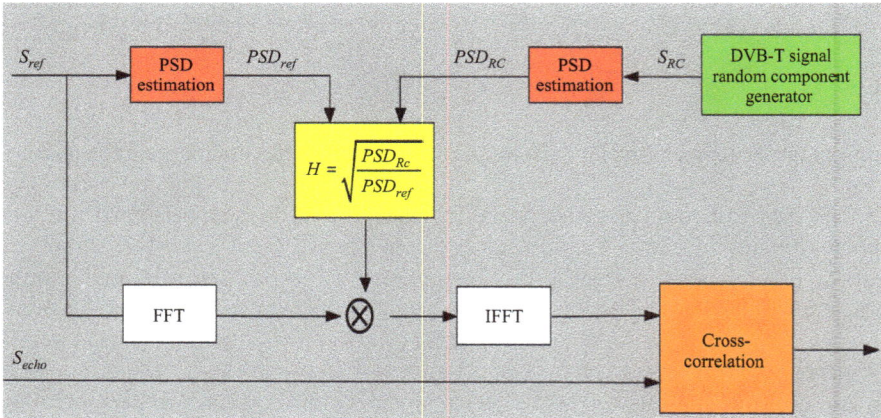

Figure 1.19   Equalisation algorithm functional block scheme

reference signal by the function **H**. An IFFT is then applied to generate the spurious-free reference signal time sequence. An example of the effect of such an equalisation is shown in Figure 1.20, where the zero-Doppler slice of the cross-ambiguity function (CAF) along the range direction is shown before (blue line) and after (red line) the equalisation process. The zero-range slice of the CAF along the Doppler direction is shown in Figure 1.21 instead by using the same colour coding. Spurious peaks are

Range profile f_d=0

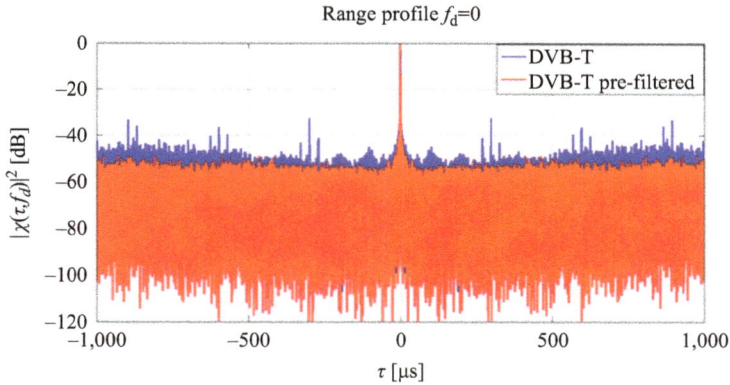

Figure 1.20 Equalisation effect CAF zero-Doppler slice along the range coordinate

Range Profile τ=0

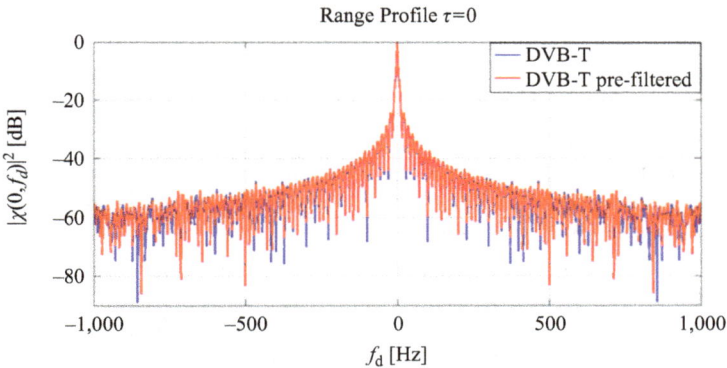

Figure 1.21 Equalisation effect CAF zero-range slice along the Doppler coordinate

visible in the zero-Doppler slice before the equalisation takes place. It is important to note that the main lobe is not distorted by this operation.

1.5 Conclusions

In this chapter, the basic principle of operation of a PCL has been described. The main characteristics related to the bistatic geometry have been introduced as well as the concepts of bistatic range and bistatic Doppler. Also the basics of passive radar signal processing have been presented, with a focus on the techniques for the calculation of the range/Doppler map. Subsequently, a section has been devoted to one of the main issues related to passive radar signal processing, that is the direct signal and multipath suppression. Some of the thematics presented in this chapter will be analysed in more detail in the following chapters.

References

[1]   Willis NJ. *Bistatic Radar*. Silver Spring, MD: Technology Service Corporation; 1995.

[2]   Howland PE, Griffith HD, and Baker CJ. Passive bistatic radar systems, in: *Bistatic Radar: Emerging Technology*. Chichester: John Wiley and Sons; 2008.

[3]   Griffith HD and Baker CJ. Passive coherent location radar systems. Part 1: performance prediction. *IEE Proceedings Radar Sonar and Navigation*. 2005;152(3):153–159.

[4]   Howland PE. Editorial: Passive radar systems. *IEE Proceedings Radar Sonar and Navigation*. 2005;152(3):105–106.

[5]   Farina A and Kuschel H. Special issue on Passive radar (Part I). *EEE Aerospace and Electronic Systems Magazine*. 2012;27(10)15–25.

[6]   Malanowski M, Mazurek G, Kulpa K, *et al*. FM-based PCL radar demonstrator. In: *Proceedings of the International Radar Symposium*; 2007.

[7]   Kuschel H, Heckenbach J, O'Hagan D, and Ummenhofer M. A hybrid multi-frequency Passive Radar concept for medium range air surveillance. In: *2011 Microwaves, Radar and Remote Sensing Symposium*; 2011, pp. 275–279, doi: 10.1109/MRRS.2011.6053653.

[8]   Griffith HD and Baker CJ. *Introduction to Passive Radar*. Artech House; 2017.

[9]   Lombardo P and Colone F. Advanced processing methods for passive bistatic radar systems, in: *Principles of Modern Radar* – Vol II. Raleigh, NC: Scitech Publishing; 2013.

[10]   Moscardini C, Petri D, Conti M, *et al*. Batches algorithm for passive radar. *IEEE Transactions on Aerospace and Electronic Systems*. 2015;52(2):1475–1487.

Chapter 2

Applications for passive radar on moving platforms

Krzysztof Kulpa[1], Piotr Samczyński[1], Damian Gromek[1], Rafał Rytel-Andrianik[1], and Gustaw Mazurek[1]

List of abbreviations

ADC analogue to digital converter
AFE analog front end
CAF cross ambiguity function
COTS commercial off the shelf
DAB digital audio broadcasting
DPCA displaced phase center array
DVB-T digital video broadcasting – terrestrial
GMTI ground moving target indication
GPS global positioning system
GSM global system for mobile communication
ISAR inverse synthetic aperture radar
NCTR non-cooperative target recognition
PAISAR passive ISAR
PASAR passive SAR
PCL passive coherent location
SAR synthetic aperture radar
STAP space–time adaptive processing
UHF ultrahigh frequency
VHF very high frequency

The chapter shows examples of current and emerging applications for passive radar on moving platforms. The application largely depends on the platform the radar is installed on, so different platforms are first considered mainly sea- and air-borne. The applications described include target detection and passive imaging of terrain or moving objects. Sample results obtained from real signals during field test campaigns are presented. Then, selected technologies applied in passive radar on mobile platforms

[1] Department of Electronics and Information Technology, Warsaw University of Technology, Poland

for multichannel signal acquisition, processing, and for system synchronization are discussed.

2.1 Introduction

Passive radar, known also as passive coherent location (PCL), has been one of the most rapidly developing fields in radar technology over the last three decades. The low cost nature of PCL has resulted in strong interest from a significant number of companies and research institutions. As a result, many passive radar demonstrators and commercial products have been developed [1–8]. Existing passive radar systems utilize various types of signals that illuminate targets of interest. The most popular signals used by modern passive demonstrators are: FM radio [2–4,6], DAB radio [5,9–11], analog and digital television (DVB-T) [12–14,17], GSM networks [16,18], and WiFi signals [15,19,21–23]. The majority of passive radars are stationary, ground-based systems dedicated to the detection and tracking of airborne targets. The last decade of PCL systems development has led this technology toward a state of relative maturity for ground-based operation. These developments have prompted the research focused specifically on passive radar on a moving platform. In the literature available on this subject, initial analyses devoted to PCL mounted on ground moving vehicles and airborne platforms can be found [24–31]. At the present moment, four different platform types are being analyzed: ground platforms (vehicles), sea platforms (ships, patrolling boats, submarines), airborne platforms (aerostats, aircrafts, helicopters, UAV), and space platforms (satellites). As it was in the past, the active radars were first deployed on the ground, and shortly were adapted to many moving platforms. So in the near future, it appears that the passive radars will also be adapted to moving platforms and play an important role in self-protection, surveillance, and reconnaissance. Placing passive radars on moving platforms not only provides many challenges to radar designers but also provides new perspectives and new capabilities. The use of mobile PCL systems can add extended functionality to the passive radar. By placing it on airborne or satellite platforms, the horizon limitation of using distant illuminators of opportunity is removed and it would be possible to detect low flying targets. The platform motion can be also exploited to perform passive imaging of the ground and create passive SAR images [7,33–37], detect slow moving objects using ground moving target indication (GMTI) techniques [38–40], and produce images of detected moving targets using the inverse synthetic aperture radar (ISAR) mode [17]. However, it also brings new challenges to radar designers. In classical ground-based passive radar, if the scene is illuminated by a stationary transmitter, the ground returns (ground clutter) have almost zero Doppler frequency. Light movement of tree branches or leaves, for example, can introduce some small Doppler spread of the clutter, but it can be easily removed by appropriate filtration. As most targets of interest have significant Doppler shift due to their motion, the detection losses caused by zero or near zero Doppler clutter cancellation are negligible. On the other hand, in the case of a moving platform PCL, the echo coming from ground clutter possesses a large Doppler frequency due to the platform motion. As the ground clutter signals arrive from different angles, one can observed the wide Doppler spread of the clutter. This Doppler frequency shift and its change as a function of time can be exploited in the passive synthetic aperture imaging of the ground. However, such clutter Doppler spread can cover echoes of slow targets which makes their detection a challenge. Of course, when only fast targets are of interest, it is possible to cancel all clutter using the extended Doppler CLEAN technique and detect the targets in a clutter-free region. The objective of signal processing in a moving platform PCL is to suppress ground clutter while preserving the echoes from moving objects. This problem is well known in classic airborne pulse Doppler radars, for which there are many solutions based on STAP or DPCA algorithms. Such algorithms for active radars have been derived using the assumption that all pulses sent by radar are the same and have the same power and waveform. It is of course not true in passive radar, where there are no pulses, and the illuminating signal is not stationary. Both the power and the waveform of the illuminating signal are functions of time, and thus more elaborate methods have to be developed.

The concept of a passive radar placed on an airborne platform is presented in Figure 2.1. The aircraft is equipped with a passive radar receiving several signals. Of main importance is the direct signal (a) used as the reference. Apart from this, several indirect signals (echoes) are received from ground objects (b), ground traffic (c), and airborne targets (d). In practical cases, airborne passive radars will observe several transmitters and many moving targets, so the scene presented in Figure 2.1 is simplified one.

To resolve all these signals, an advanced antenna array has to be used, equipped with several elementary antennas and digital receivers. The digital processing will

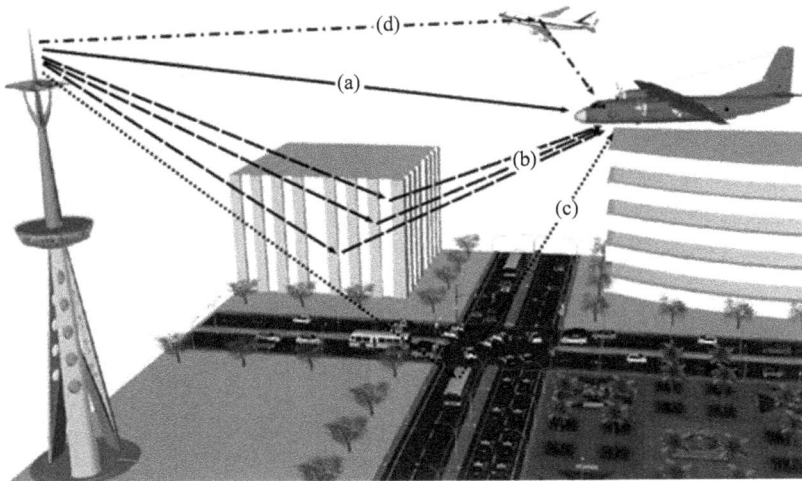

Figure 2.1 Concept of passive radar placed on airborne platform

then form several beams. One (or more) will be used as a reference signal beam, while others will be used for surveillance and imaging purposes. For the ground-based passive radar, both the illuminator of opportunity and the receiver are stationary. Thus the echoes originating from the reflection from non-moving targets (ground clutter) have no Doppler shift and can be removed using classical adaptive filtering. For a PCL installed on a moving platform, presented in Figure 2.2, the situation is much more complicated. As the receiver is in motion, all received signal components are Doppler-shifted, and the shift depends on the platform velocity as well as the angles to the transmitter and to the object of interest (or clutter point).

As a result, all targets and ground scatterers will have their own unique Doppler shift, as presented in a simplified form in Figure 2.3. Only the Doppler shift in the surveillance antenna is presented. After the calculation of the CAF, one can also see the impact of Doppler shift in the reference channel.

A typical CAF of the surveillance and reference channels in an airborne PCL, together with a situation map, are presented in Figure 2.4. The dashed lines in Figure 2.4(a) represent the theoretical limits of ground clutter Doppler spread related to platform velocity, flight altitude, and geometry. It can be seen that the clutter echoes are within the theoretical limits. The clutter in this figure has its own structure that stems from illuminated ground echoes. Such a CAF function can be easily converted into an unfocused SAR image of the ground.

PCL radar can also be placed on a ship to observe other vessels and airplanes. An example CAF obtained from a maritime PCL is presented in Figure 2.5. Again, the dashed line represents the theoretical limits of the ground (or sea) clutter, and due to the much slower platform motion it is far narrower. The small dot at the range of 2.5 km is the echo of a CASA airplane detected by PCL.

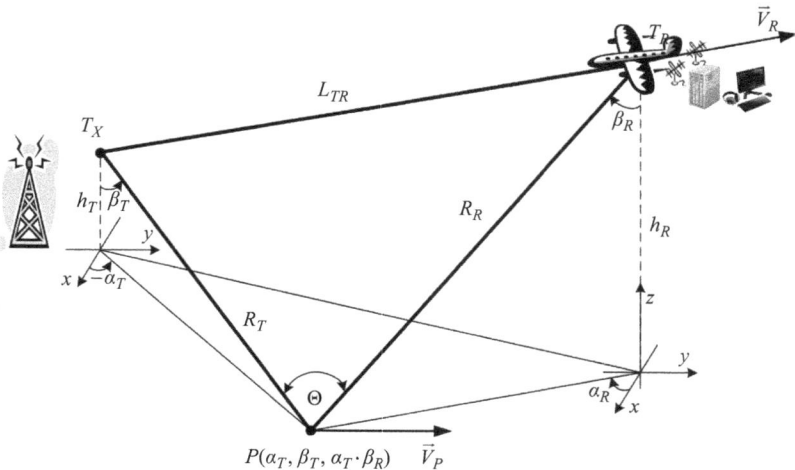

Figure 2.2   Geometry of mobile passive radar

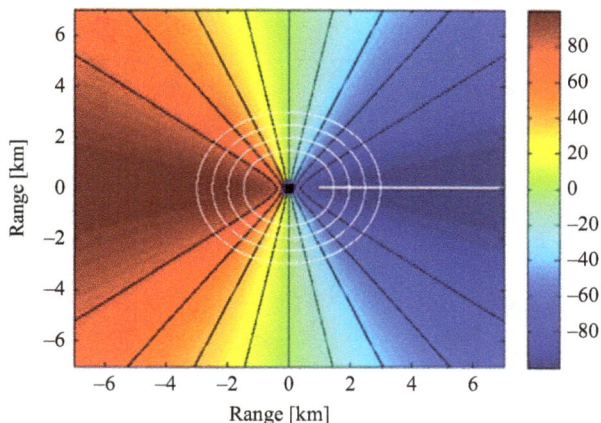

Figure 2.3 *Doppler shift of ground clutter versus clutter position, color scale in (m/s) (recalculated from Doppler shift)*

2.2 Platforms

2.2.1 Sea-borne passive radar example

A very promising platform is sea vessel. Passive radar can be deployed on a sea-borne platform operating in the vicinity of an illuminating transmitter. In most cases, a ground-based broadcast transmitter (DVB-T, DAB, FM, LTE, 5G, etc.) is considered for scene illumination, and this naturally limits the operation to coastal zones only. But in the case of small seas surrounded by ground areas, the illumination is sufficient for PCL operation over the whole area. On the other hand, the operation of a PCL radar in remote, uninhabited regions would require deploying a cooperative transmitter for scene illumination, which would then be converted to a multistatic cooperative system.

A set of experiments has been conducted to show the value of sea-borne passive radar. One of them was conducted in Poland by Warsaw University of Technology. During NATO APART-GAS (active passive radar trials ground-based, airborne, sea-borne) trials in September 2019, an experimental PCL installation was investigated on a navy ship that was sailing on the Baltic Sea, near the Polish coastline [41]. Receiving antennas were installed on the port side of the ship and on the bow of the ship, as shown in Figure 2.6.

Typical VHF and UHF antennas intended for consumer DVB-T and DAB signal reception were used. The installed antenna setup consists of three surveillance antennas fixed in a linear array on the side of the ship, with one reference antenna pointing in the opposite direction, and another reference antenna on the front of the ship. Such an antenna array was replicated in both horizontal and vertical polarization, with the exception that the front reference antenna was present only in the vertical polarization.

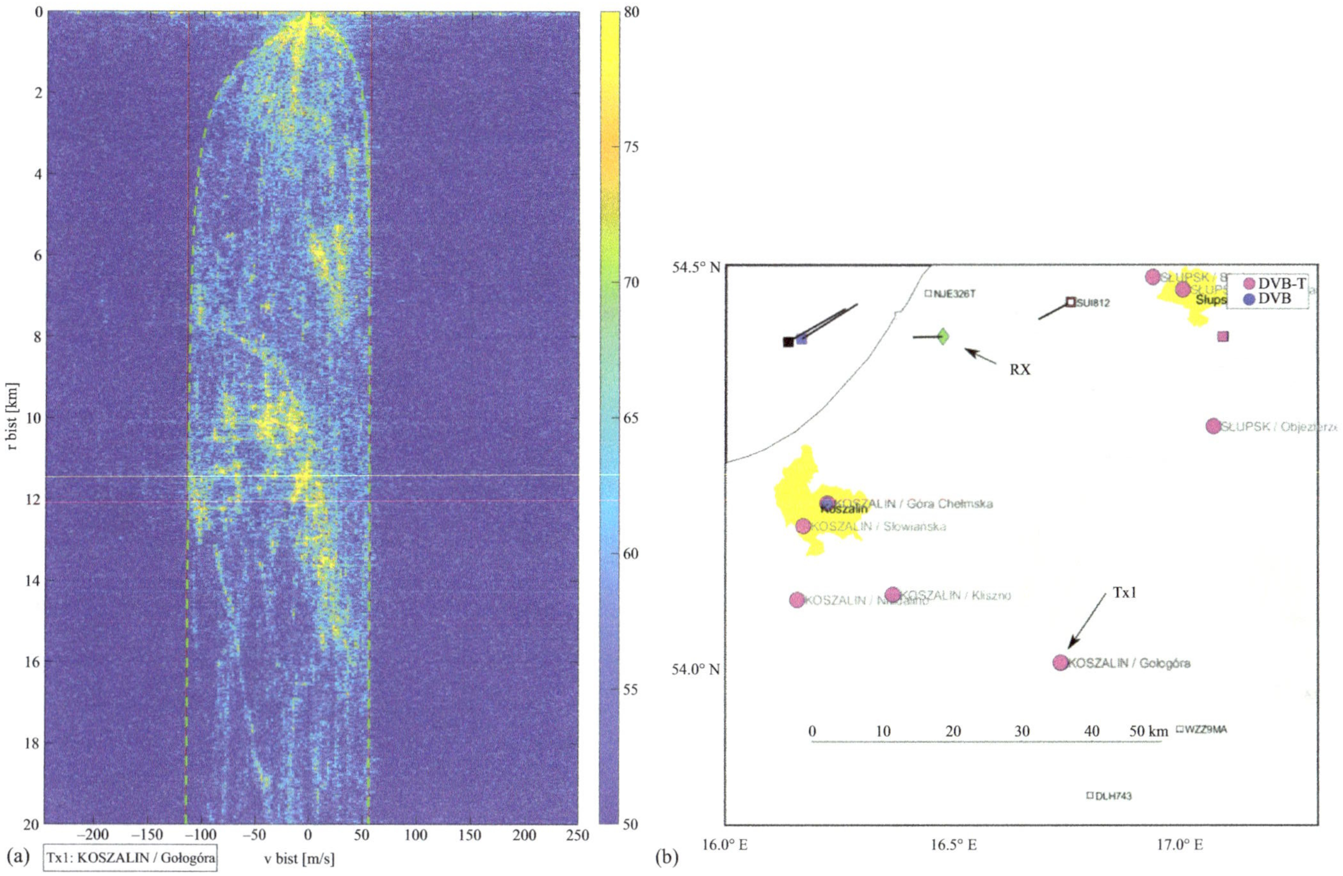

Figure 2.4 Asymmetric ground clutter for airborne passive radar

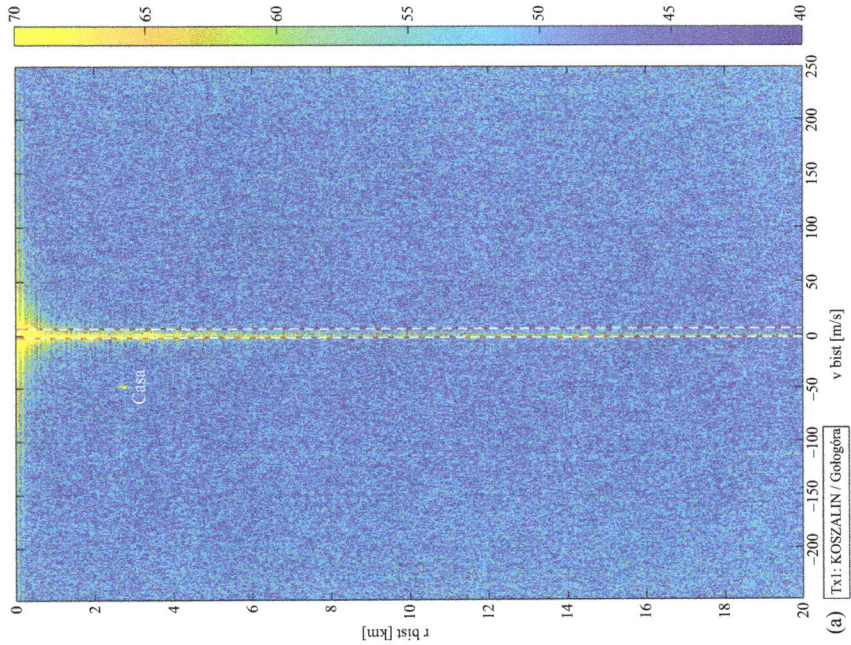

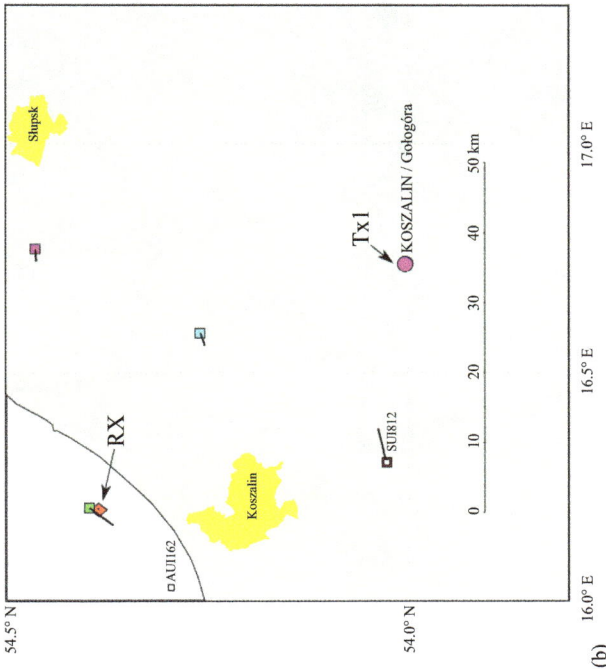

Figure 2.5   Sea clutter for ship-based passive radar

Figure 2.6   Photograph of experimental antenna array

Results of the experiments carried out with that experimental system have confirmed that it is possible to detect aircraft flying in the vicinity of the ship with a PCL receiver in the coastline patrol application, and also to detect other sea vessels. Several cooperative targets, like transport aircrafts and jet fighters, were detected using DAB (218.64 MHz) and DVB-T (191.5 MHz and 450–800 MHz band) illumination signals from nearby broadcast transmitters located in Koszalin, Poland. Example detections of these targets are shown in the bistatic velocity/range (v/R) plane in Figures 2.7 and 2.8.

It was also possible to detect cruise airplanes accidentally passing the coastline at higher altitudes during the signal acquisition. Example detections of such noncooperative targets, like THY9 (Boeing B789 Dreamliner [42]) and SAS778 (Airbus A320 [43]), are depicted in Figures 2.9 and 2.10, respectively.

The application of a sea-borne PCL to detect maritime targets, like other ships and boats, is also possible but somehow limited due to a target's low velocity and zero altitude. This results in very low power of reflected signals and low values of their Doppler shift. The echo of such a target can easily be lost in a clutter area, so advanced clutter cancelation methods, based on STAP technology, have to be used. Nevertheless, it is sometimes possible to detect targets of this class after applying longer integration periods and thus increase the Doppler resolution. In Figure 2.11, we can see the detection of a cooperative boat sailing near the ship with a PCL installation that was visualized after increasing the signal integration period to 0.5 s. This echo, however, is not as clear as in the case of airborne targets and is very hard to extract from the noise. More sophisticated signal processing algorithms, focused on maritime targets, could help to obtain higher quality pictures in such a case.

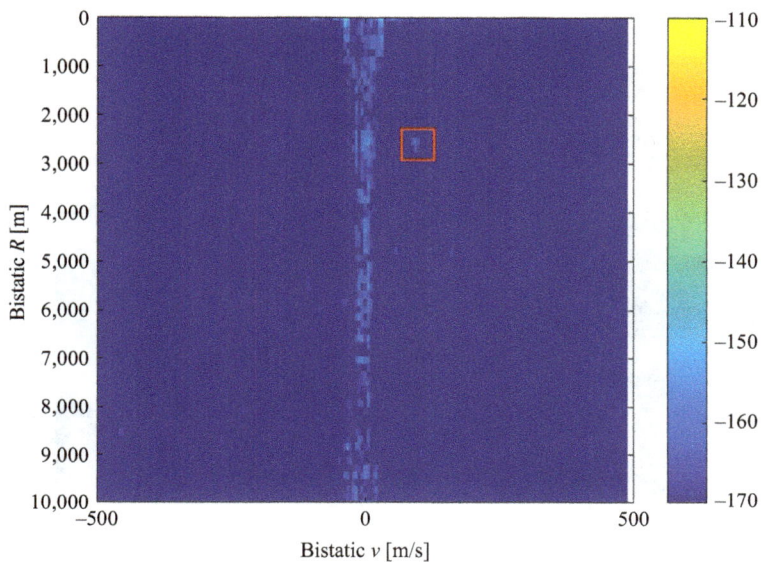

Figure 2.7   Detection of transport aircraft

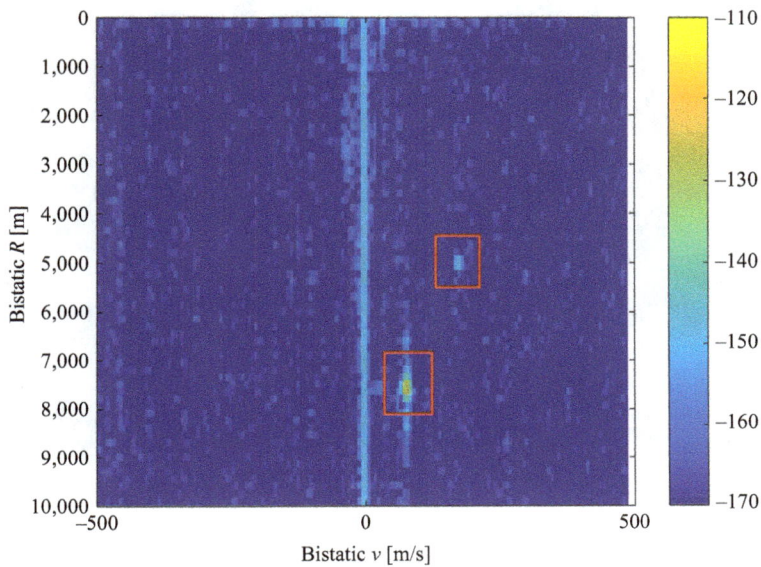

Figure 2.8   Detection of two jet fighters

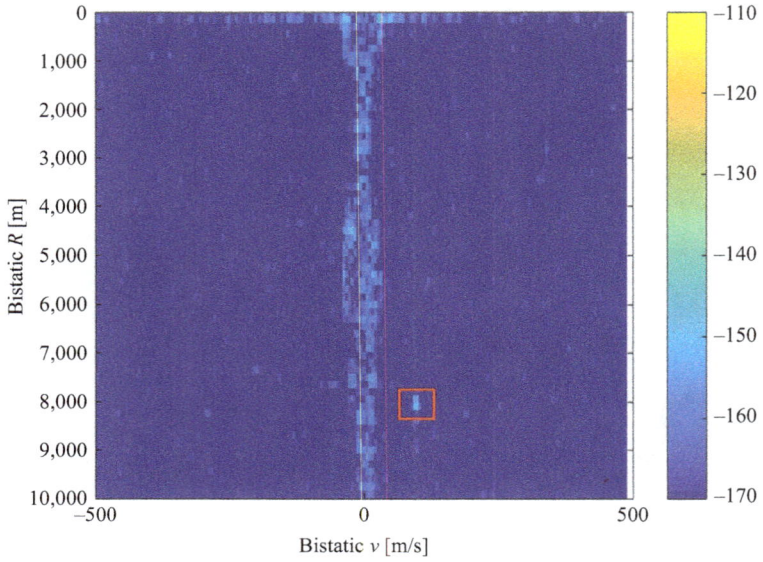

Figure 2.9   Detection of Boeing B789 (THY9)

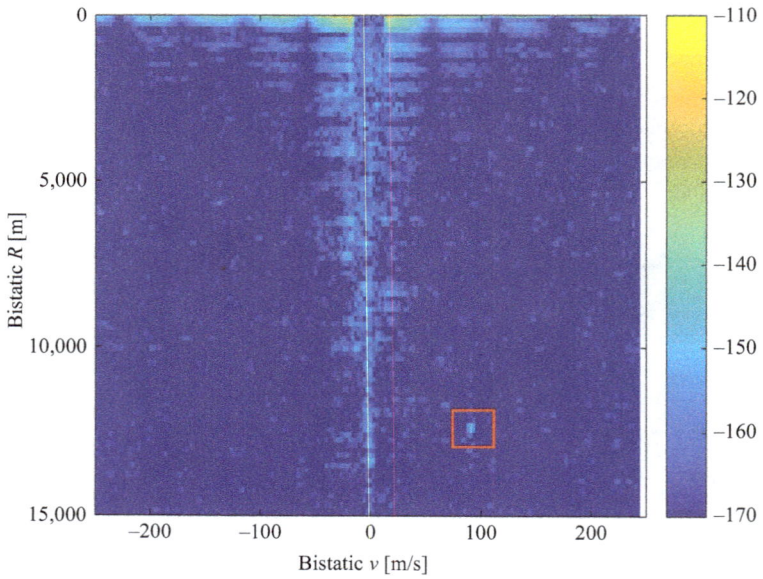

Figure 2.10   Detection of Airbus A320 (SAS778)

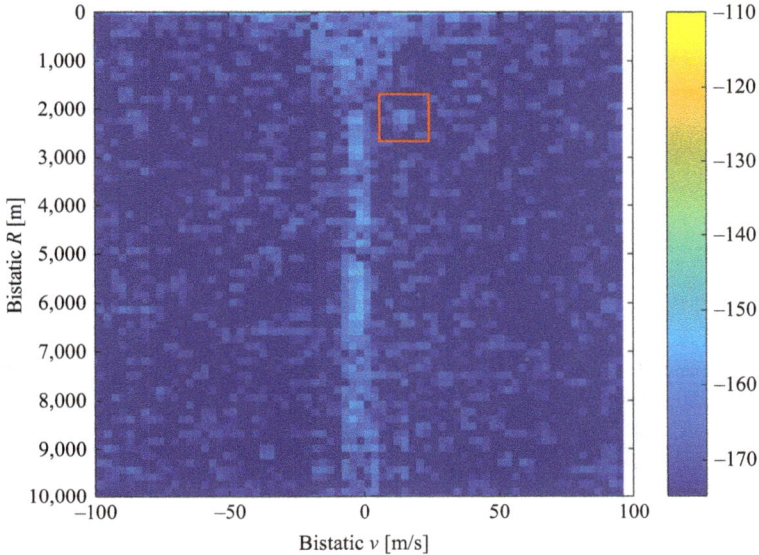

Figure 2.11   Example PCL detection of a boat

All these experiments show that passive sea-borne radar will be used in the near future and will support other sensors to increase the quality of information used for collision avoidance and warnings.

2.2.2   Air-borne passive radar example

An essential application of radar is detection of moving targets. Due to a much further horizon, airborne passive radar is theoretically able to detect distant targets even if they are low or moving on the ground. However, the movement of the radar platform causes ground clutter to spread in the Doppler dimension and possibly cover useful target echoes, which hampers their detection.

Airborne passive radars can be used for several purposes, as depicted in Figure 2.12. They can be used for aerial target detection and tracking. This feature provides additional functionality such as collision avoidance, missile detection, and drone detection. Tracking ability is obtained either if the PCL can measure precisely the angle of the echo signal arrival or if the PCL radar can generate measurements originating from different illuminators. At least three illuminators are required for precise 3D tracking. Another feature is related to the length of time on the target. Integration time is limited only by the time the target is visible (usually minutes) and when the echo is processed. The coherent integration time for detection is usually limited to a fraction of a second. By extending the processing time to seconds or even minutes it is possible to obtain the SAR and ISAR images. Moreover other techniques

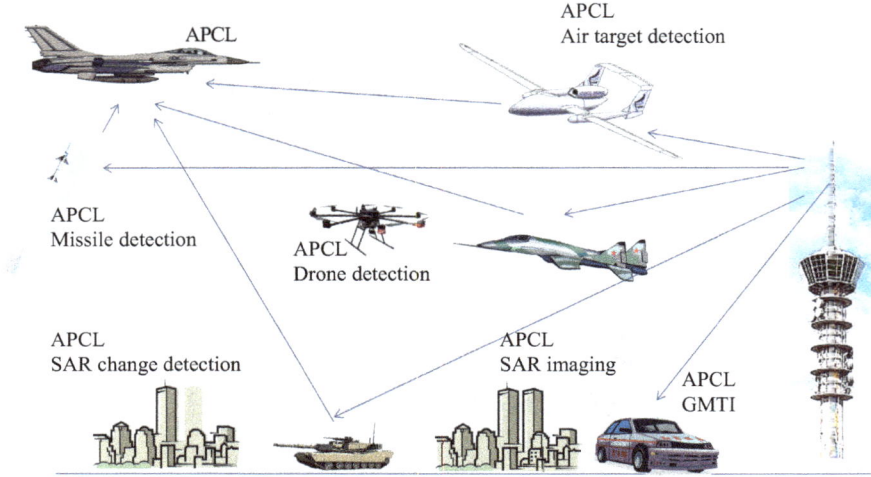

Figure 2.12   Airborne PCL applications

such as polarimetry, multistatic processing, and sparse processing can also be used for further image information enrichment.

2.2.2.1   Airborne trials

Airborne passive radar trials have already been conducted by different entities worldwide. To the author's knowledge, the first such experiment was conducted by the Warsaw University of Technology in 2008. In the airborne trial campaign, the PCL sensor was mounted to a Skytruck aircraft as depicted in Figure 2.13. The radar was equipped with two 3-element antenna arrays placed on both sides of the aircraft able to receive FM radio transmission. The antenna elements were placed inside the aircraft in window openings, as shown on the right in Figure 2.13. The signal acquisition and processing were performed by specially designed software defined radio receivers with direct sampling (middle picture). The trials were carried out over the Baltic sea close to the Polish coast.

Several processing methods have been tested to obtain air detection, including the DPCA and STAP methods. The results, depicted in Figure 2.14, were obtained using a simple DPCA method. The two antenna elements on one side of the aircraft were used as the surveillance antennas in the DPCA algorithm. In the processing, the direct signal was removed using an adaptive filter [20,32]. The Doppler spread clutter was reduced using DPCA principles and two non-cooperative targets were detected.

Other large airborne trials were conducted in Poland in 2019 (APART-GAS trials) where several airborne passive radars mounted on different aircraft were tested. The most advanced system was mounted on a CASA airplane, as depicted in Figure 2.15. It consists of two subsystems: a DVB-T-based 4 channel passive radar and a DAB-based 4 channel passive radar. Antennas were mounted at the windows on both sides of the

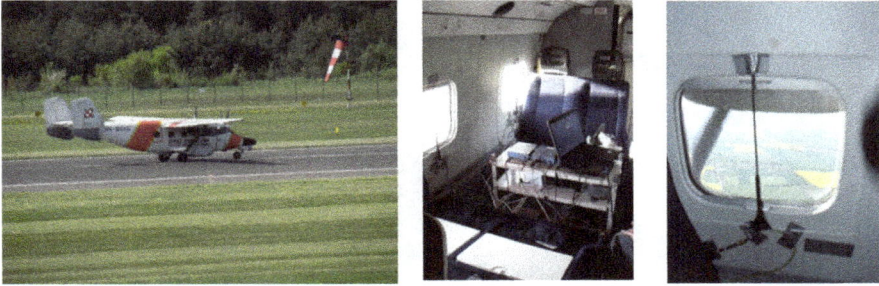

Figure 2.13   *First Airborne PCL trials POLAND 2008 – left: Skytruck platform, center: processing unit, right: antenna mounted in the window*

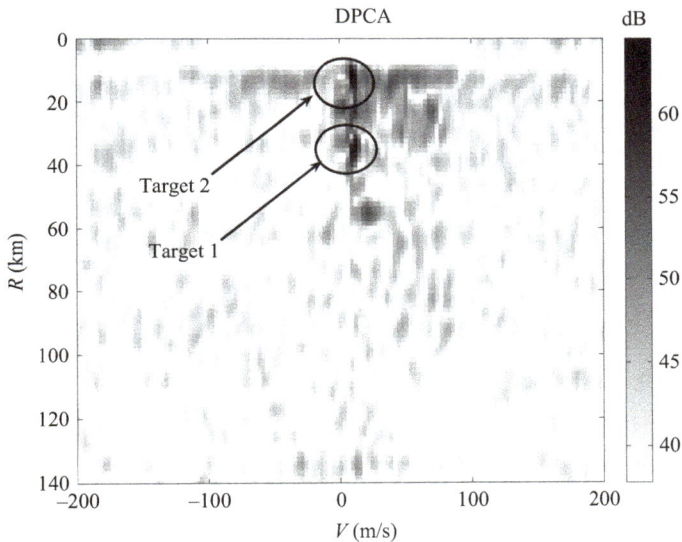

Figure 2.14   *Result of DPCA signal processing of APCL trials POLAND 2008*

plane to allow DPCA and STAP signal processing. The test flights were performed over the Baltic Sea near the Polish coast. Data was processed is several modes: detection (DPCA and STAP), GMTI, and SAR.

Example results of the ground clutter Doppler spread from the APART-GAS trials are presented in Figure 2.16a. The speed of the radar platform was 85 m/s (which defined the extent of the Doppler spread) and the speed projected on the transmitter was −26 m/s (which defined the center location of the clutter region). Targets with range and velocity parameters outside of the clutter region can be detected in a standard way (as an example, an echo of a fighter jet is clearly visible in Figure 2.16b), but

Figure 2.15   *The CASA plane with passive airborne radar working in DAB and DVB-T modes – POLAND 2019*

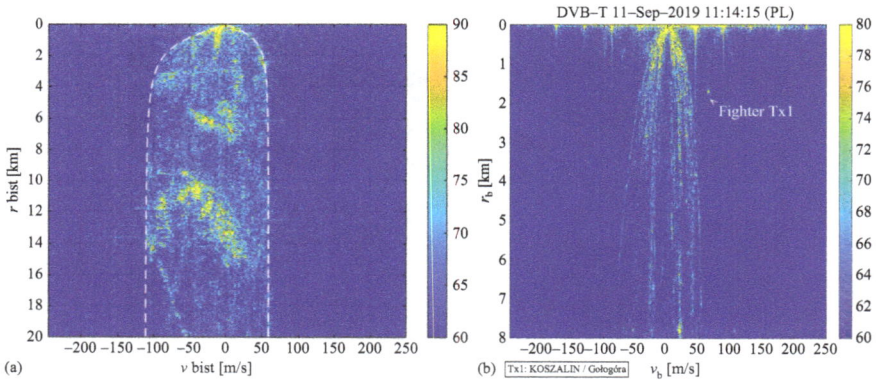

Figure 2.16   *Examples of clutter spread in Doppler on range-Doppler map: (a) ground clutter, white dotted line is analytically obtained envelope of the clutter region; (b) sea clutter, target visible outside of the clutter region*

for detection of moving targets located within a cluttered region a method of clutter reduction is necessary. This is true particularly for the ground moving target indication (GMTI) application, as echoes of such targets are most often buried in clutter.

The clutter reduction methods generally depend on the length of the surveillance receiving array. In SAR images, moving objects appear as a smeared region, which was analyzed in [40]. Such analysis can potentially lead to the estimation of moving target parameters for a bistatic radar with a single channel surveillance receiver, but does not solve the detection problem.

If two receiving channels are used in a radar with antennas placed along platform velocity vector, then the DPCA algorithm can be used [44]. In this method, the delayed signal from the trailing antenna is subtracted from the leading antenna signal. The delay corresponds to the time the platform needs to move the inter-antenna distance, thus such subtraction cancels echoes of stationary clutter. It may seem that this method suits well the character of passive radar which is typically non-expensive and lightweight because in the DPCA method, the antenna array is short (only two elements), processing is relatively simple, and the continuous nature of most signals used in passive radar makes the application of arbitrary delay easy. In practice, there are a few problems to be solved. One problem is that a transmitted signal changes with time, thus the leading signal is not the same as the delayed trailing signal (contrary to DPCA in active pulsed radar where identical pulses can be transmitted). This problem is solved by replacing matched filtering with reciprocal filtering [45], which allows one to concentrate on the impulse response of the scene and not on the signal nature. Another problem is the need to carefully calibrate receiving channels, which was dealt with in [46,47].

If more than two antennas are used in an array, then other STAP (space time adaptive processing) algorithms can be used to filter out clutter echoes [48]. Such methods are based on the observation that the bistatic velocity of a clutter patch is related to its angle of arrival, hence in the Doppler-angle map, clutter echoes appear as a well-defined ridge, while target echoes are often present in different locations.

2.2.3 Passive ground imaging (PASAR)

A promising alternative to the active SAR system is a passive one. In a passive SAR system, the target or the scene is illuminated by external illuminators of opportunity (such as DAB/DVB-T/GSM/FM). The first proof of concept was done in a similar way to active radar where passive radars were used in a ground-based configuration for air phenomena and target detection. The first passive radar experiment was conducted almost a century ago. In 1924, Appleton and Barnett measured the heights of the ionosphere using a BBC short wave broadcast transmitter operating at 770 kHz, located in Bournemouth on the south coast of England. The passive receiver was placed in Oxford at a distance of some 120 km. After that in 1935 Arnold Wilkins and Robert Watson-Watt demonstrated the detection of a bomber aircraft in Daventry, UK, based only on the illumination from a BBC short-wave broadcast station.

A well-known and popular radar technique for Earth surface imaging is the synthetic aperture radar (SAR). The SAR imaging technique has been known since 1951 when Carl Wiley from the Goodyear Aircraft Corporation proposed it for the first time [49,50]. Over the last 70 years, SAR has been intensively developed and many new concepts of image formation such as polarimetry, interferometry, and bistatic/multistatic configuration have been proposed and demonstrated [51,52]. The classical active SAR works in the monostatic configuration where the transmitter and receiver compound one radar device. Nowadays, classical active SAR is at the stage of technological maturity and it is commonly used as a sensor on satellites, aircraft, and unmanned aerial vehicles (UAVs).

But passive technology was then forgotten for several decades, as in the twentieth century only analog signal processing was possible, and the implementation of passive radar using continuous waves was impossible. Nowadays, passive technology is being rediscovered [7,35–37,53,54] due to the rapid development of digital technology and the availability of high computational power platforms. This modern technology allows the use of existing emissions (e.g. FM, DAB, GSM, DVB-T, DVB-S, among others) for object detection. The fundamental advantage of passive technology is its lack of its own emission, so it is impossible to detect a passive radar (receiver) based on radar emission. Another benefit is that there is no need to apply for a frequency allocation or a transmission license. Passive SAR technology is thus a "green" technology which does not pollute the electromagnetic environment.

The concept of passive airborne synthetic aperture radar (PASAR) is a combination of three technologies, namely synthetic aperture radar (SAR), bistatic (or multistatic) radar and passive radar. The PASAR device is an airborne radar that utilizes ground-based transmitters of opportunity for the illumination of ground targets to obtain passive SAR images of them. To obtain a good quality passive SAR image of the ground, a good PASAR illuminator should be used. In this case, the PASAR illuminator should have high power density on the ground, high bandwidth and favorable ambiguity function. A lot of commercial illuminators (FM, GSM, LTE, DVB-T, see Figure 2.17) can be used for the illumination of the ground for PASAR purposes. From Figure 2.17, the best candidate seems to be the DVB-T transmitter, which has relatively high power (up to 1 MW) and reasonable bandwidth (7.6 MHz for a single DVB-T channel), giving a good power budget and fine resolution of up to 20 m [34]. In many countries, it is possible to find illumination of a wider bandwidth, while multiple channels are often transmitted from the same location. This can improve the range resolution of the passive SAR system.

Other PASAR concept descriptions can be found in the literature [34,55], but up to now few preliminary results of experimental trials have been published in the open

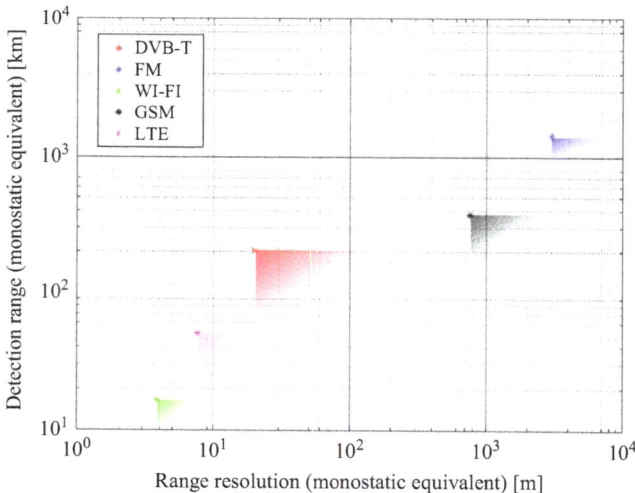

Figure 2.17   Illuminator of opportunity – range and range resolution for different types of terrestrial illuminators

literature. In [34], the authors present a range-Doppler map of the observed ground targets reflections obtained using a DVB-T based passive radar. Such a method is similar to unfocused SAR processing. In [7], the authors present the results of single channel passive SAR processing obtained from a CARABAS SAR system working in a passive mode.

An exemplary PASAR radio-location scenario is depicted in Figure 2.18. The transmitter of opportunity illuminates the scene and also an aircraft. The PASAR receiver consists of two physical radio channels, one for the reference signal and the second for collecting the echo from ground targets. Signals from both channels – reference and imaging – are converted into digital form using ADC converters and passed to the processing unit. It is assumed that both channels are fully synchronized in frequency and phase. The synchronization is not required between the passive radar and the DVB-T transmitter since the SAR image is obtained by methods relying on the correlation of the signal from the reference and surveillance channel.

A more detailed PASAR scenario operating in SAR StripMap mode is presented in Figure 2.19. The passive radar, placed on board the aircraft, consists of two or more antennas and two or more receiving channels. One antenna is directed towards the illuminator (transmitter of opportunity) and receives the reference channel signal. The second antenna is directed towards the observed scene and receives the echo signal (indirect path signal) reflected from the buildings, ground surface, and all objects illuminated by the transmitter and visible to this antenna.

As can be seen, the received signal consists of a few components; some of them are wanted and some not, such as the direct path signal, reflected echoes from stationary targets (multipath signal), echoes from moving objects, and noise. The main task of the PASAR radar is ground image formation, so the echoes from moving objects can be treated as a a clutter. Due to velocity mismatch moving targets in the final passive SAR image will be defocused and appear blurred in the image. Signal processing in PASAR radar is complex and power demanding.

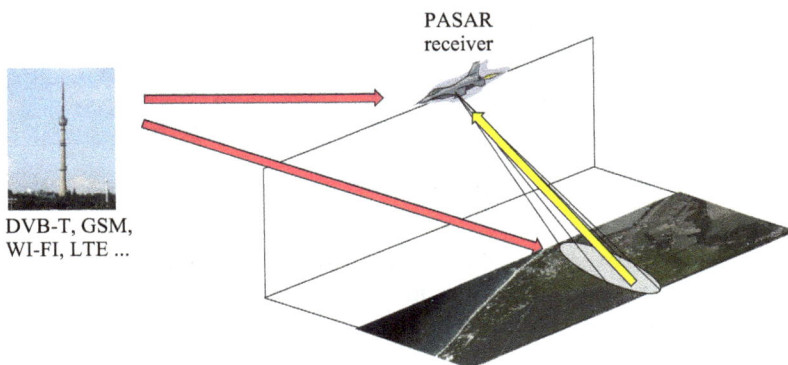

Figure 2.18 *PASAR radiolocation scenario – red arrows corresponds to direct path signal from illuminator of opportunity to PASAR receiver and observed scene, yellow arrow correspond to reflected echo from ground targets*

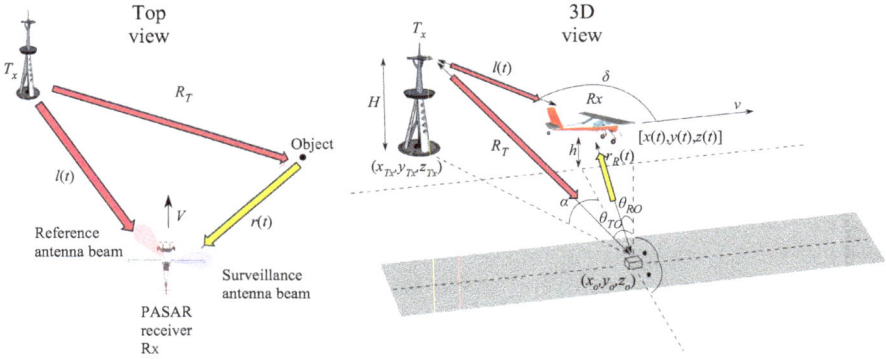

Figure 2.19   PASAR radiolocation scenario – StripMap SAR mode of operation

Figure 2.20   PASAR experiment radiolocation scenario – Płock area in Poland

Up to now some experimental work considering PASAR imaging has been done in several universities and research institutes. A few measurements campaigns have been done at the Warsaw University of Technology, Warsaw, Poland. One measurement campaign took place in December 2014. Measurements were made in the vicinity of the city of Sierpc, Poland. The DVB-T transmitter used as a illuminator of opportunity was located in Rachocin. The GPS coordinates of the transmitter are 52.890N, 19.650E. The PASAR experimental scenario and DVB-T illuminator parameters are shown in Figure 2.20.

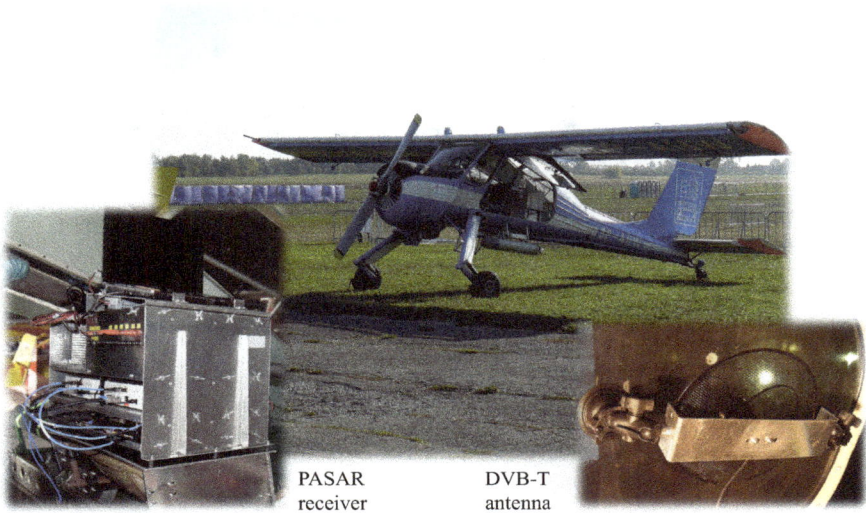

Figure 2.21   PASAR experiment radiolocation scenario – Płock area in Poland

As a passive airborne radar carrier the Polish aircraft PZL-104 called "Wilga" was used. Figure 2.21 presents the aircraft with on-board equipment. The passive radar demonstrator was constructed using COTS – (Commercial Off The Shelf) components such as DVB-T antennas, a USRP N210 as a data acquisition unit, and a portable PC for data recording and processing. As the transmitter was on the opposite side of the imaging area, two DVB-T antennas were mounted on the opposite aircraft windows. In the measurement scenario presented in Figure 2.20, the white line represents the flight path of the aircraft given by the GPS/INS device. According to the GPS device, the speed of the platform during the flight was around 40 m/s and the altitude was around 300 m above the ground. The total travelled distance (length of the red line), shown in Figure 2.20, was around 8 km (200 s × 40 m/s). The distance between the start point of the flight path and the transmitter was around 11 km. The approximate localization of the illuminator is marked in Figure 2.20. The whole area which was processed using an algorithm developed at WUT in order to obtain the SAR images is presented in Figure 2.20. The pixel size of the image was set to 3 m × 3 m. The integration time in azimuth in this case was 4 s (which is around 1.6 km of synthetic aperture) and the integration path is marked with a white line in Figure 2.20. The grazing angle of the illumination from the DVB-T transmitter was around 3.50 for the processed area presented in Figure 2.20.

The resulting passive SAR image is presented in Figure 2.22 with its corresponding Google Earth map image. Most of the targets which are visible in the Google optical image can be easily found in the passive SAR image. Structures such as: the edges of the forest as well as roads, private properties, and even a wind farm are visible. Some of the targets which are visible in the optical image have no representation

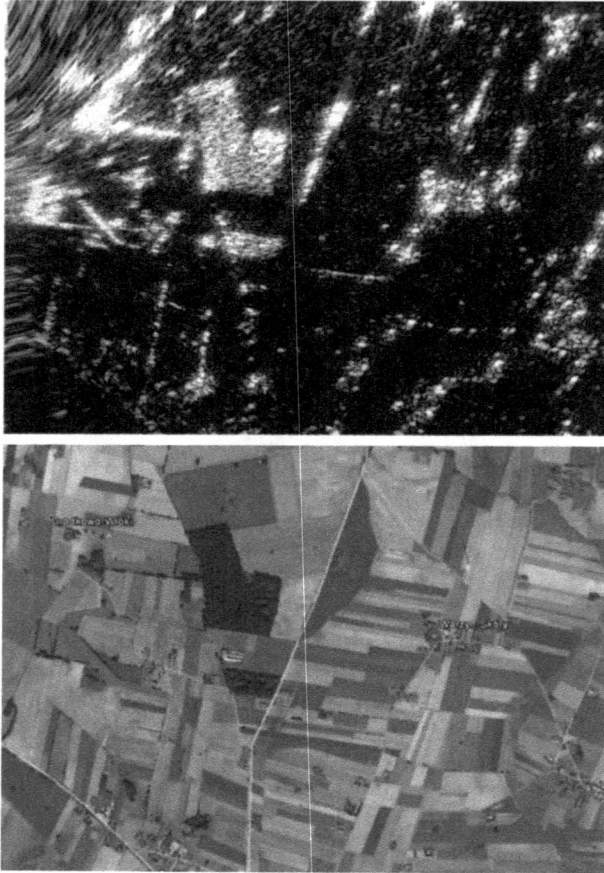

*Figure 2.22   Passive SAR image – top, and it is Google Map optical counterpart –
bottom*

in their SAR counterparts and vice versa. This may be due to various reasons such
as the differing time of the acquisition of both images. An additional reason might
be the strong influence of shadowing between objects. It has to be taken into account
that the grazing angle of the illumination is 3.5°. For such low grazing angles of the
DVB-T illumination the terrain profile also has a strong influence on SAR imaging
capability.

2.2.4   Passive imaging of moving targets (PAISAR)

An emerging application of passive radar on moving platforms is the imaging of targets. Compared to the SAR technique, where imaging is possible due to the platform's
own movement while the imaged scene is stationary, in ISAR (inverse synthetic aperture radar), imaging is possible due to the movement of the target being imaged.

This difference is a source of significant difficulties because the trajectory of the non-cooperating target is unknown so must be precisely estimated and compensated.

Waveforms utilized in passive radar are typically narrowband, so the expected imaging resolution in the range dimension is not high (20 m for a single DVB-T channel [17]). Additionally, the carrier frequencies are low, hence long coherent processing intervals (CPIs) are required to obtain satisfactory cross-range resolution, which makes target trajectory estimation even more challenging. One idea for improving the resolutions is to use a multistatic configuration and increase the bandwidth by the use of multiple transmissions. Figure 2.23 shows an example of multistatic passive ISAR imaging. Several receivers with various aspect angles were then used to form separate images which were noncoherently integrated to form the final image. An outline of the plane model corresponding to the observed target has been added to facilitate the visual identification of the characteristic elements of the image. The

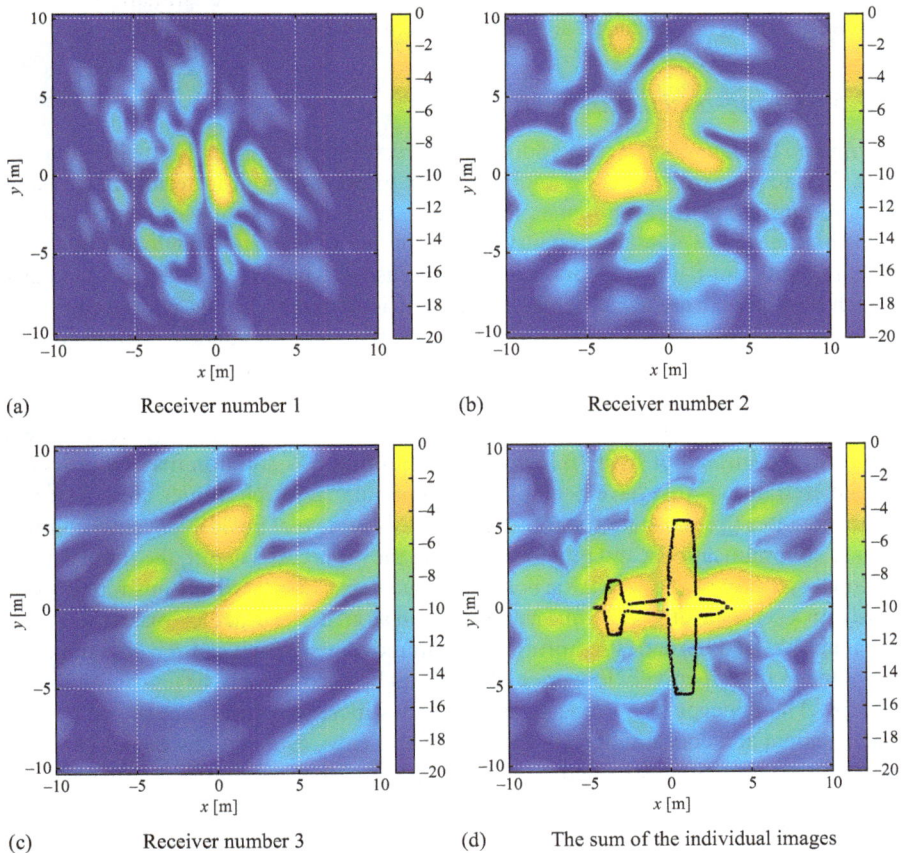

(a)   Receiver number 1

(b)   Receiver number 2

(c)   Receiver number 3

(d)   The sum of the individual images

Figure 2.23   Results of an object imaging for individual receivers and combined image

obtained range resolution of 2.5 m corresponds to 56 MHz which is seven DVB-T channels. In this example, the radar was stationary while the target was moving, but it is expected that similar results can also be obtained for moving radar platforms.

Although the expected quality of ISAR images is far from "photographic," it can be good enough for non-cooperative target recognition (NCTR). In this application, even a simple classification based on target size can be very valuable in some situations.

The literature on ISAR imaging with passive radar on moving platforms is still rather scarce. The signal processing chain of a moving radar must take into account that imaged targets, especially if they are ground targets, often fall into the region of clutter echoes. Therefore, target imaging must be preceded with clutter cancelation and moving target detection stages, hence the natural coexistence of DPCA or STAP with ISAR [56]. A single receiving channel was assumed in the context of detecting and imaging moving targets in SAR. Theoretical backgrounds were presented in [38,39]. The effects of a moving target on SAR images in bistatic radar, the estimation of target motion parameters, and target signature extraction were analyzed in [40].

2.2.5   Drone-based passive radars

Compared to planes with human pilots on board, drones are typically smaller, require less time to prepare for action and are easier to operate both in terms of cost and required infrastructure. Of course, the small size of a drone imposes severe restrictions on the size, weight, and energy consumption of the carried payloads, but such requirements suit passive radars well. Other reasons why drones and passive radars match are that both are low cost and both are difficult to detect, thus a drone equipped with a passive radar can fly unnoticed where an active radar on a bigger platform would be easily detected. Nevertheless, a drone with a passive radar installed is not totally silent due to required communications – needed both to control the drone (unless the drone is autonomous) and to transmit the results of radar operation (unless signals are recorded instead of being transmitted in real time). A solution may be provided by tethered drones that have been gaining in popularity. Such drones are physically connected to a ground station through a tether – a cable that besides physical control, allows communications, and provides power to the drone. As a result, the flight time is no longer limited by the battery on board and data does not need to be transmitted wirelessly. In a sense, a tethered drone can be viewed as an easy-to-deploy mast; therefore, it cannot be used in modes requiring platform movement, such as SAR imaging.

Due to their relatively low price, drones can be used in more or less autonomous groups. The concept of a drone swarm provides increased resistance at the cost of management and coordination complexity. In this case, even if a few devices are lost during a mission, the rest can continue their task. Another possibility is to use several coordinated drones, each with a receiver, to form a large antenna array. This would not be achievable with a single drone for FM transmission, for example, due to the small size of a drone compared to the wavelength in the VHF band [57].

Concepts of drones as a platform for passive radar have appeared in the context of various applications. The advantage of the greater coverage of a drone-borne passive

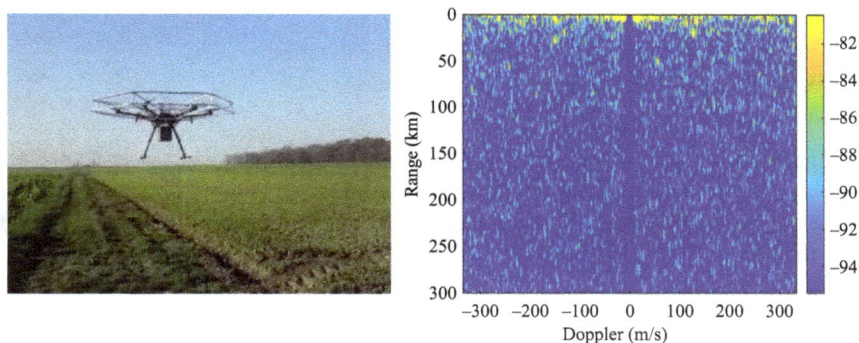

Figure 2.24 Photo of the drone used and example of obtained range-Doppler spectrum in the experiments conducted by ONERA (courtesy of Dominique Poullin)

radar, compared to a ground based one, led to the concept of using it to detect intruding drones. This was described and analyzed in [58] where positive results were obtained. The concept of a passive multistatic drone-borne weather radar was analyzed in [59], where it was shown that such a system would be able to retrieve a 3D wind field.

Despite all the advantages, there are not many working passive radar systems installed on drones and described in open literature (e.g., the concept of weather radar [59], as of today, was implemented only as ground-based). Some reasons for this were discovered during the project DORADA conducted in the French aerospace lab ONERA [57,60]. It turned out that both fixed-wing and multi-rotor drones have their disadvantages. Fixed-wing drones require runways for starting and landing and are more cumbersome to operate, particularly in the vicinity of an airport. On the other hand, in the case of easier to manage multi-rotor drones, signals received by wide-angle antennas were severely distorted. These distortions manifested themselves in strong interfering terms at the bistatic range close to zero and were likely caused by multipaths from the drone's rotating blades. For this reason, to obtain the reference signal for clutter cancellation, a separate ground-based receiver was used in the experiment. A photograph of the drone used and an example of the resulting range-Doppler map are presented in Figure 2.24. In this experiment, the drone was stationary; therefore, clutter was not widely spread in Doppler and could be removed using the Extensive Cancellation Algorithm (ECA).

2.3 Technologies

The modern passive radar is a software defined radar, a digital device with a limited front-end. Generally it consists of several major components:

1. Analog front end
2. Software defined receivers

3. High powered computer for signal processing
4. High powered computer for tracking and localization
5. Communication and synchronization system
6. Man–machine interface

The analog front-end consists of a set of receiving antennas, band-pass filters, low noise amplifiers, and a signal conditioning unit. Despite simple structure, it has to have a very low noise figure, very good linearity, and high dynamic range – preferably more than 80 dB.

The multichannel software defined receiver has to digitize the input signal at selected frequencies, so it should be equipped with high-speed ADC converters with good linearity, a high number of effective bits (preferable more than 12), extremely low jitter (below 1 ps), and high-efficiency data streaming.

The signal and data processing computers have to posses sufficiently high computational power to be able to process the signal and data in real time. Nowadays, advanced GPUs are usually used to exploit the fully parallel processing concept.

Often, several receivers work in a coherent mode, therefore adequate synchronization within the site and multisite is required. For this purpose, GPS is often used, but requirements are far beyond the GPS constraints nowadays so more advanced solutions are being developed. In this chapter, you will find considerations on contemporary multichannel receivers and synchronization means.

2.3.1 Multi-channel receivers and signal acquisition

Signal acquisition for PCL processing requires signal acquisition from at least two RF receiving channels: one for surveillance and the other for the reference path. More channels are necessary if advanced algorithms (i.e. with digital beamforming or direction of arrival estimation) are employed.

The signal acquisition subsystem for a PCL application usually consists of a host PC (which may be a typical notebook or desktop PC with sufficient performance on I/O ports) and a multichannel digital RF receiver. To realize this task without designing custom hardware, it is possible to employ a commercial off the shelf (COTS) FPGA-based board equipped with a sufficient number of analog to digital converters (ADCs) and fast enough I/O peripherals. In Figure 2.25, we can see a few examples of such COTS boards with FPGA chips:

- ADC-SoC board [61]
- Nuand BladeRF 2.0 micro [62]
- STEMLab 125-14 (Red Pitaya) [63]
- Xilinx Zynq UltraScale+ RFSoC platform [64]

This approach gives full control of the signal processing path, synchronization and data transfer by means of an FPGA configuration. However, there are also a few drawbacks that have to be pointed out. The design, implementation, and testing of the FPGA configuration in a classical way require specialized knowledge and is a time consuming process. The application of fast data transfer, for example, by means of a PCI Express connection or 10 Gbit/s Ethernet, needs either purchase of commercial

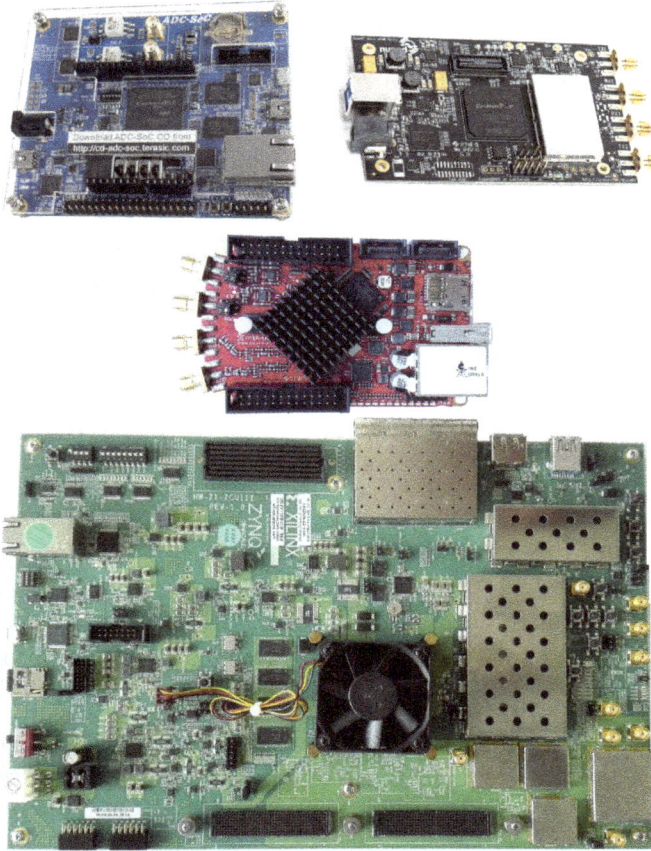

Figure 2.25 FPGA development boards suitable for PCL signal acquisition

intellectual property (IP) blocks and software drivers, or spending months designing and testing one's own solutions. Moreover, in spite of integrated ADCs, an analog front-end (AFE) usually needs to be designed and built, with the exception of only a few boards with FPGA (e.g., the Nuand BladeRF 2.0 micro board). This functional block is required to allow the digitizing of weak RF signals from the antennas and has high requirements in terms of analog performance (i.e., the noise level, gain, linearity, crosstalk, etc.)

It is also possible to use COTS software-defined radio (SDR) receivers for signal acquisition. This approach relies on already designed and tested functional blocks and drivers. It does not give the potential of full control of the signal processing pipeline, but it reduces to the minimum the costs and workload required for the implementation of AFE and fast data transfer to the host PC.

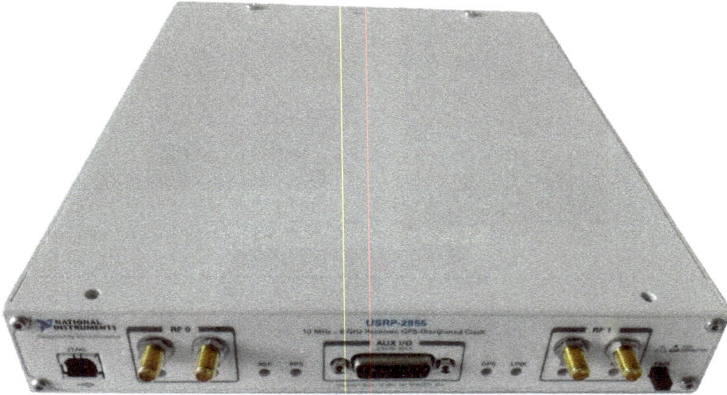

Figure 2.26   USRP-2955: four-channel SDR receiver

For high-performance applications, where the broadband signal from four or more channels needs to be acquired, the USRP-2955 SDR platform from National Instruments (NI) can be employed. The USRP-2955, depicted in Figure 2.26, is a scalable platform for designing and deploying wireless communications systems [65]. This device is based on a four-channel NI X310 mainboard and provides four SDR receivers operating in the frequency range from 10 MHz to 6 GHz with 80 MHz of available bandwidth. The maximum sample rate for each channel is 100 MHz with a 14-bit ADC resolution. The receiving channels are equipped with variable-gain low-noise amplifiers (LNAs) with a gain range selectable from 0 to 95 dB. Only external bandpass filters (to suppress out-of-band interference) may be required to obtain the signal acquisition directly from the antennas. All the channels can operate synchronously. It is possible to achieve a four-channel coherent signal acquisition with a sustainable sample rate of 50 Msamp/s (i.e. 200 Msamp/s in total) provided that the connection with the host PC is based on PCIe x4 or 10 Gbit Ethernet. Such a system could support different frequency bands with signals selected for PCL scene illumination (e.g., FM, DVB-T, DAB, 3G, LTE, 5G).

A general block diagram of a four-channel PCL system with an SDR receiver is shown in Figure 2.27. It must be noted, however, that in the simplest configuration of a PCL system, only two receiving channels are sufficient (i.e., the reference and the surveillance channel). All tuners of the receiver operate synchronously at the same center frequency. The RF inputs of the receiver are connected to the receiving antennas via AFE blocks that contain the LNAs and bandpass filters suitable for the frequency and bandwidth of the illuminating signal.

To achieve coherent signal acquisition from all the RF input channels, additional connections may be necessary to be made to share the local oscillator (LO) signal among all tuners in the receiver. The digital port of the receiver is connected to the host PC via PCI-Express, Gigabit Ethernet, or a USB port, depending on the SDR

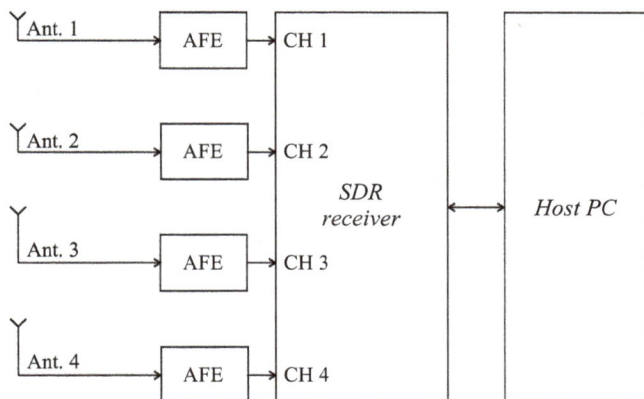

Figure 2.27   Simplified architecture of four-channel PCL receiver

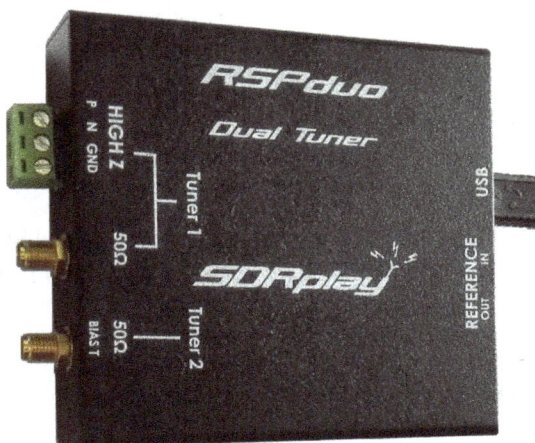

Figure 2.28   RSPduo: two-channel USB-connected SDR receiver

receiver's architecture. The host machine is responsible for the configuration of the receiver, acquisition of the digital streams, and storage of the incoming data on SSD drives (the total bandwidth of the acquired signals can be balanced among several drives of the host).

When a PCL system has to be installed on a space-limited platform, for example in a drone, smaller SDR receivers need to be utilized. An example device of this class is the RSPduo [66] from SDRplay, shown in Figure 2.28. The RSPduo is small-sized, low-power, two-tuner SDR receiver with integrated complete AFE for each tuner. The frontends include selectable bandpass filters, optional notch filters, and variable-gain LNAs. The receiver can operate with 14-bit performance, provided that the sample rate is limited to 2 MHz, or with 8-bit performance and the sample rate elevated to 10 MHz. The connection with a host PC is provided by a USB 2.0 port for both data transfer and power supply. Due to low power operation and low weight, this kind of SDR receiver suits exceptionally well the needs of small, mobile PCL platforms [67]. These devices have additional reference IN/OUT ports for sharing the clock and a PLL reference, in case more than two receiving channels are necessary. However, due to the lack of a common trigger signal and the uncertainty of the initial phase shift between the channels [67], additional hardware and software routines are needed for synchronization [68].

2.3.2   Signal synchronization

When acquiring signals from several SDR receivers in multi-channel mode, the need for the channels' synchronization arises. The signals from all channels in the PCL system need to be synchronized in time to make possible clutter cancelation, cross-ambiguity function calculation and, finally, PCL detection of the targets. The time synchronization of the signals received in the same device is usually granted by its hardware design [65,67]. However, when more signals, coming from different devices, need to be processed, some means of synchronization are necessary. We can point to three basic methods applied in this area:

- use of GPS receivers as the reference source,
- correlation of the signals with a crosstalk component,
- training sequences injection to the received signals.

A GPS-disciplined clock option is available in the advanced SDR receivers [65] to obtain the reference Pulse Per Second (PPS) and 10 MHz signals coherent with the atomic reference clock of the GPS. This method is performed prior to signal acquisition. It allows one to synchronize the sampling rate, PLL reference, and the trigger signals of different receivers, also installed in distant locations. It requires that all synchronized receivers are supported by additional GPS antennas and can receive signals from the GPS satellites.

The correlation method may be applied when the receiving antennas are installed relatively close to each other. It is performed after signal acquisition. When the signal propagation delay between the antennas is much smaller than the sampling interval of the receivers, it may be assumed that the acquired signals should be time-aligned in the digital domain, i.e. their time delay is equal to zero samples. If all the antennas receive the signal component coming directly from the illuminating transmitter, the correlation between these signals allows one to find their real time shift [67]. After that, the shift can be corrected in the digital domain to obtain time-aligned signals. This method, however, does not allow one to compensate phase offsets introduced in several channels of the receiver [68]. This may be essential in some applications, for example, in direction of arrival (DoA) estimation.

The last of the considered methods needs additional hardware for periodic generation of the special training sequences and for injection of these synthetic signals to the RF inputs of the receiver. An example solution is depicted in Figure 2.29. In

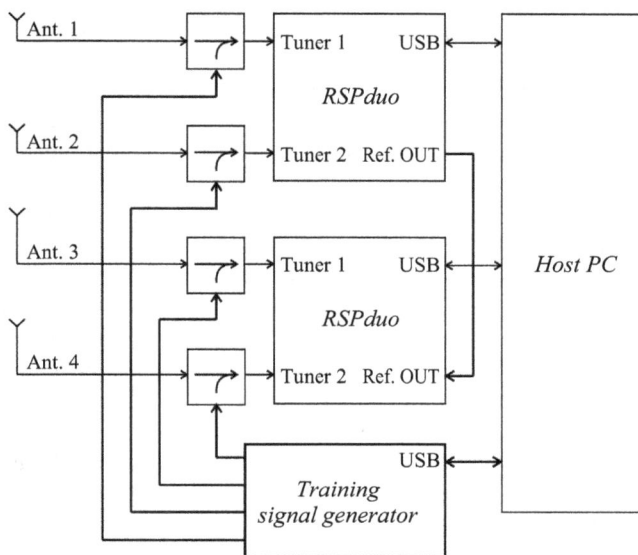

Figure 2.29   Four-channel PCL signal acquisition system with two RSPduo devices

this system, two RSPduo devices are connected to the same host PC and share the reference clock to synchronize PLLs and sample rate generators. However, no trigger signal sharing is possible in this architecture [67]. Therefore, an additional training signal generator is employed and coupled with the receivers' inputs via directional couplers (in a PCL application, it is essential to minimize the crosstalk between the receiving channels). The synchronization is performed after signal acquisition. The training signals are processed in the digital domain, just like in all modern digital communication systems, which allows one to estimate and compensate time delays and even phase offsets between the channels [68].

The receivers in this setup do not rely on GPS signals that make the system more reliable. However, the main drawback of this approach is the periodical interference caused by the injection of the training sequences which spoils the PCL detection images during short time intervals, as demonstrated in [41]. To avoid this interference, the training sequences should be removed from the acquired signals by means of adaptive filtration. However, when the training sequence repeat interval is much longer than the sequence duration, the interferences are sparse in time and thus may be neglected in basic applications [41], by skipping the distorted images of the PCL detection.

2.4   Conclusions

Mobile passive radar is an emerging technology which should be further explored in the next decade. The passive radar does not emit anything and relies on existing emissions. The world of communication is changing rapidly, and radar technology has followed the new developments. It is obvious that analog TV and AM radio are obsolete technologies; however, the present technology of FM, DAB, DVB-T, and GSM may also become obsolete soon. Unfortunately, the 5G and 6G technology is no longer attractive for long-range mobile passive radar as it is unpredictable and has low power. However, for short-range applications, the 5G/6G illumination can give mobile and ground-based passive radar new opportunities for high-resolution passive SAR/ISAR imaging due to the wide bandwidth of transmitters [69–73]. The application for mobile platforms might be small PCL receivers mounted on small drones for surveillance in short ranges in urban areas. For future mobile passive radar applications, the new illuminations can be studied. One such illumination is a STARLINK satellites constellation [74–76]. This illuminator can give future global coverage for mobile passive radar operating in short and medium ranges. Another global sources of illumination signals are GNSS (Global Navigation Satellite Systems) [77,78]. However, the authors strongly believe that classical broadcasting will stay with us sufficiently long enough that mobile passive radar technology will be introduced in the near future to increase safety in civil applications and military superiority in the field of defence.

References

[1]   Gould D, Pollard R, Sarno C, *et al.* A multiband passive radar demonstrator. In: *2006 International Radar Symposium.* Krakow, Poland: IEEE; 2006. p. 1–4. Available from: http://ieeexplore.ieee.org/ document/4338144/.

[2]   Howland PE, Maksimiuk D, and Reitsma G. FM radio based bistatic radar. *IEE Proceedings – Radar, Sonar and Navigation.* 2005;152(3):107. Available from: https://digital-library.theiet.org/ content/journals/10.1049/ip-rsn_20045077.

[3]   Di Lallo A, Farina A, Fulcoli R, *et al.* Design, development and test on real data of an FM based prototypical passive radar. In: *2008 IEEE Radar Conference.* Rome, Italy: IEEE; 2008. p. 1–6. Available from: http://ieeexplore.ieee.org/ document/4720985/.

[4]   Malanowski M, Kulpa K, and Misiurewicz J. PaRaDe – PAssive RAdar DEmonstrator family development at Warsaw University of Technology. In: *2008 Microwaves, Radar and Remote Sensing Symposium.* Kiev, Ukraine: IEEE; 2008. p. 75–78. Available from: http://ieeexplore.ieee.org/ document/4669549/.

[5]   Poullin D and Flecheux M. Recent progress in Passive Coherent Location (PCL) concepts and technique in France using DAB or FM broadcasters. In: *2008 IEEE Radar Conference.* Rome, Italy: IEEE; 2008. p. 1–5. Available from: http://ieeexplore.ieee.org/document/ 4721009/.

[6]   Griffiths H and Baker C. The signal and interference environment in passive bistatic radar. In: *2007 Information, Decision and Control.* Adelaide, Australia: IEEE; 2007. p. 1–10. Available from: http://ieeexplore.ieee.org/ document/4252468/.

[7]   Ulander LMH, Frölind PO, Gustavsson A, *et al.* VHF/UHF bistatic and passive SAR ground imaging. In: *2015 IEEE Radar Conference (RadarCon)*; 2015. p. 669–673.

[8]   Berger CR, Demissie B, Heckenbach J, *et al.* Signal processing for passive radar using OFDM waveforms. *IEEE Journal of Selected Topics in Signal Processing.* 2010;4(1):226–238. Available from: http://ieeexplore.ieee.org/document/5393298/.

[9]   Coleman C and Yardley H. DAB based passive radar: performance calculations and trials. In: *2008 International Conference on Radar.* Adelaide: IEEE; 2008. p. 691–694. Available from: https://ieeexplore.ieee.org/document/4654009.

[10] Yardley H and Coleman C. Passive bistatic radar based on target illuminations by digital audio broadcasting. *IET Radar, Sonar & Navigation.* 2008;2(5):366–375. Available from: https://digital-library.theiet.org/content/journals/10.1049/iet-rsn_20080019.

[11] Coleman CJ, Watson RA, and Yardley H. A practical bistatic passive radar system for use with DAB and DRM illuminators. In: *2008 IEEE Radar Conference.* Rome, Italy: IEEE; 2008. p. 1–6. Available from: http://ieeexplore.ieee.org/document/4721007/.

[12] Howland PF. Target tracking using television-based bistatic radar. *IEE Proceedings – Radar, Sonar and Navigation.* 1999;146(3):166. Available from: https://digital-library.theiet. org/content/journals/10.1049/ip-rsn_19990322.

[13] Cai M, He F, and Wu L. Application of UKF algorithm for target tracking in DTV-based passive radar. In: *2009 2nd International Congress on Image and Signal Processing.* Tianjin, China: IEEE; 2009. p. 1–4. Available from: http://ieeexplore.ieee.org/document/5304230/.

[14] Saini R and Cherniakov M. DTV signal ambiguity function analysis for radar application. *IEE Proceedings – Radar, Sonar and Navigation.* 2005;152(3):133. Available from: https://digital-library.theiet.org/content/journals/10.1049/ip-rsn_20045067.

[15] Guo H, Coetzee S, Mason D, *et al.* Passive radar detection using wireless networks. In: *IET International Conference on Radar Systems 2007.* Edinburgh, UK: IEE; 2007. p. 167–167. Available from: https://digital-library.theiet.org/content/conferences/ 10.1049/cp_20070643.

[16] Hongbo Sun, Tan DKP, Yilong Lu, *et al.* Applications of passive surveillance radar system using cell phone base station illuminators. *IEEE Aerospace and Electronic Systems Magazine.* 2010;25(3):10–18. Available from: http://ieeexplore.ieee.org/document/5463951/.

[17] Bączyk MK, Samczyński P, and Kulpa K. Passive ISAR imaging of air targets using DVB-T signals. In: 2014 *IEEE Radar Conference*; 2014. p. 502–506.

[18] Tan DKP, Sun H, Lu Y, *et al.* Passive radar using Global System for Mobile communication signal: theory, implementation and measurements. *IEE Proceedings – Radar, Sonar and Navigation.* 2005;152(3):116. Available from: https://digital-library.theiet.org/content/journals/10.1049/ip-rsn_20055038.

[19] Chetty K, Smith G, Guo H, *et al.* Target detection in high clutter using passive bistatic WiFi radar. In: *2009 IEEE Radar Conference.* Pasadena, CA: IEEE; 2009. p. 1–5. Available from: http://ieeexplore.ieee.org/document/4976964/.

[20] Baker CJ, Griffiths HD, and Papoutsis I. Passive coherent location radar systems. Part 2: waveform properties. *IEE Proceedings – Radar, Sonar and Navigation*. 2005;152(3):160. Available from: https://digital-library.theiet.org/content/journals/10.1049/ip-rsn_20045083.

[21] Rzewuski S, Wielgo M, Kulpa K, *et al*. Multistatic passive radar based on WIFI – results of the experiment. In: *2013 International Conference on Radar*. Adelaide, Australia: IEEE; 2013. p. 230–234. Available from: http://ieeexplore.ieee.org/document/6651990/.

[22] Falcone P, Colone F, and Lombardo P. Localization of moving targets with a passive radar system based on WiFi transmissions. In: *IET International Conference on Radar Systems (Radar 2012)*. Glasgow, UK: Institution of Engineering and Technology; 2012. p. 14. Available from: https://digital-library.theiet.org/content/conferences/ 10.1049/cp.2012.1561.

[23] Falcone P, Colone F, and Lombardo P. Potentialities and challenges of WiFi-based passive radar. *IEEE Aerospace and Electronic Systems Magazine*. 2012;27(11):15–26. Available from: http://ieeexplore.ieee.org/document/6380822/.

[24] Brown J, Woodbridge K, Stove A, *et al*. Air target detection using airborne passive bistatic radar. *Electronics Letters*. 2010;46(20):1396. Available from: https://digital-library.theiet.org/content/journals/10.1049/el.2010.1732.

[25] Dawidowicz B and Kulpa KS. Experimental results from PCL radar on moving platform. In: *2008 International Radar Symposium*. Wroclaw, Poland: IEEE; 2008. p. 1–4. Available from: http://ieeexplore.ieee.org/document/4585777/.

[26] Tan DKP, Lesturgie M, Sun H, *et al*. Target detection performance analysis for airborne passive bistatic radar. In: *2010 IEEE International Geoscience and Remote Sensing Symposium*. Honolulu, HI: IEEE; 2010. p. 3553–3556. Available from: http://ieeexplore.ieee.org/ document/5652159/.

[27] Gromek D, Kulpa K, and Samczyński P. Experimental results of passive SAR imaging using DVB-T illuminators of opportunity. *IEEE Geoscience and Remote Sensing Letters*. 2016;13(8):1124–1128. Available from: http://ieeexplore.ieee.org/document/7492308/.

[28] Cetin M and Lanterman AD. Region-enhanced passive radar imaging. *IEE Proceedings – Radar, Sonar and Navigation*. 2005;152(3):185. Available from: https://digital-library.theiet.org/ content/journals/10.1049/ip-rsn_20045019.

[29] Wu Q, Zhang YD, Amin MG, *et al*. High-resolution passive SAR imaging exploiting structured Bayesian compressive sensing. *IEEE Journal of Selected Topics in Signal Processing*. 2015;9(8):1484–1497. Available from: http://ieeexplore.ieee.org/document/7270255/.

[30] Willis NJ. *Bistatic Radar*. Edison, NJ: SciTech Publishing; 2005.

[31] Dawidowicz B, Kulpa KS, and Malanowski M. Suppression of the ground clutter in airborne PCL radar using DPCA technique. In: *2009 European Radar Conference (EuRAD)*. 2009; p. 306–309.

[32] Malanowski M. Comparison of adaptive methods for clutter removal in PCL radar. In: *2006 International Radar Symposium*. Krakow, Poland: IEEE; 2006. p. 1–4. Available from: http://ieeexplore.ieee.org/document/4338044/.

[33] Gustavsson A, Frölind PO, Hellsten H, *et al.* The airborne VHF SAR system CARABAS. In: *Proceedings of IGARSS '93 – IEEE International Geoscience and Remote Sensing Symposium*, vol. 2; 1993. p. 558–562.

[34] Gromek D, Samczyński P, Kulpa K, *et al.* Initial results of passive SAR imaging using a DVB-T based airborne radar receiver. In: *2014 11th European Radar Conference*; 2014. p. 137–140.

[35] Evers A and Jackson JA. Experimental passive SAR imaging exploiting LTE, DVB, and DAB signals. In: *2014 IEEE Radar Conference*; 2014. p. 0680–0685.

[36] Ma H, Antoniou M, and Cherniakov M. Passive GNSS-based SAR resolution improvement using joint Galileo E5 signals. *IEEE Geoscience and Remote Sensing Letters*. 2015;12(8):1640–1644.

[37] Krieger G, Fiedler H, Houman D, *et al.* Analysis of system concepts for bi- and multi-static SAR missions. In: *IGARSS 2003. 2003 IEEE International Geoscience and Remote Sensing Symposium. Proceedings* (IEEE Cat. no. 03CH37477). vol. 2; 2003. p. 770–772.

[38] Wacks S and Yazıcı B. Passive synthetic aperture hitchhiker imaging of ground moving targets – Part 1: image formation and velocity estimation. *IEEE Transactions on Image Processing*. 2014;23(6):2487–2500.

[39] Wacks S and Yazıcı B. Passive synthetic aperture hitchhiker imaging of ground moving targets – Part 2: performance analysis. *IEEE Transactions on Image Processing*. 2014;23(9):4126–4138.

[40] Duman K and Yazıcı B. Moving target artifacts in bistatic synthetic aperture radar images. *IEEE Transactions on Computational Imaging*. 2015;1(1): 30–43.

[41] Mazurek G, Kulpa K, Malanowski M, *et al.* Experimental seaborne passive radar. *Sensors*. 2021;21(6). Available from: https://www.mdpi.com/1424-8220/21/6/2171.

[42] Turkish Airlines 9 (THY9). Accessed: 2021-03-25. https://flightaware.com/live/flight/THY9.

[43] SAS 778/SK778. Accessed: 2021-03-25. https://flightaware.com/live/flight/SAS778.

[44] Dawidowicz B, Kulpa KS, Malanowski M, *et al.* DPCA detection of moving targets in airborne passive radar. *IEEE Transactions on Aerospace and Electronic Systems*. 2012;48(2):1347–1357.

[45] Wojaczek P, Colone F, Cristallini D, *et al.* Reciprocal-filter-based STAP for passive radar on moving platforms. *IEEE Transactions on Aerospace and Electronic Systems*. 2019;55(2):967–988.

[46] Wojaczek P, Cristallini D, O'Hagan DW, *et al.* A three-stage inter-channel calibration approach for passive radar on moving platforms exploiting the minimum variance power spectrum. *Sensors*. 2021;21(1). Available from: https://www.mdpi.com/1424-8220/21/1/69.

[47] Blasone GP, Colone F, Lombardo P, *et al.* Passive radar DPCA schemes with adaptive channel calibration. *IEEE Transactions on Aerospace and Electronic Systems*. 2020;56(5):4014–4034.

[48] Blasone GP, Colone F, Lombardo P, *et al.* Passive radar STAP detection and DoA estimation under antenna calibration errors. *IEEE Transactions on Aerospace and Electronic Systems.* 2021;57(5):2725–2742.

[49] Willey C. Synthetic aperture radars. *IEEE Transactions on AeroSpace and Electronic Systems.* 1985;3:440–443.

[50] Carrara WG, Goodman RS, and Majewski RM. In: *Spotlight Synthetic Aperture Radar.* Artech House; 1995.

[51] Papathanassiou K, Kugler F, and Hajnsek I. Exploring the potential of Pol-InSAR techniques at X-band first results and experiments from TanDEM-X. In: *2011 3rd International Asia-Pacific Conference on Synthetic Aperture Radar (APSAR)*; 2011. p. 1–2.

[52] Griffiths HD, Baker C, Baubert J, *et al.* Bistatic radar using satellite-borne illuminators. In: *RADAR 2002.* 2002. p. 1–5.

[53] Prati C, Rocca F, Giancola D, *et al.* Passive geosynchronous SAR system reusing backscattered digital audio broadcasting signals. *IEEE Transactions on Geoscience and Remote Sensing.* 1998;36(6):1973–1976.

[54] Gutierrez Del Arroyo JR, and Jackson JA. WiMAX OFDM for passive SAR ground imaging. *IEEE Transactions on Aerospace and Electronic Systems.* 2013;49(2):945–959.

[55] Kulpa K, Malanowski M, Samczyński P, *et al.* The concept of airborne passive radar. In: *2011 Microwaves, Radar and Remote Sensing Symposium*; 2011. p. 267–270.

[56] Gelli S, Bacci A, Martorella M, *et al.* Clutter suppression and high-resolution imaging of noncooperative ground targets for bistatic airborne radar. *IEEE Transactions on Aerospace and Electronic Systems.* 2018;54(2):932–949.

[57] Gabard B, Casadebaig L, Deloues T, *et al.* A UAV airborne passive digital radar for aerial surveillance. In: *2018 15th European Radar Conference (EuRAD)*; 2018. p. 162–165.

[58] Vinogradov E, Kovalev DA, and Pollin S. Simulation and detection performance evaluation of a UAV-mounted passive radar. In: *2018 IEEE 29th Annual International Symposium on Personal, Indoor and Mobile Radio Communications (PIMRC)*; 2018. p. 1185–1191.

[59] Byrd A, Palmer R, and Fulton C. Concept for a passive multistatic UAV-borne weather radar. In: *2018 IEEE Radar Conference (RadarConf18)*; 2018. p. 845–850.

[60] Gabard B, Wasik V, Rabaste O, *et al.* Airborne targets detection by UAV-embedded passive radar. In: *2020 17th European Radar Conference (EuRAD)*; 2021. p. 346–349.

[61] ADC-SoC User Manual. Accessed: 2021-03-25. https://www.terasic.com.tw/cgi-bin/page/archive_download.pl?Language=English&No=1061&FID=217 0285f16d7e05879607fccd2f31574.

[62] bladeRF 2.0 micro. Accessed: 2021-03-25. https://www.nuand.com/bladerf-2-0-micro.

[63] Red Pitaya STEMlab board 125-14. Accessed: 2021-03-25. https://www.redpitaya.com/194/Red%20Pitaya%20STEMlab%20board%20125-14.

[64] Zynq UltraScale+ RFSoC. Accessed: 2021-03-25. https://www.xilinx.com/products/silicon-devices/soc/rfsoc.html.

[65] USRP-2955 10 MHz to 6 GHz Tunable RF Receiver. Accessed: 2021-03-25. https://www.ni.com/pdf/manuals/376707a.pdf.

[66] RSPduo Dual Tuner 14-bit SDR. Accessed: 2021-03-25. https://www.sdrplay.com/wp-content/uploads/2018/05/RSPduoDatasheetV0.6.pdf.

[67] Mazurek G. Lightweight SDR platform for passive coherent location applications. In: *2019 Signal Processing Symposium (SPSympo)*; 2019. p. 39–44.

[68] Mazurek G and Rytel-Andrianik R. Pilot-based calibration of dual-tuner SDR receivers. In: *2020 28th European Signal Processing Conference (EUSIPCO)*; 2021. p. 1971–1975.

[69] Samczyński P, Abratkiewicz K, Płotka MP, *et al.* 5G network-based passive radar. *IEEE Transactions on Geoscience and Remote Sensing*. 2022;60;1–9.

[70] Kanhere O, Goyal S, Beluri M, and Rappaport TS. Target localization using bistatic and multistatic radar with 5G NR waveform. In: *2021 IEEE 93rd Vehicular Technology Conference (VTC2021-Spring)*, 2021. p. 1–7.

[71] Baquero Barneto C, Riihonen T, Turunen M, *et al.* Full-duplex OFDM radar with LTE and 5G NR waveforms: challenges, solutions, and measurements. *IEEE Transactions on Microwave Theory and Techniques*. 2019;67(10):4042–4054.

[72] Ai X, Zhang L, and Zheng Y. Passive detection experiment of UAV based on 5G new radio signal. In: *2021 Photonics and Electromagnetics Research Symposium (PIERS)*, 2021. p. 2124–2129.

[73] Maksymiuk R, Abratkiewicz K, Samczyński P, and Płotka M. Renyi entropy-based adaptive integration method for 5G-based passive radar drone detection. *Remote Sensing*. 2022;14(23):6143. https://www.mdpi.com/2072-4292/14/23/6146.

[74] Sayin A, Cherniakov M, and Antoniou M. Passive radar using Starlink transmissions: a theoretical study. In: *2019 20th International Radar Symposium (IRS)*, 2019. p. 1–7.

[75] Blázquez-García R, Ummenhofer M, Cristallini D, and O'Hagan D. Passive radar architecture based on broadband LEO communication satellite constellations. In: *2022 IEEE Radar Conference (RadarConf22)*, 2022. p. 1–6.

[76] Gomez-Del-Hoyo P, Gronowski K, and Samczyński P. The STARLINK-based passive radar: preliminary study and first illuminator signal measurements. In: *2022 23rd International Radar Symposium (IRS)*, 2022. p. 350–355.

[77] Gronowski K, Samczyński P, Stasiak K, and Kulpa K. First results of air target detection using single channel passive radar utilizing GPS illumination. In: *2019 IEEE Radar Conference (RadarConf)*, 2019. p. 1–6.

[78] Gomez-del Hoyo P, del Rey-Maestre N, Jarabo-Amores M-P, Mata-Moya D, and Benito-Ortiz M-d-C. Improved 2D ground target tracking in GPS-based passive radar scenarios. *Sensors*. 2022;22(5):1724. https://www.mdpi.com/1424-8220/22/5/1724.

Chapter 3
Passive radar illuminators of opportunity
Stephen Paine[1] and Christof Schüpbach[2]

3.1 General transmitter characteristics

There are many different illuminators of opportunity (IoO)s that can be exploited for use in passive radar (PR). These IoOs include but are not limited to terrestrial transmitters such as amplitude modulated (AM) radio, frequency modulated (FM) radio, analogue and digital TV, WiFi nodes and cellular base stations, as well as space-borne transmitters such as satellite television broadcasts, global positioning system (GPS) satellites and various other communication satellites such as Starlink. While there are many different transmitters, not all of them are necessarily considered desirable for use in PR.

For an IoO to be considered "desirable," three main conditions need to be met [1]:

- The transmit power and antenna pattern must be sufficient for the desired coverage.
- The modulation bandwidth of the illuminating signal should be sufficient to meet the desired range resolution.
- The signal structure should not induce range or Doppler ambiguities.

3.1.1 Power and coverage

The coverage of a PR can be inferred by observing the bi-static radar equation for computing the expected signal-to-noise ratio (SNR) under the assumption of free-space propagation:

$$\text{SNR} = \frac{P_R}{P_n} = \frac{P_T G_T}{4\pi R_T^2} \cdot \sigma_b \cdot \frac{1}{4\pi R_R^2} \cdot \frac{G_R \lambda^2}{4\pi} \cdot \frac{G_P}{kT_0 BF}, \tag{3.1}$$

where P_R, P_n and P_T denote the received signal power, the receiver noise power and the transmit power, G_T and G_R the antenna gains of the transmitter and the receiver, σ_b the bi-static radar cross section, λ the wavelength of the signal, k the Boltzmann constant,* T_0 the noise temperature of the receiver, B the bandwidth of the signal and F the noise factor of the receiver. The factor G_P is the processing gain resulting from

[1]University of Cape Town, South Africa
[2]armasuisse Science and Technology, Switzerland
*$1.38064852 \times 10^{-23}$ m$^2 \cdot$ kg $\cdot$ s$^{-2} \cdot$ K^{-1}

integrating over a processing interval. The term R_R denotes the distance between the target and the receiver and R_T is the distance between the transmitter and the target. Since the SNR is proportional to the factor $1/(R_R^2 R_T^2)$, it has a minimum value when both distances are equal and becomes high when the target is either close to the transmitter or the receiver.

For reliable detection, the target signal would have to be approximately $10-15$ dB above the noise. To get a rough estimate of the detection range, we can thus set $R_T = R_R = R$ and solve the radar equation for R:

$$R = \sqrt[4]{\frac{P_T G_T}{SNR} \cdot \frac{\sigma_b}{4\pi} \cdot \left(\frac{\lambda}{4\pi}\right)^2 \cdot \frac{G_R G_P}{kT_0 BF}}. \tag{3.2}$$

Keeping everything at the receiver-end constant, it becomes clear that the most significant parameter for the achievable coverage is the transmitter's effective radiated power (ERP) which is a result of the product $P_T G_T$. Note that this also implies that the coverage of a PR is also dependant on the antenna pattern G_T.

3.1.2   Bandwidth and delay-resolution

Additional examination of the radar equation (3.1), indicates that the modulation bandwidth, B, of a signal also has an effect on system coverage. However, signal bandwidth has a more significant role to play than that for a different reason. The *bandwidth of a signal determines the delay resolution* and therefore also the range resolution of a PR. The delay resolution is determined as:

$$\Delta\tau = \frac{c}{2B\cos\theta}. \tag{3.3}$$

where θ is the bi-static angle between the transmitter, target and receiver. Due to the bi-static geometry of a typical PR deployment, the resulting range resolution in space is dependent on the relative location of the target to the transmitter–receiver pair [2]. For the purpose of comparing the difference in system performance when making use of different IoOs, geometrical dependency is not typically considered since the geometrical restrictions are the same for all illuminator types.

3.1.3   Doppler resolution and integration time

Another important factor that one must consider when working with PR, and in fact radar systems in general, is the Doppler frequency induced in the return echo as a result of the motion of a target. Again, the bi-static geometry plays an important role as the magnitude of the Doppler shift is not only dependent on the target's velocity but also on its location relative to the transmitter–receiver pair. Remember, the induced Doppler shift of a target is due to the targets *radial* velocity relative to the transmitter–receiver pair. The relative Doppler shift of a target is therefore determined as:

$$f_D = (\vec{u}_t + \vec{u}_r) \cdot \frac{\vec{v}}{c} \cdot f_c, \tag{3.4}$$

with $\vec{u}_t$ and $\vec{u}_r$ denoting the unit vectors pointing from the transmitter to the target and the receiver to the target. The target's velocity vector is denoted by $\vec{v}$. Note that the first dot in (3.4) denotes the inner product between the vectors while the second is a scalar multiplication.

Another factor to consider when observing the induced Doppler shift caused by a targets' motion is the carrier frequency, f_c, of the transmitted signal. The Doppler shift, f_D, for a given geometrical configuration, is directly proportional to the carrier frequency. This is to say that at higher carrier frequencies, even slow moving targets generate higher Doppler shifts than at low carrier frequencies and as such, the signal frequency poses a restriction on the acceptable Doppler resolution to avoid what is known as Doppler-walk. Doppler-walk occurs when the target migrates across multiple Doppler bins during a single integration period. This migration manifests as a loss of integration gain as the energy from the target echo is no longer coherently integrated within a single bin.

The Doppler resolution itself, that is how wide in Hz a single Doppler bin is, is inversely proportional to the integration time, T, used during a single integration interval or coherent processing interval (CPI). Again, for the purpose of comparing different illuminator signals, we ignore the geometrical dependency as it is consistent across the different illuminators and instead we work with:

$$\Delta f_D = \frac{1}{T}. \tag{3.5}$$

With the Doppler resolution of a PR inversely proportional to the integration time we can observe that for a CPI of 2 s, the resultant Doppler resolution is ($f_D = \frac{1}{T} = \frac{1}{2s} = 0.5\,\text{Hz}$). In addition to determining the Doppler resolution, the integration time also greatly affects the achievable processing gain as the processing gain is determined by the time-bandwidth product:

$$P_G = B \cdot T \tag{3.6}$$

Looking at (3.6), we can theoretically expect arbitrarily large values if we were able to integrate for long enough. For a moving target, however, there is a practical limit on how long we can integrate since the target moves (or migrates) out of its Doppler and range bins after some time. Once either range or Doppler migration occurs, it needs to be corrected for or non-coherent integration will occur. Usually, however, these restrictions are not prohibitive for practical radar applications.

3.1.4 Ambiguity functions

So far we have, apart from bandwidth and frequency, ignored the signal characteristics for our considerations. While this is perfectly suitable for defining the basic parameters of a radar's performance, there is clearly more to consider. Since PR does not typically use cooperative transmitters, provided the power and bandwidth considerations are met, there are two major considerations that need to be made when selecting a suitable IoO.

The first consideration is that a PR needs to be able to obtain a "good enough" reference signal that it can use for the matched filter processing stage where the reference signal is cross-correlated with a surveillance signal. While not strictly necessary

under all circumstances, this means that a PR usually operates with at least two channels, one for the reference and another for the surveillance signal. Range and Doppler
processing is then performed by computing the cross-ambiguity function also called
delay-Doppler or amplitude-range-Doppler processing.

This is achieved by repeatedly cross-correlating the surveillance signal with the
reference signal across different Doppler shifts of the reference signal. This process
is mathematically defined as:

$$\psi(f_D, \tau) = \int_T r(t) \cdot e^{2\pi i f_D t} \cdot s(t + \tau)dt, \tag{3.7}$$

where $r(t)$ denotes the reference and $s(t)$ the surveillance signal. In practice, the direct
implementation of (3.7) is computationally expensive and rarely done exactly. Most
methods use a combination of some approximation and the fast Fourier transform
(FFT) as illustrated in [3,4].

To measure the delay and Doppler frequency of a target echo, the location of the
corresponding peak on the range-Doppler map is extracted. The height of the peak
above the noise floor of the ambiguity function determines whether a target can be
detected or not. To have an unambiguous measurement, it is important that there is
only one perfect peak for any single target. This is where the time and frequency characteristics of the signal determine how well-suited they are for radar operation. If there
are periodic components or if there are e.g. pilot tones at different sub-frequencies or
guard intervals (cyclic components) between different components within a signal,
secondary peaks can be generated that create ambiguities or even mask other target
peaks.

An example of a signal with an ideal ambiguity function with a so-called
tumbtack-shaped response is a white (Gaussian) noise signal. Due to its random
nature, it only correlates with itself when it is perfectly aligned at zero-delay and
zero-Doppler. The amount of randomness in a signal usually provides a good indication of how well suited it is for radar purposes. Conversely, deterministic parts usually
impose some restrictions for the processing parameters such as integration time or
pulse duration for pulse-Doppler processing. In some cases, such as with digital
broadcast signals, the reference can be altered to mitigate some of these effects [4].

3.1.5 Digital vs. analogue waveforms

Traditional radar signal processing requires a reference signal of sufficient quality for
cross-correlating with the echo signal. As mentioned previously, the reference signal
for a PR is commonly acquired by utilising a dedicated reference channel that has
sufficient antenna gain in the direction of the transmitter. For analogue signal-based
systems, the use of a dedicated reference channel is critical; however, in the case of digital signals, it is often possible to extract a good reference signal by de-modulating the
signal and applying error correction before re-modulating the reconstructed, error-free
signal to obtain a perfect or near-perfect representation of the originally transmitted
signal. This process is commonly referred to as demod-remod.

Table 3.1 *Typical transmitter parameters*

Typical IoO parameters	FM radio (VHF)	DAB (VHF)	DVB-T2 (UHF)	WiFi (2.4 GHz)	WiFi (5 GHz)
Carrier frequency	88–108 MHz	30–300 MHz	470–870 MHz	2.4–2.5 GHz	5–6 GHz
Channel bandwidth	200 kHz	1.537 MHz	1.7–10 MHz	20 MHz or 40 MHz	20–160 MHz
ERP	10–100 kW	1–10 kW	5–50 kW	100 mW–1 W	100 mW–1 W
Range resolution	1.5 km	200 m	30 m	15 m	4 m

Depending on the robustness of the signal standard to be demodulated without bit errors, the lower SNR of a surveillance channel could already be sufficient to reproduce a good reference signal. This is e.g. the case for digital audio broadcast (DAB) radio signals and was demonstrated in [5], where targets were detected with only one receiver channel.

There are many ways to classify and categorise IoOs. The most basic but important characteristics are whether a transmitter is stationary or mobile and the size of the area it is intended to serve since this is usually directly coupled to its transmission power. Especially for the application of passive radar on moving platforms, the coverage has to be large enough so that the receiver is not leaving the covered area too quickly.

For airborne receivers, the mode of operation of an illuminator network is very important as well. In many cases, a network of transmitters has some level of frequency re-use such as in FM-radio where the same frequency can be used by two transmitters that are either sufficiently far apart or somewhat separated by topography. In some cases, the transmitters even operate on the same frequency with either the same, synchronised signal content such as e.g. in DAB single-frequency network (SFN) or with different signals that are especially tailored for each user such as in long-term evolution (LTE) networks.

While a ground-based passive radar receiver will most likely only receive a limited number of illuminators at the same time, an airborne one will be exposed to many more that can potentially be interfering or introducing too many ambiguities.

Table 3.1 shows an overview of different IoOs and their characteristics.

3.2 Analogue IoO

3.2.1 Analogue TV transmitters

Along with FM radio transmitters, analogue TV transmitters were some of the earliest to be used as an IoO for PR [6–8]. Analogue TV transmitters typically broadcast a high power ($\approx$ 250 kW per channel), wide bandwidth ($\approx$ 8 MHz per channel) signal.

This allows for extremely long-range target detections at a relatively high-range reso-lution. In addition to this, analogue TV transmitters typically operate in the ultra high frequency (UHF) band from 300 MHz to 900 MHz.

While analogue TV transmitters meet two of the three main conditions required for them to be considered a desirable IoO, they have a significant disadvantage in that they have sync pulses embedded within the signal that results in significant autocorrelation issues. If we take a look at the experiments performed in 1986 [9], Griffiths *et al.* used the Crystal Palace Transmitter in South London as an IoO to detect aircraft taking off and landing at Heathrow Airport. In these experiments, the receiver was placed 11.8 km away from the transmitter that had an omni-directional antenna with an ERP of 1 MW. The system operated on TV channels between 487.25 and 567.25 MHz with each separate channel having a channel bandwidth of 8 MHz, resulting in an ERP of 250 kW per channel.

It was found that the autocorrelation function of the sync-plus-white waveform of the analogue TV signal exhibited high sidelobes with poor range resolutions and severe ambiguities at 9,600 m and integer multiples thereof. It was therefore concluded at the time that analogue TV signals were not a good IoO due to the sync pulses. It must be noted that at the time of the experiments, analogue-to-digital converters (ADC) technology was limited to 8 bits with an even lower effective number of bits (ENOB), resulting in a maximum of 48 dB of dynamic range.

In another experiment by Howland in 1999 [10], it was shown that even though the signal exhibited severe ambiguities, Doppler and bearing information could be extracted from the target echos. This was achieved by extracting Doppler information using a stable single carrier within the transmitted signal, thereby bypassing the waveform modulation effects experienced in earlier experiments.

While it is likely possible that one could use modern processing techniques to remove these ambiguities inherent in the signal, most analogue TV broadcasts around the world have been decommissioned in favour of satellite and Internet protocol television (IPTV), making research into the use of analogue TV-based IoOs no longer feasible.

3.2.2   FM radio transmitters

Along with analogue TV transmitters, FM radio transmitters were some of the earliest to be used as an IoO for PR [6–8]. While analogue TV transmitters are no longer used around the world, FM radio transmitters have been shown to be an extremely useful IoO.

3.2.2.1   Signal structure

The mathematical description of broadcast FM radio signals is developed here. The mathematical description follows that by Stremler in [11]. Frequency modulation is a form of analogue modulation where the baseband information carrying signal, typically called the message signal, $m(t)$, modulates the frequency of the carrier wave. Broadcast FM radio signals are generated by applying a message signal to a voltage-controlled oscillator (VCO). The output of the VCO is a constant amplitude sinusoidal carrier wave whose frequency is a function of the control voltage, $m(t)$. When no message signal exists, the carrier wave is simply at its centre frequency, f_c. When a message signal exists, the instantaneous output signal varies about the carrier frequency as expressed by:

$$f_i(t) = f_c + K_{VCO} \times m(t) \tag{3.8}$$

where K_{VCO} is the voltage-to-frequency gain of the VCO, expressed in units of Hz/V. The resultant output of $K_{VCO} \times m(t)$ is the instantaneous frequency deviation, Δf. The instantaneous phase of the signal is equal to 2π multiplied by the integral of the instantaneous frequency, giving:

$$
\begin{aligned}
\theta_i(t) &= \int_0^t f_i(t)dt \\
&= \int_0^t 2\pi f_c dt + \int_0^t 2\pi K_{VCO} \times m(t)dt \\
&= 2\pi f_c t + 2\pi K_{VCO} \int_0^t m(t)dt
\end{aligned}
\tag{3.9}
$$

The output FM waveform, $X_{FM}(t)$, is therefore represented by:

$$X_{FM}(t) = A_c \cos(\theta_i(t)) \tag{3.10}$$

If the initial phase is assumed to be zero for simplicity, the output FM signal becomes:

$$X_{FM}(t) = A_c \cos\left[2\pi f_c t + 2\pi K_{VCO} \int_0^t m(t)dt \right] \tag{3.11}$$

where the frequency modulated output, $X_{FM}(t)$, has a non-linear dependence on the message signal, $m(t)$, making it difficult to analyse the exact properties of an FM signal. The baseband message signal, $m(t)$, can be modelled mathematically as a sum of each channel component:

$$
\begin{aligned}
m(t) = \ &C_0[L(t) + R(t)] \\
&+ C_1 \cos(2\pi \times 19 \text{ kHz} \times t) \\
&+ C_0[L(t) - R(t)] \cos(2\pi \times 38 \text{ kHz} \times t) \\
&+ C_2 RDS(t) \cos(2\pi \times 57 \text{ kHz} \times t)
\end{aligned}
\tag{3.12}
$$

where C_0, C_1 and C_2 are the gains used to scale the amplitudes of the left channel audio, the right channel audio, the 19 kHz pilot tone, and the radio data system (RDS) subcarrier, respectively, to generate the appropriate modulation index, β. For simplicity, the FM signal bandwidth can be estimated by representing the message signal, $m(t)$, by a single tone:

$$m(t) = A_m \cos(2\pi f_m t) \tag{3.13}$$

where A_m is the amplitude of the message signal and f_m is the message tone frequency. Substituting (3.13) into (3.11) gives the following:

$$
\begin{aligned}
X_{FM}(t) &= A_c \cos\left(2\pi f_c t + \frac{K_{VCO}A_m}{f_m} \sin(2\pi f_m t)\right) \\
&= A_c \cos\left(2\pi f_c t + \frac{\Delta f}{f_m} \sin(2\pi f_m t)\right) \\
&= A_c \cos(2\pi f_c t + \beta \sin(2\pi f_m t))
\end{aligned} \tag{3.14}
$$

The peak frequency deviation, $\Delta f = K_{VCO}A_m$, is a result of the message amplitude and gain of the VCO. The ratio of the peak frequency deviation, Δf, and the message signal frequency, f_m, is known as the modulation index, β. The number of significant sidebands in the output spectrum is a function of the modulation index. This can be seen by representing the FM output signal in terms of n-th-order Bessel functions of the first kind:

$$
X_{FM}(t) = A_c \sum_{n=-\infty}^{\infty} J_n(\beta) \cos(2\pi (f_c + n f_m)t) \tag{3.15}
$$

By taking the Fourier transform of (3.15), the discrete FM spectrum can be represented in magnitude by coefficients as a function of β:

$$
X_{FM}(f) = \frac{A_c}{2} \sum_{n=-\infty}^{\infty} J_n(\beta) \left[\delta(f - f_c - n f_m) + \delta(f + f_c + n f_m)\right] \tag{3.16}
$$

The number of sidebands of an FM signal and its associated magnitude coefficient can be found with the help of Bessel function tables [12,13].

The average power envelope of an FM signal is constant and can be determined using the Bessel functions as:

$$
P_{ave} = \frac{1}{2}A_c^2 \sum_{n=-\infty}^{\infty} J_n^2(\beta) \tag{3.17}
$$

As the bandwidth of the message signal reduces, the number of significant sidebands required for transmission reduces. This leads to an increased level of the carrier component, J_0. As the bandwidth of the message signal increases, however, the number of significant sidebands required for transmission increases, i.e. J_n for $n > 0$ increases, and the level of the carrier tone, J_0, subsequently decreases.

Figure 3.1 illustrates the FM radio station spectrum output for high, medium and low bandwidth message signals. Each signal is normalised to the maximum level within each plot. From (3.17), the average power envelope across the channel must remain constant. It is therefore clear by comparing the high bandwidth signal (left) to the low bandwidth signal (right) that the carrier tone levels are significantly higher for a lower valued β than for high values. In the context of using FM radio signals as IoOs, it is important to note that the integration gain achieved through matched filtering is directly proportional to the instantaneous signal bandwidth. If the message bandwidth occupied the maximum permissible FM radio station bandwidth [14], the

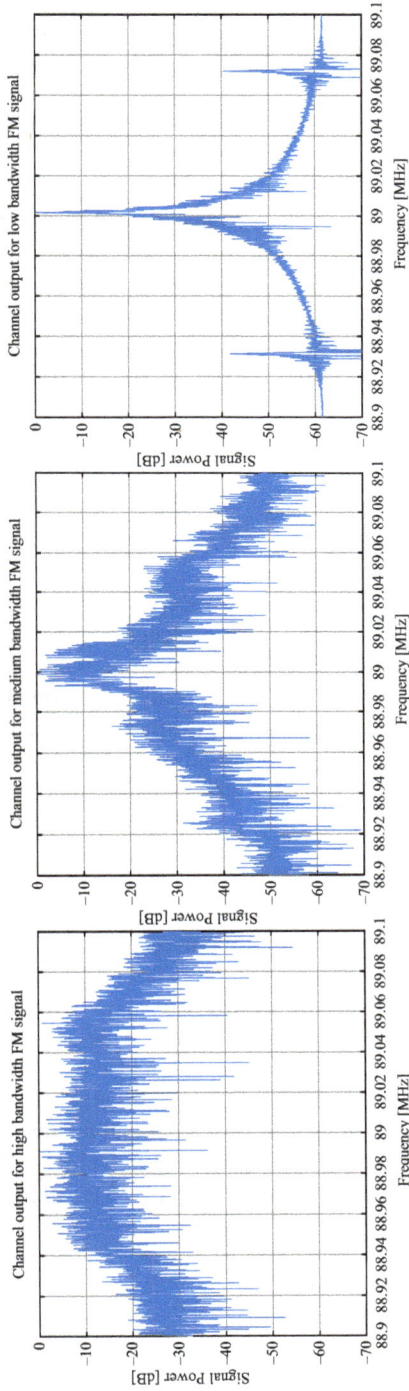

Figure 3.1   FM channel output for high, β = 4 (left), medium β = 2 (middle) and low β = 0.25 (right) bandwidth message signals [4]

corresponding maximum integration gain would be 59 dB (assuming for the time being an integration time of 4 s), as calculated in (3.18)

$$G_{int} = t_{int} \cdot \text{B}$$
$$= 10 \log (4 \text{ s} \cdot 200 \text{ kHz}) \qquad (3.18)$$
$$= 59.03 \text{ dB}$$

A 59 dB integration gain can only be achieved when the entire modulation bandwidth is filled for the duration of the CPI. Realistically, this will only be the case when applying broadband noise jamming to the PR receiver. A more realistic integration gain for the FM based PR receiver under normal operation is determined by the average modulation bandwidth across the CPI. This results in an integration gain of approximately 55 dB as shown in (3.19):

$$G_{int} = t_{int} \cdot \text{B}$$
$$= 10 \log (4 \text{ s} \cdot 30 \text{ kHz}) \qquad (3.19)$$
$$= 55.05 \text{ dB}$$
$$\therefore 55 \text{ dB} < G_{int} < 59 \text{ dB} \qquad (3.20)$$

As with almost all FM PR, there are two receive channels, one for reference signal and another for surveillance signal. The reference channel is directed towards the transmitter of opportunity and is used to record the illuminating signal. This is done in order to obtain a clean reference signal. The surveillance channel is directed away from the illuminating transmitter and towards the desired coverage area. The raw data is sampled and filtered, usually using a front-end pre-select filter stage to remove out-of-band interference. It is important that the two channels are synchronised to allow for coherent processing.

3.2.2.2 Range and Doppler resolution

As mentioned, the range and Doppler resolution for an FM band PR is determined by the signal bandwidth and the integration time used. With FM-based systems, the maximum range resolution is determined using (3.3) to be 1.5 km with a maximum channel bandwidth of 200 kHz. In typical FM PR systems with typical bandwidth fluctuations, this range resolution is commonly found to be anywhere from 5 km to 10 km.

As you can see, the range resolution of FM PR systems is relatively poor; however, this allows for longer integration times while ensuring that the target remains within a single range bin. The integration time for an FM PR is typically 1–4 s, resulting in a relatively high Doppler resolution of between 1 and 0.25 Hz in real systems.

3.3 Digital IoO

Digital IoOs differ from analogue IoOs in that unlike analogue waveforms whose signal content and bandwidth vary over time depending on content, digital waveforms maintain a constant signal structure and bandwidth regardless of signal content. Here we explore the two most common types of IoOs used in airborne PR platforms, digital video broadcast terrestrial 2 (DVB-T2) and digital audio broadcasting (DAB). Before going into the specifics of both DVB-T2 and DAB signals, we first need to understand the backbone of most commercial digital transmissions, orthogonal frequency division multiplexing (OFDM).

3.3.1   OFDM-based signals

OFDM-based signal protocols have the advantage that they have high and constant bandwidth compared to traditional analogue systems. The wider bandwidths provided by digital broadcast services yield improved PR range resolution. Couple this with the ability to demodulate and remodulate the reference signal to create a perfect, noise-free reference signal, you can see why these types of signals would be of interest to PR engineers.

In addition to their popularity for use as IoOs for PR, of the many reasons modern broadcast systems are moving towards digital OFDM-based standards is that the communication links can be made extremely robust against multi-path propagation and selective fading while also allowing for operation in a single frequency network (SFN) configuration. In this deployment configuration, several transmitters use the same transmission frequency and transmit the signal in a synchronised manner. As long as the differences in propagation time between the different transmitters and the receiver do not exceed what is known as the guard interval, the receiver has no problem decoding the signal. In this way, networks can be set up with a more powerful transmitter at a central location and many smaller transmitters at a distance to either extend coverage or fill difficult to reach gaps like valleys.

To keep the propagation time differences below the duration of the guard interval, fill-in transmitters usually transmit their signal with a so-called network delay. This is a parameter set by the network operator and is usually not published. For a PR, this poses the problem that the arrival times of the signal components from the different transmitters no longer correspond to their relative location. The operator of the PR therefore first needs to find out the different delays used which is not always straight forward. In [15], the authors demonstrate how the network-delays can be measured using a mobile receiver. Other methods use targets of opportunity or the direction of arrival of the direct signal for associating the transmitter delays.

Apart from the problem with the network delays, the SFN configuration also poses additional challenges in target association since any given range-Doppler map would contain several transmitters and a target detection can therefore not unambiguously be associated to a transmitter. This makes the subsequent tracking and location stages more involved because multiple hypotheses need to be tracked.

With that said, the use of OFDM-based IoOs for a PR has many significant advantages and as such, been investigated by numerous researchers, most notably [16–21]. There are two common approaches to processing OFDM-based signals in PR systems, namely mismatched filtering as proposed in [16,17] and further explored in [19,20] and inverse filtering, as shown by [18] and expanded on by [21].

Mismatched filtering is a traditional cross-correlation based approach which involves demodulating the reference signal and then 'remodulating' it to obtain a clean, slightly modified reference. The modifications are needed to remove pilot ambiguities in the range-Doppler map. This new reference signal is then used as a 'mismatched' filter when performing the range-Doppler processing.

The so-called inverse filtering is a process whereby the signal undergoes demod–remod to produce a noise-free reference signal. This noise-free reference signal is then used to perform the range-Doppler processing through element-wise division of the OFDM symbols in the surveillance channel by the same OFDM symbols in the remodulated reference channel. A 2-D FFT is then applied to produce a range-Doppler map. This, in effect, normalises the carriers and removes direct signal clutter from the range-Doppler map. This will be covered in additional detail in Chapter 7.

3.3.2   Digital video broadcast transmitters

The first generation of digital video transmitters was known as digital video broadcast terrestrial (DVB-T). DVB-T is the European standard for the broadcast of digital terrestrial television and first appeared in 1997 [22] with its first broadcast being achieved in Singapore in February 1998 [23,24]. DVB-T is designed to allow commercial terrestrial broadcasters to transmit compressed digital audio, digital video and other data in an MPEG transport stream using OFDM or coded orthogonal frequency division multiplexing (COFDM) modulation techniques.

DVB-T is capable of carrying data encoded at quadrature phase shift keying (QPSK), 16-QAM and 64-QAM and while these encodings are perfectly suitable for standard definition (SD) television, the move to high definition (HD) television in recent times has driven the need for additional throughput. As such, DVB-T has been further developed into a newer, higher data-rate DVB-T2 which was initially finalised in August 2011.

Since being finalised in 2011, DVB-T2 has since become the standard for HD television broadcasts and has largely pushed the original DVB-T out of service. This section therefore focuses exclusively on the use of DVB-T2 as an IoO for use in PR.

3.3.2.1   Signal structure

As with most digital broadcast standards, the DVB-T2 standard utilises OFDM with a pre-defined, open signal structure where the number of sub-carriers, bandwidth and encoding technique is dependent on the operating mode [25]. The DVB-T2 signal has three levels of abstraction, a super frame which carries multiple smaller T2 frames, each of which contain symbols carrying content data.

The maximum length of a super frame is 63.75 s if future extension frames (FEFs) are not used which corresponds to 255 T2 frames of 250 ms each. If FEFs are used, each super frame is 127.5 s long. The number of data symbols in each T2 frame depends on the length of each symbol. If, for example, each symbol is 32K (32,768 samples) long, then a total of 60 useful symbols can fit into each frame [25]. Data

P1	G$_{P2}$	P2	G$_1$	Data Symbol 1	G$_2$	Data Symbol 2	•••	G$_{n-1}$	Data Symbol n-1	G$_n$	Data Symbol n

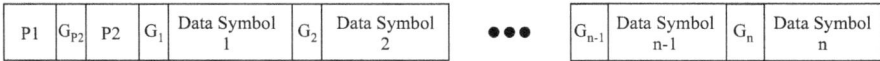

*Figure 3.2   Single T2 frame structure illustrating the position of the P1 symbol,
            followed by the guard symbol and the P2 symbol before the data
            symbols*

is then encoded onto each symbol carrier using a variation of quadrature amplitude modulation (QAM).

In order to work with a DVB-T2 signal, it is important to understand how each T2 frame is constructed. The basic time domain structure of a T2 frame is demonstrated in Figure 3.2.

The frame begins with the P1 symbol. The P1 symbol is used to speed up channel search and allows for basic timing and coarse frequency offset correction between the receiver and the transmitter. The P1 symbol is a 1K (1,024 samples) OFDM symbol that consists of three distinct parts A, B and C. The P1 symbol also contains information concerning the Fourier transform (FFT) size (32K, 16K, 8K, 4K, 2K and 1K) of the remaining OFDM symbols (P2 and data) within the frame and whether either multiple input single output (MISO) or single input single output (SISO) transmission is being used.

After the P1 symbol, a P2 symbol and a number of data symbols follow. The number of data symbols within the frame is dependent on the FFT size specified by P1. The P2 and data symbols contain pilot tones which are known reference values spaced at different frequency carriers and symbol indices within each symbol. Pilot tones are used by a receiver for channel estimation, timing and fine frequency offset correction.

The P2 symbols contain the relevant layer-1 (L1) signalling, split into L1-pre signaling and L1-post signaling, as well as a highly dense pilot pattern. The number of pilot tones within a P2 symbol is much higher than the number of pilot tones in a data symbol and is dependent on the parameters given by P1. The L1 signalling provides the means to demodulate the DVB-T2 frame and extract information. This information includes the number of data symbols, the constellation size as well as the pilot pattern present within the data symbols.

As with most high bandwidth OFDM-based communication protocols, pilot carriers are utilised which have pre-defined patterns. The pilot carriers are required in high bandwidth OFDM waveforms as they are used to correct for and equalise out various channel effects that occur due to things like multipath and selective fading.

The data symbols contain the data and pilots grouped into physical layer pipes (PLP). The pilots can be further divided into scattered pilots (SP) and continual pilots (CP). The SP are spread in symbol and frequency bins in one of eight pre-defined patterns using a binary phase shift keying (BPSK) mapping scheme (specified by the L1 information) as highlighted in Table 3.4 in Section 3.3.2.2. The CP are placed at the same sub-carrier index for all data symbols in a frame. The sub-carrier index of the CP is dependent on the scattered pilot pattern in use and the FFT size of each symbol.

Each PLP in the data symbol contains their own constellation mapping (QPSK, 16-QAM, 64-QAM or 256-QAM), where the number and configuration of each PLP is given by the L1 information. Any remaining cells in the DVB-T2 frame are filled with either auxiliary streams, which are transmitter specific and are not required for demodulation, or dummy cells, which are empty cells.

3.3.2.2   Summary of pilot signal parameters

Tables 3.2–3.5 provide a brief summary of the various different DVB-T2 pilot signal parameters.

3.3.2.3   Range-Doppler ambiguities

While pilot carriers within high bandwidth OFDM signals are essential to the signal demodulation process, they also lead to ambiguities within the range-Doppler map

Table 3.2   DVB-T2 signal parameters used in the Cape Town Area

Parameters	Value
FFT size	32,768
Active carriers	27,841
Pilot pattern	PP4
T_u	3,584 μs
T_g	224 μs (1/16)
Subcarrier spacing	279 Hz
Bandwidth	7.77 MHz
P2 pilot amplitude	$\sqrt{37}/5$
Continual pilot amplitude	8/3
Scattered pilot amplitude	7/4
Scattered pilot: separation of pilot bearing carriers	12
Scattered pilots: number of symbols forming one scattered pilot sequence	2
P2 encoding	64 QAM
Data encoding	256 QAM
Number of continual pilots	178
Scattered pilots per symbol	1,159+
Total pilots per symbol[a]	1,196

[a]Where the scattered pilots fall on the same frequency bin as the continual pilots, the amplitude is set to the scattered pilot amplitude.

Table 3.3   DVB-T2 continual and P2 pilot parameters for different FFT sizes

Pilot amplitude	FFT size					
	1K	2K	4K	8K	16K	32K
A_{CP}	4/3	4/3	$(4\sqrt{2})/3$	8/3	8/3	8/3
A_{P2}	$\sqrt{31}/5$	$\sqrt{31}/5$	$\sqrt{31}/5$	$\sqrt{31}/5$	$\sqrt{31}/5$	$\sqrt{37}/5$

Table 3.4   DVB-T2 scattered pilot pattern parameters

Scattered pilot (SP) pattern	Separation of pilot-bearing carriers (P_s)	No. of symbols forming one SP sequence (C_s)	Amplitude (A_{SP})	Equivalent boost (dB)
PP1	3	4	4/3	2.5
PP2	6	2	4/3	2.5
PP3	6	4	7/4	4.9
PP4	12	2	7/4	4.9
PP5	12	4	7/3	7.4
PP6	24	2	7/3	7.4
PP7	24	4	7/3	7.4
PP8	6	16	7/3	7.4

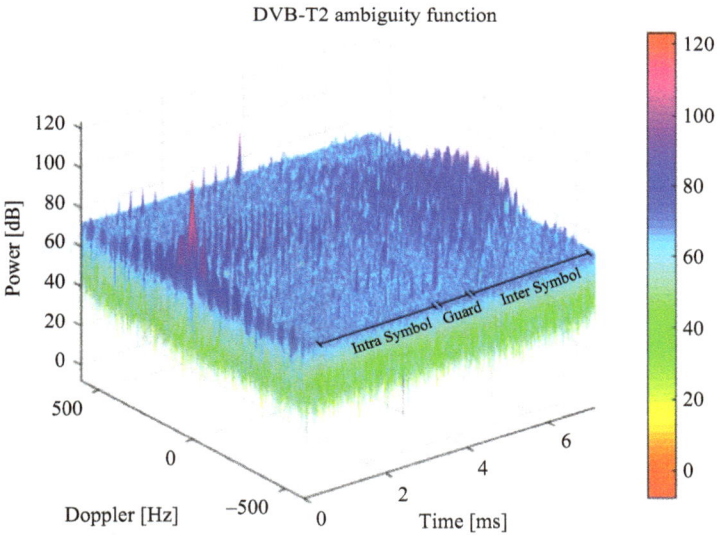

Figure 3.3   3D AF of a frame within the DVB-T2 signal [4]

due to their periodicity and raised amplitude levels, as highlighted in Tables 3.3 and 3.4. To evaluate the effect that these pilots have on range-Doppler ambiguities, the auto ambiguity function (AF) of a single DVB-T2 frame containing 32K symbols is observed in Figure 3.3. The AF is a two-dimensional function of time delay and Doppler shift and is described mathematically using (3.21):

$$\chi(\tau, v) = \int_{-\infty}^{\infty} r(t)r^*(t-\tau)e^{-j2\pi vt}dt \qquad (3.21)$$

Table 3.5   Summary of active and pilot carriers for different DVB-T2 FFT sizes

		1K	2K	4K	8K N	8K E	16K N	16K E	32K N	32K E
Total no. active carriers		853	1,705	3,409	6,817	6,913	13,633	13,921	27,265	27,841
PP1	CP	20	45	45	45	45	89	93		
	SP	71-	142-	284-	568-	576-	1,136-	1,160-		
	Total*	76	164	306	590	598	1,178	1,206		
PP2	CP	20	42	44	46	50	87	89	175	177
	SP	70+	141+	283+	567+	575+	1,135+	1,159+	2,271+	2,319+
	Total*	75	158	302	588	600	1,167	1,193	2,307	2,357
PP3	CP	45	42	43	43	45	87	89		
	SP	35+	71-	142-	284-	288-	568-	580-		
	Total*	40	90	162	304	310	608	622		
PP4	CP	20	43	44	46	48	90	92	176	178
	SP	35	70+	141+	283+	287+	567+	579+	1,135+	1,159+
	Total*	39	88	160	304	310	600	614	1,170	1,196
PP5	CP	19	42	45	46	46	90	92		
	SP	17++	35+	71-	142-	144-	284-	290-		
	Total*	22	52	89	161	163	315	323		
PP6	CP						88	90	176	180
	SP						283+	289+	567+	579+
	Total*						314	322	598	614
PP7	CP		45	50	53	58	88	91	180	182
	SP		17++	35+	71-	72-	142-	145-	284-	290-
	Total*		40	60	97	103	171	177	317	325
PP8	CP				47	52	86	89	175	181
	SP				71-	72-	142-	145-	284-	290-
	Total*				103	109	181	187	332	344

*Note that some of the pilots fall on the same carriers and therefore they do not add up as expected.

where τ and v represent the time delay and the Doppler shift, respectively. The peaks seen in Figure 3.3 are due to the boosted CP and SP within the frame (pilot pattern (PP) in this instance). The CP are on the same frequency bins throughout the DVB-T2 frame which creates low amplitude ridges across the range profile at fixed Doppler frequencies. While these ridges are not clearly visible above the noise floor in Figure 3.3, they are illustrated clearly in Figure 3.4 where the cross-ambiguity function (CAF) of the CP is shown. The SP can be seen scattered around the range-Doppler map in a pattern defined by the cyclic nature of their positioning within the DVB-T2 frame. The mathematical representation for the positions of the ambiguities is described in detail in [16,17,19,20]. The ambiguities and their influence on target detection depends heavily on the transmit mode and pilot pattern used. The three ambiguity regions are demonstrated in Figure 3.3 and highlighted as:

• Intra-symbol ambiguity region

- Guard ambiguity region
- Inter-symbol ambiguity region

A. Intra-symbol ambiguities

Intra-symbol ambiguities refer to ambiguities corresponding to $0 < \tau < T_u$. This leads to ambiguities in both range and Doppler. These peaks are due to the boosted pilot amplitudes relative to the data symbols as highlighted in Table 3.4. The intra-symbol ambiguities need to be addressed regardless of the symbol length as they are always present in the system unlike the other ambiguities which may or may not appear depending on the symbol length.

Range ambiguities

The SP pattern causes peaks in fast-time that repeat every $\frac{1}{P_s \cdot C_s}$ seconds, where P_s is the number of carriers between each pilot and C_s is the carrier spacing.

Doppler ambiguities

Since the CP pattern repeats every N symbols, there is also periodicity over slow-time. The symbol length, T_u, corresponds to a periodicity every $\frac{1}{N \cdot T_u}$ Hz in Doppler.

B. Guard interval ambiguities

Guard interval ambiguities are a result of the guard intervals at delays equal to T_u. The guard intervals between symbols are repetitions of the end of each succeeding symbol. These repetitive symbols result in ambiguities that only appear at regular intervals equal to T_u as seen in Figure 3.3 where $\tau = 3.584$ ms for a 32K symbol with a 2K guard [26]. In this case, the guard interval ambiguities appear well outside the detection range of the system at greater than 1,000 km.

C. Inter-symbol ambiguities

Inter-symbol ambiguities refer to ambiguities corresponding to $\tau > T_u$. Inter-symbol ambiguities arise from SP pattern repetition every N symbols, which results in additional ambiguities appearing at a delays of $n \cdot N \cdot T_u$ seconds. For systems utilising 1K, 2K or 4K modes, this potentially falls within the detection range however, when 8K, 16K or 32K modes are used, these ambiguities fall well outside the detection range of a typical system as highlighted in Table 3.6 where the guard interval range represents the bistatic range at which guard interval and inter-symbol ambiguities will appear.

3.3.2.4 Quantifying range-Doppler ambiguities

Since the creation of an AF is a linear process, it can be broken down into the individual contributions of each component. To get a comprehensive understanding of the individual effects of each component within the DVB-T2 signal, the AF of each component including the data symbols, the CP, the SP and the P2 symbol pilots is calculated.

Continual pilot effects

Figure 3.4 illustrates the CAF of the CP after demod–remod where all other signal components are set to zero and the original demod–remod signal. The CP pattern consists of raised carriers appearing on the same carriers across time. This causes ridges that appear along delay. Where a SP falls on the same carrier as the CP, the CP

Table 3.6   DVB-T2 OFDM symbol guard intervals for different FFT sizes

FFT size	T_u (ms)	Guard interval range (km)	Guard interval (ms)						
			1/128	1/32	1/16	19/256	1/8	19/128	1/4
1k	0.112	34	NA	NA	0.007	NA	0.014	NA	0.028
2k	0.224	67	NA	0.007	0.014	NA	0.028	NA	0.056
4k	0.448	134	NA	0.014	0.028	NA	0.056	NA	0.112
8k	0.896	268	0.007	0.028	0.056	0.0665	0.112	0.133	0.224
16k	1.792	537	0.014	0.056	0.112	0.133	0.224	0.266	0.448
32k	3.584	1075	0.028	0.112	0.224	0.266	0.448	0.532	NA

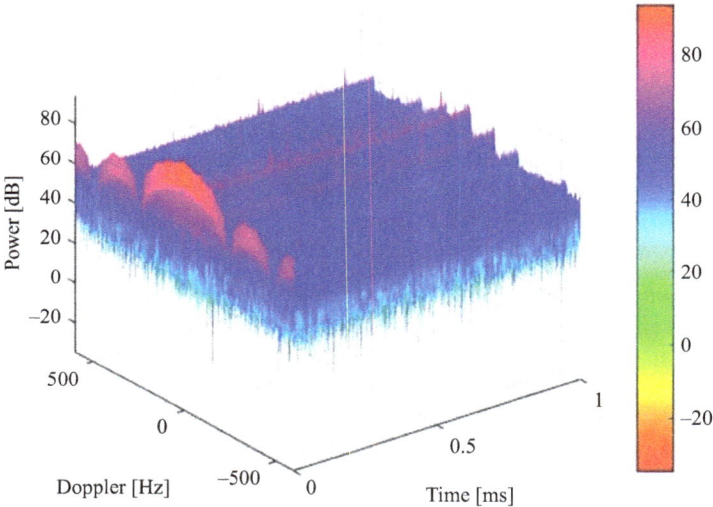

Figure 3.4   Continual pilot ambiguity function [4]

is set to the same amplitude as the SP. This results in a pulse train along the carrier where the two pilots combine, leading to an additional peak forming on top of the CP ridge in the same arrangement as the SP peaks in Figure 3.5. The contribution of the CP ridges to the AF is relatively small since they appear only 2 dB above the noise floor. The peaks caused by the SP appearing on the same carriers as the CP, however, appear 13 dB above the noise floor. The peak levels are governed by the amplitude level of the pilots which change depending on the FFT size.

Scattered pilot effects
Figure 3.5 illustrates the CAF of the SP where all other carriers are set to zero with the original demod–remod signal. As with the AF demonstrated in Figure 3.3, the SP

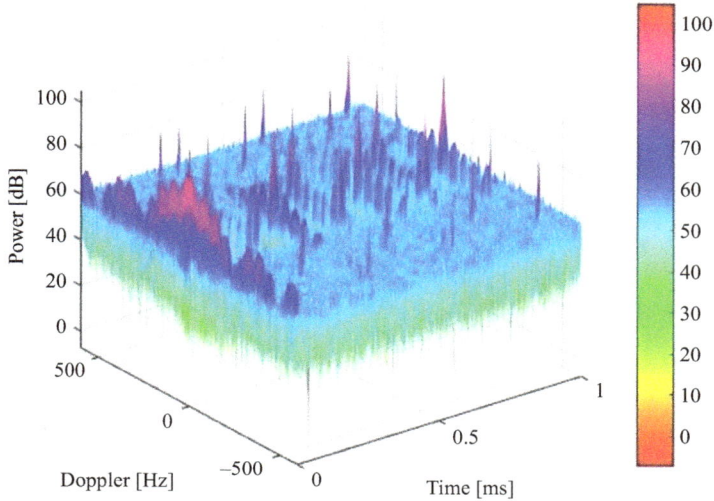

Figure 3.5   Scattered pilot ambiguity function [4]

AF exhibits peaks which are periodic in both range and Doppler. The SP contribute significant power to the AF as the peaks appear 33 dB above the noise floor, compared to 13 dB for the CP.

Along with the guard interval ambiguities (which are not shown here due to their delay being impractical for a real 32K system), the final contribution to the AF is the P2 pilot pattern.

P2 pilot effects
Figure 3.6 illustrates the CAF of the P2 symbol pilots after demod–remod where all other components set to zero with the original demod–remod signal. It is clear from Figure 3.6 that the P2 symbol contributes almost nothing to the complete AF and can therefore be ignored for the purposes of this work.

3.3.2.5 Ambiguity removal

As demonstrated, the ambiguities within the DVB-T2 AF are caused as a direct result of the periodic elements within the signal structure. These ambiguities can largely be removed from the resultant range-Doppler map with the removal process depending on the range-Doppler processing technique employed.

The process required to remove the ambiguities within each of the three ambiguity regions is described by Gao [17] and Harms [19] and summarised below. It is noted that the removal of ambiguities using the techniques discussed below are valid for time-domain based correlation processing only and may need to be slightly modified to remove ambiguities when using frequency-domain-based processing techniques.

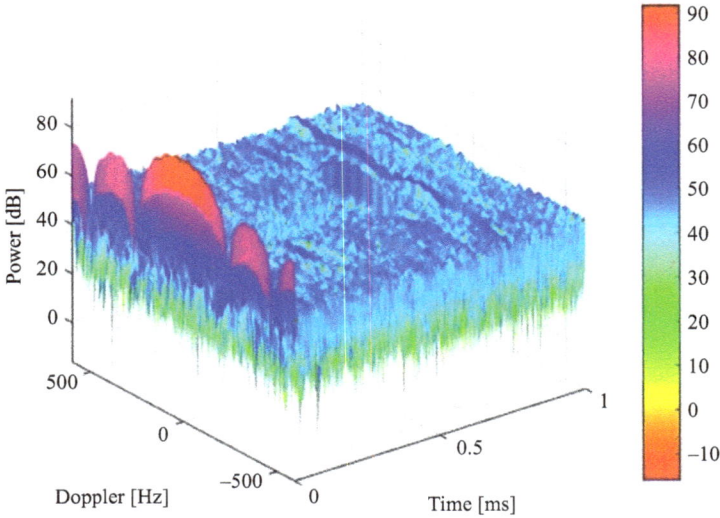

Figure 3.6   P2 pilot ambiguity function [4]

Intra-symbol ambiguity removal

It is mentioned in [26] and demonstrated in [27] that these ambiguities can be removed by blanking the pilot carriers, however, since the ambiguities in the intra-symbol region are due to the relative amplitude differences between the pilot and the data carriers, as highlighted in Table 3.4, they are therefore removed through a normalisation process. This normalisation is achieved by modifying the pilot levels in the remodulating phase to $1/A_{SP}$. This then normalises the pilots to the background carrier levels during the correlation process, resulting in the removal of the intra-symbol pilot ambiguities. Normalising the pilots in such a way results in a slight (approximately 0.44 dB) decrease in integration gain (in a 32K system) as shown in (3.22).

$$G_{loss} = 10 \cdot \log_{10} \left(\frac{\text{Active carriers} - \text{Pilot carriers}}{\text{Active carriers}} \right)$$

$$= 10 \cdot \log_{10} \left(\frac{27841 - 1196}{27841} \right) \tag{3.22}$$

$$= -0.44 \text{ dB}$$

Setting the pilot amplitudes to zero, as suggested in [26,27], can be modeled as an inverted pulse train, leading to additional ambiguities. As these ambiguities are originally a result of the amplitude differences between the pilot and data carriers, setting them to zero once again results in ambiguities within the range-Doppler map as described by (3.25).

Assuming the spectrum of a single symbol has no zero-value carriers, i.e. contains only data carriers and pilots, the blanking of these pilots can be modeled by

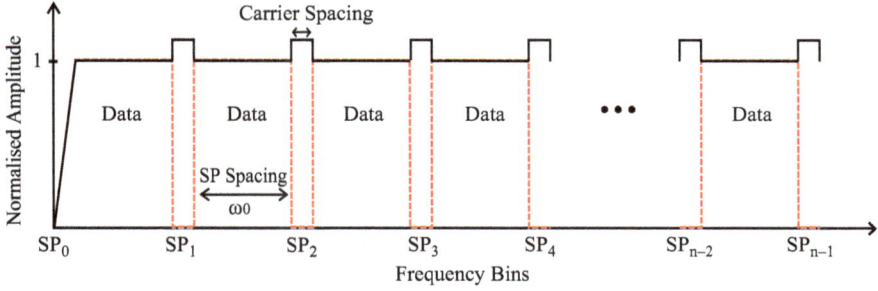

Figure 3.7 OFDM signal structure in frequency domain [4]

multiplying the spectrum by a pulse train of zeros and ones which follows the SP pattern as demonstrated in Figure 3.7. Blanking the carriers at regular time intervals corresponding to the SP pattern is the same as multiplying the spectrum by a pulse train in the frequency domain. This is shown in (3.23) where the pulse period is $\frac{1}{(P_s)(C_s)}$ with P_s representing the pilot spacing and C_s representing the carrier spacing:

$$S_{out}(f) = S_{data}(f) \times S_{blank}(f) \tag{3.23}$$

where $S_{blank}(f)$ is described as:

$$
\begin{aligned}
S_{blank}(f) &= 1 - \omega_0 \sum_{n=0}^{N_{SP}-1} \delta(\omega - n\omega_0) \\
&= 1 - 2\pi(P_s)(C_s) \sum_{n=0}^{N_{SP}-1} \delta(2\pi f_0 - n2\pi(P_s)(C_s))
\end{aligned}
\tag{3.24}
$$

$$S_{blank}(t) = \delta(t) - \underbrace{\sum_{n=0}^{N_{SP}-1} \delta\left(t - n \cdot \frac{1}{(P_s)(C_s)}\right)}_{\text{Pulse train with period } \frac{1}{(P_s)(C_s)}} \tag{3.25}$$

The pulse period, T, which defines where the peaks will appear is therefore:

$$T = \frac{1}{(P_s)(C_s)} \tag{3.26}$$

The ambiguities arising from blanking the pilots will therefore be located in the same position in the range-Doppler map as the original boosted pilot ambiguities.

Guard interval ambiguity removal
Unlike the intra-symbol ambiguities that arise due to the relative level differences between the pilot and data carriers, the guard interval ambiguities arise due to the cyclic repetition of the guard symbols. As a result, these ambiguities can be removed by blanking the guard intervals in the remodulation step as demonstrated in [17,19,20].

This however, comes with a loss of integration gain relative to the length of the guard interval itself. Table 3.6 highlights the bistatic ranges at which the guard interval ambiguities are present for a given symbol length.

Inter-symbol ambiguity removal
As with the ambiguities caused by the guard intervals, ambiguities appearing in the inter-symbol region are a result of their repetition over different symbols and as such, can be removed by blanking the pilots. In most scenarios, as with the guard intervals, the inter-symbol ambiguities will fall well outside the instrumented range of the radar and can therefore be ignored.

Since the process of blanking the pilots and normalising the pilots are counter-active, two parallel processes need to be initialised whereby one normalises and the other blanks the pilots. Each ambiguity region then needs to be calculated in parallel using the appropriately modified reference function with the two results stitched together to form an ambiguity-free range-Doppler map as demonstrated in [17].

3.3.2.6   Range and Doppler resolution
In Section 3.1, it was shown that the range and Doppler resolution for a PR is determined by the signal bandwidth and the integration time used. Unlike analogue signal-based systems where the signal bandwidth fluctuates from CPI to CPI depending on the signal content, leading to range resolution fluctuations, the bandwidth of a digital OFDM-based signal is constant regardless of signal content. This results in a predictable, range resolution regardless of signal content. Depending on the exact signal structure used in the DVB-T2 transmission, range resolutions of up to 30 m can be achieved.

Typically the integration time used with DVB-T2-based PR is in the order of a single OFDM frame and while longer integration times are achievable, this is typically not required in real systems. Using a single frame of 250 ms results in a Doppler resolution of 4 Hz.

3.3.3   *Digital audio broadcast transmitters*
Digital audio broadcasting is an European Telecommunications Standards Institute (ETSI) standard [28] and has been developed as a digital successor of FM radio. In contrast to other digital successor technologies to the FM waveform, DAB is not designed to be downward compatible with FM. It was designed to be used for both mobile and stationary receivers, to be operated in a single-frequency network configuration and to be robust even in severe multi-path propagation environments. Given these requirements, and like many of the digital broadcast standards, the OFDM multi-carrier waveform was chosen for the physical layer. The spectrum of a DAB signal is shown in Figure 3.9.

Since its inception, DAB has been shown to be a very useful IoO for PR applications for both medium range air surveillance and even detection of micro unmanned aerial vehicles (UAV).

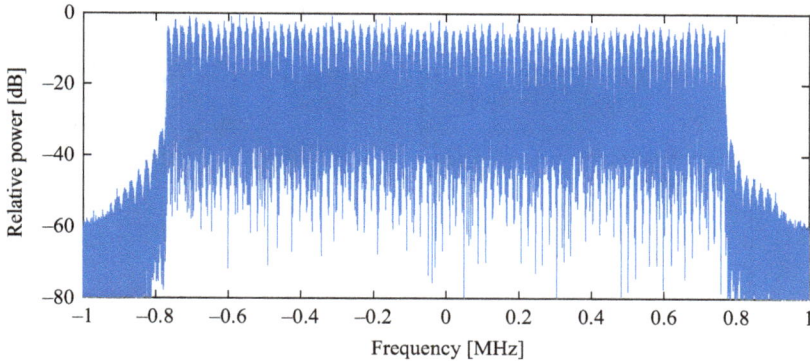

Figure 3.8 Spectrum of a DAB signal

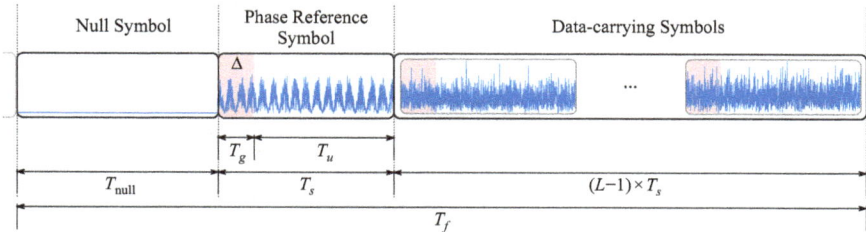

Figure 3.9 Illustration of a DAB transmission frame [29]

3.3.3.1 Signal structure

The standard for DAB was developed by the European Union project Eureka-147 from 1987 to 2000 and is now available under the code EN 300 401 from the European Telecommunications Standards Institute. Today, it is widely adopted as a successor of the analogue FM waveform used for broadcasting terrestrial audio content such as commercial radio stations. Even though most countries with regular DAB services also still maintain the legacy FM systems, many of them have plans to stop analogue broadcasting altogether.

As with the DVB-T2 signal, the DAB signal is organised into different symbols within a single frame as depicted in Figure 9. Every frame starts with the Null-symbol that allows for coarse synchronisation, followed by what is known as the phase reference signal (PRS) which is used for fine synchronisation as well as a phase reference for the differential demodulation of the following data-carrying symbols.

The exact shape of the PRS is defined within the DAB standard and is the same for every frame. The carrier modulation used for the data symbols is differential quadrature phase shift keying (D-QPSK). In contrast to traditional or absolute QPSK,

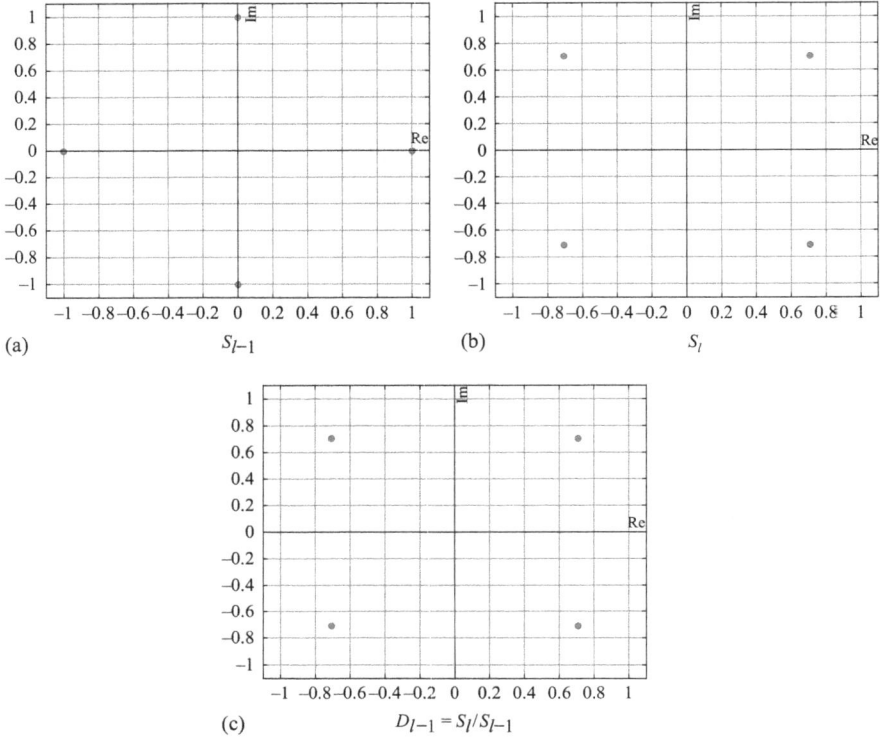

Figure 3.10   *Plots shown on the complex plane of various data derived from a perfect DAB signal [29]*

the information is not encoded in the absolute position of the carrier value in the I-Q complex plane, but in the phase difference of two consecutive symbols.

Figures 3.10a and 3.10b illustrate the perfect carrier values of two consecutive OFDM symbols. This shows how the carrier values are offset by a factor of $\pm\pi/4$. In the demodulation process, the values of the carriers of one OFDM symbol is divided by the values of the previous symbol. Figure 3.10c is the result of the division of the two symbols, leading to a perfect demodulation of the resulting data symbol.

Applying this process to real data results in the constellation plots as shown in Figure 3.11 where all the values of the carriers in Figure 3.11c lie in one of the four quadrants, allowing for the carriers to be used for decoding the content of the signal. This process of dividing by the previous symbol's values naturally requires the knowledge of the first symbol in the frame, which is, in our case, given by the phase reference symbol.

The differential modulation has several advantageous side-effects. First, the division by the previous symbol will automatically take care of frequency equalisation

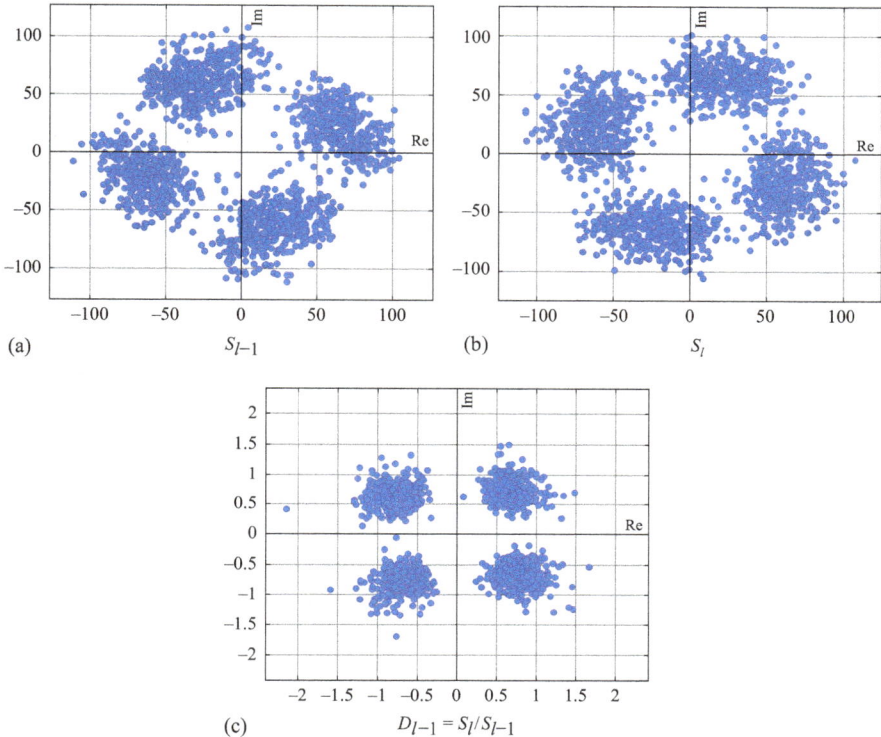

Figure 3.11 *Plots shown on the complex plane of data derived from a real-world DAB signal [29]*

since two consecutive symbols will experience a similar frequency response within the same physical channel. This means that carriers that are attenuated by the channel will be boosted by the division and are thus also normalised. Compared to pilot-based equalisation where the frequency response for the data-carriers can only be estimated by interpolating between two pilot carriers, the differential modulation allows for having an equalisation for every carrier without having to sacrifice any carriers for pilot slots. Second, slow changes in the channel characteristics, i.e. in the order of the symbol duration are also taken care of which makes the reception with a moving mobile receiver possible. For PR processing, the lack of pilot carriers is a great advantage since there are no additional ambiguities introduced by the deterministic parts of the signal as is the case with DVB-T2-based systems.

Unfortunately, while there are no pilot ambiguities present when using a DAB signal for PR, there are still other sources of periodicity that lead to ambiguities. Like with most OFDM-based systems, including DVB-T2, in order to mitigate inter-symbol interference, that is, interference caused by delayed copies of the signal due to multi-path, a guard interval (also known as a *cyclic prefix*) is added to the beginning of every symbol. This way the receiver can delay the process decoding until all the received signal components belonging to the current symbol are received. Thanks to the definition of the carriers in the frequency domain, the symbol is cyclic in the time-domain. This is why it makes sense to use a cyclic extension of the time signal as a signal for the guard interval. In practise, this is achieved by adding part of the end of each symbol in the time domain to the beginning of the same symbol, hence the name cyclic prefix. The duration of the guard interval determines what the maximally tolerable delay-spread is before multi-path components cause interference.

The mathematical representation of the DAB signal is given in the DAB standard as follows:

$$s(t) = \text{Re}\left(e^{2j\pi f_c t} \sum_{m=-\infty}^{\infty} \sum_{l=0}^{L} \sum_{k=-K/2}^{K/2} z_{m,l,k} \cdot g_{k,l}(t - mT_F - T_{NULL} - (l-1)T_S) \right) \quad (3.27)$$

The various parameters are described as follows:

L is the number of OFDM symbols per transmission frame without the Null symbol

K is the number of transmitted carriers

T_F is the transmission frame duration

T_U is the inverse of the carrier spacing

Δ is the duration of the time interval called guard interval or cyclic prefix

T_S is the duration of the OFDM symbols ($T_U + \Delta$) other than the Null symbol ($l = 1, 2, \ldots, L$)

$z_{m,l,k}$ are the complex D-QPSK symbols of carrier k, OFDM symbol l in transmission frame m. The central carrier, i.e. the DC-carrier is not transmitted and therefore $z_{m,l,0} = 0$.

f_c is the carrier frequency of the transmitted signal

Looking closely at (3.27), you will notice that it contains three summations. The first one occurs over all the transmission frames, the second one over the symbols within the frame and, finally, the third one is over all the carriers in a symbol. The summation over the carriers is performed over $k = -K/2 \ldots K/2$. This simply re-arranges the carriers and leaves out the ones that are set to 0 anyway, i.e. the guard carriers – not to be confused with the guard symbol.

The terms within the argument of the function g(.) account for the different time shifts that are necessary to make sure there are no overlaps of the symbols. The term T_{NULL} is necessary because the Null symbol is slightly longer than the other symbols. The different parameters were conceived for operation in four different modes and are summarised in Table 3.7. The duration is given as multiples of the OFDM clock period $T = 1/2048000s$. Depending on the choice of parameters, different network configurations and deployments are possible.

For use in PR, the most relevant parameters are the symbol duration and the number of active carriers K as these are the parameters that influence the maximum unambiguous Doppler and the range resolution if frequency-domain symbol-wise

Table 3.7   Definition of parameters for the different transmission modes

Parameter	Mode I	Mode II	Mode III	Mode IV
L	76	76	153	76
K	1,536	384	192	768
T_F	19,6608T	49,152T	49,152T	98,304T
	96 ms	24 ms	24 ms	48 ms
T_{NULL}	2,656T	664T	345T	1,328T
	1.297 ms	324 ms	168 ms	623 ms
T_S	2,552T	638T	319T	1,276T
	1.246 ms	312 ms	156 ms	623 ms
T_U	2,048T	512T	256T	1,024T
	1 ms	250 ms	125 ms	500 ms
Δ	504T	126T	64T	252T
	246 ms	62 ms	31 ms	123 ms

correlation is applied. These parameters have slightly less importance when traditional time-domain-based correlation is used for creating the range-Doppler maps.

As discussed previously, DAB has been deployed in several countries as a successor to FM-radio. Being a European initiative, most European countries have adopted regular DAB service; however, most of them also maintain the analog FM service. An updated list of countries using DAB can be found in [30].

Mode I is conceived for terrestrial broadcasting in very high frequency (VHF) Band III. This frequency band is administrated differently depending on the region of the world, but in most cases, the frequencies between 174 and 230 MHz are used in this mode. This is the most prevalent mode and very well suited for passive radar applications. All the other modes, i.e. II–IV are conceived for both terrestrial and satellite-based use for frequencies below 3 GHz.

In terms of transmit power, DAB is usually operated at lower power than FM transmitters serving a similar area. As a rule of thumb, one can expect roughly 10 dB less transmit power than from a comparable FM transmitter. In addition to the use of lower power transmitters, digital systems also typically utilise higher frequencies to take advantage of the increased bandwidth. This leads to increased propagation attenuation and therefore detection ranges for PR using DAB are in general shorter than with FM radio.

3.3.3.2   Range-Doppler ambiguities

The ambiguity function of a DAB signal has a perfect thumb-tack response with no delay-Doppler coupling and very good zero-Doppler rejection. If the correlation is done over more than one OFDM symbol, there are secondary inter-symbol ambiguities – peaks at a delay corresponding to the length of the useful part of the symbol T_U. This is due to the cyclic extension for the guard interval that naturally correlates with the end of the symbol. This is, however, usually not a problem since in mode I $T_U = 1$ ms which corresponds to 300 km of range.

The peak-to-sidelobe ratio (PSLR) of targets within the range-Doppler map can be significantly improved by only correlating in a symbol-wise fashion and only using the cyclic, i.e. the useful part of the symbol of duration T_U. In this way, the theoretical self-ambiguity function is indeed perfect, i.e. the PSLR becomes infinite. In practise, channel effects and various imperfections in the transmit and receive chain such as frequency errors limit the maximum possible PSLR to 80–90 dB.

3.3.3.3   Range and Doppler resolution

As with all radar systems, the range and Doppler resolution is determined by the signal bandwidth and the integration time. Like DVB-T2-based PR, the range and Doppler resolution of a DAB PR remains constant and independent of signal content. Depending on the exact signal structure used in the DAB transmission, range resolutions of up to 150 m can be achieved.

Typically the integration time used with DAB-based PR is in the order of a single OFDM frame and while longer integration times are achievable, this is typically not required in real systems.

References

[1]   D.W. O'Hagan. *Passive Bistatic Radar Performance Characterisation Using FM Radio Illuminators of Opportunity*. PhD Thesis, University College London, UK, April 2009.

[2]   H. Griffiths and C. Baker. An *Introduction to Passive Radar*, Artech House, 2017.

[3]   C.A. Tong. *A Scalable Real-time Processing Chain for Radar Exploiting Illuminators of Opportunity*. PhD Thesis, University of Cape Town, South Africa, December 2014.

[4]   S. Paine. *Electronic Countermeasures Applied to Passive Radar*. PhD Thesis, University of Cape Town, 2019.

[5]   D. Moser, G. Tresoldi, C. Schüpbach, and V. Lenders. Design and evaluation of a low-cost passive radar receiver based on iot hardware. In: *2019 IEEE Radar Conference (RadarConf)*, pp. 1–6, 2019.

[6]   P.E. Howland, D. Maksimiuk, and G. Reitsma. FM radio based bistatic radar. *IEE Proceedings – Radar, Sonar and Navigation*, 152(3):107–115, 2005.

[7]   H. Kuschel, M. Ummenhofer, D. O'Hagan, and J. Heckenbach. On the resolution performance of passive radar using DVB-T illuminations. In *11th International Radar Symposium*, pp. 1–4, June 2010.

[8]   H. Kuschel. Approaching 80 years of passive radar. In *2013 International Conference on Radar*, pp. 213–217, September 2013.

[9]   H.D. Griffiths and N.R.W. Long. Television-based bistatic radar. *Communications, Radar and Signal Processing, IEE Proceedings F*, 133(7):649–657, 1986.

[10] P.E. Howland. Target tracking using television-based bistatic radar. *IEE Proceedings – Radar, Sonar and Navigation*, 146(3):166–174, 1999.

[11] F.G. Stremler. *Introduction to Communication Systems*, 3rd ed. Pearson, 1990.

[12]   S.S. Haykin and M. Moher. *Communication Systems*. Wiley, 2010.
[13]   F.E. Relton. Applied Bessel functions. *Dover Books on Intermediate and Advanced Mathematics*. Dover, 1965.
[14]   M. Inggs, C. Tong, D. O'Hagan, *et al.* Noise jamming of a FM band commensal sensor. *IET Radar, Sonar Navigation*, 11(6):946–952, 2017.
[15]   C. Schüpbach, C. Patry, and A. Jaquier. Passive radar illuminator identification in single frequency networks. In *2019 IEEE Radar Conference (RadarConf)*, pp. 1–5, 2019.
[16]   R. Saini and M. Cherniakov. DTV signal ambiguity function analysis for radar application. *IEE Proceedings – Radar, Sonar and Navigation*, 152(3):133–142, 2005.
[17]   Z. Gao, R. Tao, Y. Ma, and T. Shao. DVB-T signal cross-ambiguity functions improvement for passive radar. In *2006 CIE International Conference on Radar*, pp. 1–4, October 2006.
[18]   C.R. Berger, B. Demissie, J. Heckenbach, P. Willett, and S. Zhou. Signal processing for passive radar using OFDM waveforms. *IEEE Journal of Selected Topics in Signal Processing*, 4(1):226–238, 2010.
[19]   H.A. Harms, L.M. Davis, and J. Palmer. Understanding the signal structure in DVB-T signals for passive radar detection. In *2010 IEEE Radar Conference*, pp. 532–537, May 2010.
[20]   J.E. Palmer, H.A. Harms, S.J. Searle, and L. Davis. DVB-T passive radar signal processing. *IEEE Transactions on Signal Processing*, 61(8):2116–2126, 2013.
[21]   L. Fang, X. Wan, G. Fang, F. Cheng, and H. Ke. Passive detection using orthogonal frequency division multiplex signals of opportunity without multipath clutter cancellation. *IET Radar, Sonar Navigation*, 10(3):516–524, 2016.
[22]   European Telecommunications Standards Institute. *ETSI EN 300 744 – Digital Video Broadcasting (DVB)*.
[23]   Singapore Infocomm Media Development Authority. *Dataone Limited Response to Consultation Paper on Datacasting*. PhD Thesis, 1998.
[24]   Advent Television. *The Future is in Digital Broadcasting and that future is with Advent Television*. PhD Thesis, 2001.
[25]   European Telecommunications Standards Institute. ETSI TS 102 831-Digital Video Broadcasting (DVB); Implementation guidelines for a second generation digital terrestrial television broadcasting system (DVB-T2).
[26]   D.W. O'Hagan, M. Setsubi, and S. Paine. Signal reconstruction of DVB-T2 signals in passive radar. In *2018 IEEE Radar Conference (RadarConf18)*, pp. 1111–1116, April 2018.
[27]   D.A. Kovalev and V.I. Veremyev. Correction of DVB-T2 signal cross-ambiguity function for passive radar. In *2014 International Radar Conference*, pp. 1–4, October 2014.
[28]   ETSI. Radio broadcasting systems; digital audio broadcasting (DAB) to mobile, portable and fixed receivers. Technical Report, 2017.
[29]   C. Tilbury. *DAB Processing Chain for Passive Radar Applications*. Master's Thesis, University of Cape Town, 2020.
[30]   Wikipedia Contributors. Countries using DAB/DMB—Wikipedia, the free encyclopedia, 2022. Accessed 23-October-2022.

Chapter 4

Land clutter statistics in passive bistatic radar

Luke Rosenberg[1] and Vichet Duk[2]

Passive operation is an important operational concept for future airborne radar platforms. Successful target detection requires an understanding of the underlying clutter statistics in order to set the detection threshold accurately. In addition, clutter statistics are important for modelling of passive radar performance and simulation of passive bistatic radar to aid in the development of new target detection algorithms. In this chapter, the clutter statistics of data from an airborne platform are studied over a range of bistatic geometries and receive polarisations. The illuminator is a digital video broadcast-terrestrial (DVB-T) station and the collected data spans both ground and maritime regions. The key contributions in this chapter include a study of amplitude statistics using a number of common distributions, understanding the impact of the reference signal quality and an analysis showing the distribution accuracy as the integration time and bistatic range varies.

Acronyms

3MD	tri-modal
APART-GAS	Active Passive Radar Trials-Ground-borne Airborne Stationary
BD	Bhattacharyya distance
CCDF	complementary cumulative distribution function
CNR	clutter to noise ratio
CPI	coherent processing interval
DSTG	Defence Science and Technology Group
DVB-T	digital video broadcast-terrestrial
ECA+	extensive cancellation algorithm plus
FFT	fast Fourier transform
FM	frequency modulation
GIS	geographic information system
GNSS	global navigation satellite system
I	in-phase
IFFT	inverse fast Fourier transform

[1] Defence Science and Technology Group, Australia
[2] Fraunhofer FHR, Germany

K+N	K+noise
K+R	K+Rayleigh
LN	log-normal
MRC	maximal ratio combining
NRL	Naval Research Laboratory
OFDM	orthogonal frequency-division multiplexing
P+N	Pareto+noise
PDF	probability density function
Q	quadrature
RF	radio frequency
SAR	synthetic aperture radar
TE	threshold error

List of symbols

$a_n, \mathbf{a}$	3MD-distribution coefficients
$a_{Tx}(\cdot)$	complex magnitude for the transmitted signal
a_q	complex magnitude for the qth scatterer
b_k	K-distribution scale parameter
b_p	pareto-distribution scale parameter
b_r	K+Rayleigh-distribution scale parameter
$c_n, \mathbf{c}$	3MD-distribution coefficients
$\mathcal{C}$	clutter-to-noise ratio
$\mathcal{C}_r$	clutter-to-noise plus Rayleigh ratio
$\delta(\cdot)$	delta function
d_r	parameter used in $z\log z$ estimator
d_z	parameter used in $z\log z$ estimator
D	number of elements in the 3MD-distribution
D_{BD}	Bhattacharyya distance metric
D_{TE}	threshold error metric
$E(\cdot)$	generalised exponential integral function
f_D	Doppler frequency
$f_{D,q}$	Doppler frequency for the qth scatterer
f_S	sample frequency
$h(\cdot)$	reciprocal filter
k_r	K+Rayleigh-distribution Rayleigh to clutter ratio
K_{int}	number of histogram samples
m_{LN}	log normal mean
M	number of pulses in the coherent integration time
ν_k	K-distribution shape parameter
ν_p	pareto-distribution shape parameter
ν_r	K+Rayleigh-distribution shape parameter
$n(\cdot)$	thermal noise and RF interference
N_c	number of scatterers
ϕ	angular direction

ϕ_q angular location for the qth scatterer
ϕ_{Tx} angular location for the transmitter
Φ_q angular region for the qth scatterer
p_c clutter mean power
p_n noise mean power
$P(\cdot)$ probability density function
ρ_c normalised clutter mean power
ρ_n normalised noise mean power
σ_{LN} log-normal standard deviation
$s_{\text{Rx}}(\cdot)$ received signal
$s_{\text{Tx}}(\cdot)$ transmitted signal
t time
τ speckle mean power/texture
τ_r speckle mean power/texture for the K+Rayleigh-distribution
τ_q delay for the qth scatterer
τ_{Tx} delay for the transmitter
$x(\cdot)$ processed signal after reciprocal filter
z intensity signal

4.1 Introduction

Over the past decade, there has been significant research into the exploitation of passive bistatic radar systems for imaging, detection and tracking of ground, maritime and airborne targets [1]. The attraction of passive radar is its covert operation, wide coverage area and reduced cost. This is due to the use of an illuminator of opportunity which can either be terrestrial or space based. Some examples include WiMAX [2], frequency modulation (FM) radio [3], digital video broadcast-terrestrial (DVB-T) [4], global navigation satellite system (GNSS) [5], Iridium and INMARSAT communication satellites [6] and Sirius-XM digital radio [7,8]. In this chapter, the characteristics of ground clutter from an airborne platform are studied. This will aid with target detection, the modelling of passive radar performance and simulation of passive bistatic radar to aid in the development of new target detection schemes.

There has been a significant amount of work understanding the statistics of monostatic clutter [9,10] and to a lesser extent active bistatic clutter [11,12]. The majority of the published work on bistatic land clutter has focussed on the mean radar cross section with comprehensive tables provided in [11]. When it comes to characterising the amplitude statistics, the research has focussed on sea-clutter with an extensive study using the NetRAD multistatic radar system [12,13]. The key take away from this work is that the bistatic mean reflectivity and amplitude statistics vary with the bistatic angle, typically having a lower mean backscatter and are less spiky than the monostatic data [12].

Target detection in passive bistatic radar is typically performed in the range/Doppler domain where the extent of the clutter dominated delay bins depends on the system geometry, the strength of the illuminator and the sensitivity of the radar receiver. Malanowski *et al.* [3] investigated clutter distributions of land and sea clutter using three different ground-based FM transmitters. The authors showed that the in-phase (I) and quadrature (Q) components of the clutter were broadly Gaussian distributed, but had extended tails. Recently, the statistical characteristics of sea clutter have been studied using the Sirius-XM and Optus D-3 illuminators [7,14]. In these papers, the spectrogram of the sea clutter was extracted and the amplitude and Doppler characteristics were studied. The K+Rayleigh amplitude distribution was used to model the data over a range of integration times and bistatic range resolutions.

When radar signals are collected from a moving platform, there is a broadening of the Doppler spectrum due to the aircraft motion. More specifically, the receive beampattern is convolved with the Doppler spectrum of the clutter scene [15] with the result being that detection of ground moving targets becomes more challenging as the spread clutter can confuse and mask slow moving targets. Duk *et al.* [16] investigated the statistical distribution of land clutter but from an airborne platform using a DVB-T transmitter as an illuminator of opportunity. For this illuminator, the authors showed the distributions of the I and Q clutter components were not Gaussian, with a log normal distribution giving a good statistical representation for the dataset.

In this chapter, the clutter statistics of data from an airborne platform are studied over a range of bistatic geometries and receive polarisations. The data collection trial and signal processing are first presented in Section 4.2, with the different amplitude distributions given in Section 4.3 along with two goodness of fit metrics. Section 4.4 then describes the three datasets studied in this chapter along with a statistical analysis using the amplitude distributions. Also this section is a study looking at the impact of the reference signal quality and an analysis showing the distribution accuracy as the integration time and bistatic range varies. A summary of the key results is then given in Section 4.5.

4.2   Trial data and signal processing

4.2.1   APART-GAS trial

The 'Active Passive Radar Trials-Groundborne Airborne Stationary' (APART-GAS) measurement campaign was conducted in September 2019 by two NATO research groups dedicated to passive radar. During this trial, data was collected using an airborne passive radar system developed by the Fraunhofer FHR. This radar system consisted of a dual super-heterodyne receiver with four spatial channels and a sampling frequency of 64 MHz for the in-phase component of each radio frequency (RF) channel. The receiver hardware was mounted in the cockpit of the ultra light aircraft *Delphin* and four dual-polarized Yagi-Uda antennas were used as an along track linear array inside of a wing mounted pod as shown in Figures 4.1 and 4.2. During the trial, these were configured using different combinations of vertical and horizontal polarisations.

The trial took place in the Northern part of Poland close to the Baltic Sea with the *Koszalin-Gologora* DVB-T transmitter serving as the illuminator of opportunity. It broadcasts with five DVB-T channels, a horizontal polarisation and an effective

Figure 4.1   The airborne receiver 'Delphin' with the pod mounted below the right wing

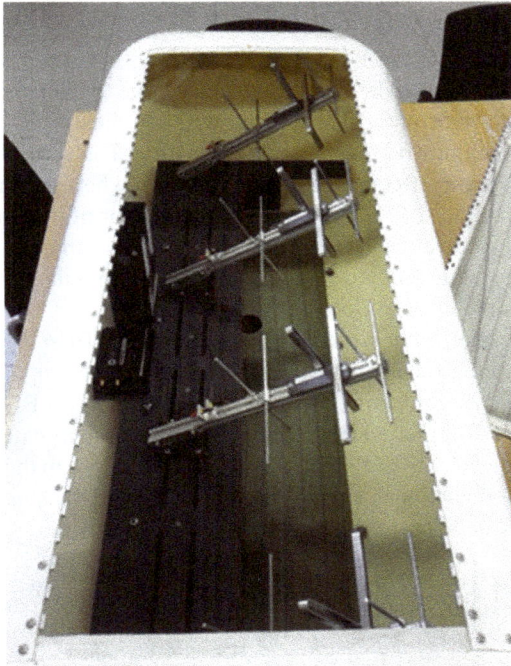

Figure 4.2   Yagi-Uda antennas shown inside the wing mounted pod

radiated transmit power of 100 kW. However, only two of these channels were within the effective bandwidth of the receiver and able to be used as illumination signals. The centre frequencies correspond to 658 MHz and 682 MHz with each channel having a bandwidth of 7.61 MHz. Also, during the trial a ground-based receiver was setup near the coast to record a clean copy of the transmitted signal.

4.2.2 Signal processing

The DVB-T signal is broadcast widely using an orthogonal frequency-division multiplexing (OFDM) waveform which is well suited for passive radar due to its wider bandwidth when compared to FM radio. To isolate the desired DVB-T channel, the airborne receiver samples the spectrum and applies a bandpass filter that matches the channel bandwidth, resulting in a pixel spacing of 39.4 m for the bistatic range. At time t, the received signal, $s_{Rx}(t)$ can be described by summation of the direct transmitted signal $s_{Tx}(t)$, the reflected signals from the illuminated scene (natural and man-made) and the thermal noise and RF interference, $n(t)$:

$$s_{Rx}(t) = a_{Tx}s_{Tx}(t - \tau_{Tx}) \exp{(j2\pi f_D(\phi_{Tx})t)}$$

$$+ \sum_{q=1}^{N_c} \int_{\Phi_q} a_q(\phi)s_{Tx}(t - \tau_q) \exp{(j2\pi f_{D,q}(\phi)t)} \, d\phi + n(t). \qquad (4.1)$$

The first term refers to the direct path signal where a_{Tx} is the complex amplitude and τ_{Tx} and $f_D(\phi_{Tx})$ are the time delay and Doppler shift due to the radial motion between the receiver and the transmitter respectively with ϕ_{Tx} being the angle between the receiver platform heading and the transmitter location. The second term refers to the clutter echoes, which are reflected from N_c range cells, where each range cell consists of multiple clutter patches. The azimuthal angular sector of the illuminated scene is defined by Φ_q and is dependent on the antenna beampattern of the receiver. For the qth range cell, $a_q(\phi)$ is the complex amplitude from the angular direction ϕ, τ_q is the delay and $f_{D,q}(\phi)$ is the bistatic Doppler shift.

Range compression is achieved using a reciprocal filter $h^{(m)}(t)$ [17,18], in a batch-wise correlation process for each slow-time instance m. This filter removes any unwanted underlying information from the transmitted DVB-T signal and hence reduces sidelobes in Doppler which could potentially contaminate the target's range cell. In this process, each DVB-T symbol, $m = 0, \ldots, M - 1$ in the coherent processing interval (CPI) is regarded as a single slow-time instance or 'pulse'. The reciprocal filtering approach can be performed in the frequency domain,

$$x^{(m)}[l] = \text{IFFT} \left\{ \frac{\text{FFT}\{s_{Rx}^{(m)}[l]\}}{\text{FFT}\{h^{(m)}[l]\}} \right\} \qquad (4.2)$$

where (I)FFT is the (Inverse) Fast Fourier Transform and $l = tf_S$ represents samples acquired with sampling frequency f_S, thus representing discretised fast-time. To be able to apply the reciprocal filter effectively, the transmitted signal must have been reconstructed using a demodulation/remodulation process. This is achieved using the maximal ratio combining (MRC) signal [19] which yields an optimum combination of the receiving channels. The next processing step then removes the direct signal interference using the extensive cancellation algorithm plus (ECA+) algorithm [20], before a final Doppler processing step converts the data into the range/Doppler domain that is used for the clutter analysis in this chapter.

4.3 Amplitude statistics

In this section, the amplitude models are presented for the log-normal (LN), K+noise (K+N), K+Rayleigh (K+R), Pareto+noise (P+N) and the (3MD)-distributions. As passive bistatic radar data is typically characterised in the range/Doppler domain, the clutter regions have undergone coherent integration on the order of seconds and hence only single look distributions are assumed. In the descriptions below, the estimation approach for the parameters in each model are described. The final section then describes the goodness of fit metrics.

4.3.1 Distribution models

Log-normal: The LN distribution is typically described in terms of intensity, z and arises if the logarithm is normally distributed with mean m_{LN} and variance σ_{LN}^2. It is given by [21],

$$P(z) = \frac{1}{z\sqrt{2\pi\sigma_{LN}^2}} \exp\left[-\frac{(\ln z - m_{LN})^2}{2\sigma_{LN}^2}\right], \quad z \geq 0. \tag{4.3}$$

To estimate the two parameters, one method is to measure the moments from the real data and apply them to the theoretical moments. In this case, the mean and variance of the logarithm of the intensity can be estimated by,

$$\hat{m}_{LN} = \ln \langle z \rangle - 0.5 \ln\left(1 + \frac{\langle z^2 \rangle}{\langle z \rangle^2}\right),$$

$$\hat{\sigma}_{LN}^2 = \ln\left(1 + \frac{\langle z^2 \rangle}{\langle z \rangle^2}\right) \tag{4.4}$$

where the kth order moment is given by

$$\langle z^k \rangle = \exp\left[km_{LN} + \frac{k^2\sigma_{LN}^2}{2}\right]. \tag{4.5}$$

K+noise-distribution: The most commonly used probability density function (PDF) model in both real and synthetic aperture radar (SAR) is the K-distribution. This model was originally developed using a large quantity of recorded data which showed that the speckle had Gaussian statistics and the texture was a good fit to the gamma family of distributions [22]. It can be formulated as a compound

Gaussian model, with the PDF of the intensity of the clutter returns found by averaging the speckle PDF, $P(z|\tau)$ over all possible values of τ so that

$$P(z) = \int_0^\infty P(z|\tau)P(\tau)d\tau. \tag{4.6}$$

One common extension of the compound model is the inclusion of the noise mean power, p_n. The noise has the same Gaussian PDF as the speckle and the mean power of speckle-pulse-noise is now $\tau + p_n$ and the PDF of the intensity of speckle-plus-noise is given by

$$P(z|\tau) = \frac{z}{\tau + p_n} \exp\left[-\frac{z}{\tau + p_n}\right]. \tag{4.7}$$

For the K-distribution, the texture is modelled by a gamma distribution,

$$P(\tau) = \frac{b_k^{\nu_k}}{\Gamma(\nu_k)} \tau^{\nu_k - 1} \exp[-b_k\tau], \qquad \nu_k, b_k > 0 \tag{4.8}$$

with shape given by ν_k and scale, $b_k = \nu_k/p_c$, where p_c is the mean clutter power. If thermal noise is included, the K+N distribution, $P(z)$ has no closed form solution.

For this distribution, the shape and scale are estimated using the $z\log z$ parameter estimation. This was first proposed by Blacknell and Tough in [23] for the single look case and $p_n = 0$. More generally, the shape estimate is found by numerically solving [24],

$$\frac{\langle z\log z \rangle}{\langle z \rangle} - \langle \log z \rangle = d_z \tag{4.9}$$

where d_z is given by,

$$d_z = \frac{1}{1 + (1/C)} \exp\left[\frac{\hat{\nu}_k}{C}\right] E_{\hat{\nu}_k + 1}\left(\frac{\hat{\nu}_k}{C}\right) \tag{4.10}$$

and C is the clutter to noise ratio (CNR) with $E_{\nu_k}(\cdot)$ being the generalised exponential integral function.

K+Rayleigh-distribution: The dominant Rayleigh component of the received backscatter has been presumed to arise from additive white receiver noise. The K+R distribution was formalised in [25] after observations by Sletten [26] and Lamont-Smith [27] who found evidence of a further Rayleigh component beyond what is captured by speckle and thermal noise. It is defined by explicitly separating the speckle mean into two components, $\tau = \tau_r + p_r$, where the extra Rayleigh component, p_r is modelled in the same fashion as the thermal noise.

The K+R model uses a gamma distribution for the texture,

$$P(\tau_r) = \frac{b_r^{\nu_r}}{\Gamma(\nu_r)} \tau_r^{\nu_r - 1} \exp[-b_r\tau_r], \qquad 0 \leq \tau_r \leq \infty \tag{4.11}$$

where ν_r is the shape and $b_r = \nu_r/p_c$ is the scale. To calculate the compound integral in (4.6), the integration is now performed with the modified speckle mean level, τ_r instead of τ. If the noise mean power is known, then the influence of the extra Rayleigh component can be measured by the ratio of the mean of the Rayleigh component to the mean of the gamma distributed component of the clutter, by $k_r = p_r/p_c$.

Parameter estimation for the K+R follows the K-distribution with the shape estimate $\hat{v} \rightarrow \hat{v}_r$ and the noise mean power, $p_n \rightarrow p_n + p_c$. For the $z\log z$ estimator, the expression for d_z in (4.10) is given by [24],

$$d_z = \frac{1}{1 + 1/\mathcal{C}_r} \exp\left[\frac{\hat{v}_r}{\mathcal{C}_r}\right] E_{\hat{v}_r+1}\left(\frac{\hat{v}_r}{\mathcal{C}_r}\right) \tag{4.12}$$

where the clutter to noise plus Rayleigh ratio is given by

$$\mathcal{C}_r = \frac{p_c}{p_n + p_r} = \frac{1}{1/\sqrt{\hat{v}_r d_r} + 1} \tag{4.13}$$

and

$$d_r = \frac{\langle z^2 \rangle}{2 \langle z \rangle^2} - 1. \tag{4.14}$$

Pareto+noise-distribution: The Pareto model was first used for modelling clutter by Balleri *et al.* [28], Fayard and Field [29] and later by others at US Naval Research Laboratory (NRL) [30] and Defence Science and Technology Group (DSTG) [31,32]. For this compound Gaussian model, the texture has an inverse gamma distribution

$$P(\tau) = \frac{b_p^{v_p}}{\Gamma(v_p)} \tau^{-v_p-1} \exp\left[-b_p/\tau\right], \qquad v_p > 1, b_p > 0 \tag{4.15}$$

where v_p is the shape and $b_p = p_c(v_p - 1)$ is the scale. When thermal noise is included, the P+N distribution from (4.6) again has no closed form.

For the $z\log z$ estimator, the shape parameter is found by numerically solving the equality in (4.9) using the following expression [24]:

$$d_z = \frac{1}{1 + (1/\mathcal{C})} \left(\frac{1}{\hat{v}_p - 1} - \exp\left[(\hat{v}_p - 1)\mathcal{C}\right] E_{\hat{v}_p+1}\left((\hat{v}_p - 1)\mathcal{C}\right)\right). \tag{4.16}$$

Tri-modal discrete (3MD)-distribution: The compound models in the literature all assume a continuous texture distribution function which suggests a small probability of texture values approaching infinity. This is not physically sound as it cannot be measured by any real radar system. The 3MD model [33,34] instead proposes the use of a discrete texture model that assumes that the clutter consists of a finite number of distinct modes or scatterer types, D. This implies that the scatterers in the observed scene are realisations from homogeneous clutter random variables with different texture values. In the original work, it was found that $D = 3$ modes were sufficient to model distributions from the spaceborne SAR imagery, hence the tri-modal in the name. However subsequent analysis has found that up to 5 modes are required to accurately model bistatic sea clutter [35]. One of the consequences of this discretisation is that spatial and long-time correlation cannot be modelled as part of the texture, and hence the model is less suitable for clutter simulation. The texture PDF is given by

$$P(\tau) = \sum_{n=1}^{D} c_n \delta(\tau - a_n), \quad \sum_{n=1}^{D} c_n = 1, \quad a_n, c_n > 0 \tag{4.17}$$

where $\delta(\,\cdot\,)$ denotes the delta-function, $\mathbf{a} = [a_1, \ldots, a_D]$ are the discrete texture intensity levels and $\mathbf{c} = [c_1, \ldots, c_D]$ are the corresponding weightings. The continuous distribution is then given by

$$P(z) = \sum_{n=1}^{D} c_n \frac{\exp\left(-\dfrac{z}{\rho_c a_n^2 + \rho_n}\right)}{(\rho_c a_n^2 + \rho_n)} \tag{4.18}$$

with $\rho_c + \rho_n = \dfrac{p_c^2}{p_c^2 + p_n^2} + \dfrac{p_n^2}{p_c^2 + p_n^2} = 1$.

To determine the model parameters, there are a number of different approaches that have been proposed [35–37]. In this work, a least squares minimisation of the log complementary cumulative distribution function (CCDF) has been used to estimate the model parameters. To determine the minimum model order, D, a single mode ($D = 1$) is first assumed and the threshold error is compared with a value of 0.5 dB. If the error is greater than this threshold, the parameters for a bi-modal fit ($D = 2$) are then estimated and only if this error is greater than the threshold value, are the parameters estimated for the full 3MD model ($D = 3$).

4.3.2 Goodness of fit measures

Assessing the accuracy of a model is important to determine its suitability for representing the radar data. In the statistical literature there are many tests and measurement techniques for assessing models such as the chi-square and Kolmogorov–Smirnov tests and the Bhattacharyya distance (BD) [38]. These are all based on measuring differences between histograms of the model and the data and require a range of intensity values with uniform bin spacings. In this work, the Bhattacharyya distance and the threshold error are used to assess the model fits.

The BD measures the similarity between two distributions [39]. It is given by

$$D_{\mathrm{BD}} = -\ln\left(\sum_{k=1}^{K_{\mathrm{int}}} \sqrt{P_1(z_k)P_2(z_k)}\right) \tag{4.19}$$

where K_{int} is the number of samples. The BD varies between 0 and ∞, where equal distributions have a distance measure of 0. A good match results in a low value of the BD and hence the BD is typically given in dBs.

The threshold error (TE) focusses on the tail of the distribution by considering the threshold difference between two CCDFs [35]. The CCDF is also known as the probability of false alarm in a detection scheme and is typically measured in the log domain. The threshold error is determined by measuring the absolute difference between the CCDFs of the model and data at a fixed CCDF value. If the threshold values at the desired CCDF levels are z_1 and z_2, then

$$D_{TE} = |z_1 - z_2|. \tag{4.20}$$

4.4 Clutter analysis

In this section, three datasets are described in terms of their collection location and the types of clutter present in the scene. A summary of the model fits for each dataset is then presented with appropriate goodness of fit metrics. The next section studies the impact of the reference signal quality, before a comprehensive study is presented showing the distribution accuracy as the integration time and bistatic range varies.

4.4.1 Data sets

Three datasets are considered in this study and summarised in Table 4.1. This includes one where only the vertical polarisation was received and two with both vertical and horizontal polarisations. The geographic region containing the aircraft and transmitter locations is shown in Figure 4.3 with geographic information system (GIS) identifiers overlaid for buildings, developed areas, natural features, railroads, roads, water and wind turbines [40]. The DVB-T transmitter is shown with a magenta box, the aircraft locations and headings that are shown with red boxes and for all datasets, the antenna was looking to the right of the aircraft.

For the first dataset 533, the aircraft had a heading of 203.4° with respect to North and hence the dominant scattering would be backscatter with the bistatic angles over the collection geometry being quite small. The processed range/Doppler image and the GIS overlay is shown in Figure 4.4. There are some interesting observations that can be made by comparing these two images. First, the upside down 'V' in the middle of the image corresponds to an area of water that has a much lower reflectivity. The other areas of strong reflectivity are then a mix of buildings, roads and developed areas. The vertical lines present in the first 15 km are due to poor reconstruction of the reference signal and the apparent target at 22 km is an ambiguity due to co-channel interference from another DVB-T broadcast station.

For the second dataset 816, the aircraft had a heading of 234.6° with respect to North and the geometry is similar to dataset 533 with the dominant reflections

Table 4.1 Dataset overview

Dataset	Receive polarisation	Aircraft location	Altitude (m)	Heading (deg)	Freq. (MHz)
533	Vert.	53.9759°N, 16.6346°E	558.3	203.4°	658
816	Vert./Horiz.	54.4070°N, 16.6703°E	618.1	234.6°	682
241	Vert./Horiz.	54.4340°N, 16.3220°E	496.2	27.1°	658

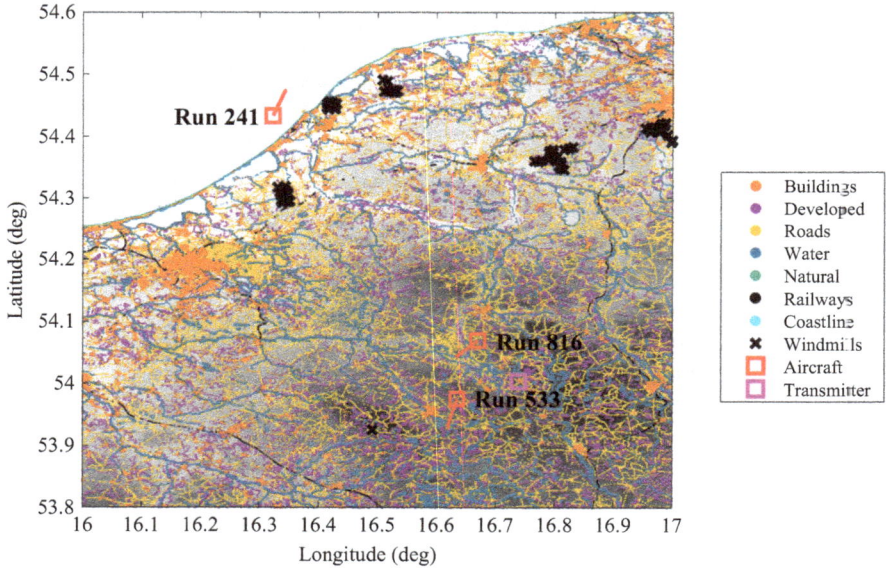

Figure 4.3   Geographic area of the trial showing the dataset locations and GIS features

being backscatter. Figure 4.5 shows the range/Doppler images for the vertical and horizontal polarisations and the GIS overlay. In this scene, the dominant scattering would be from buildings, roads and developed areas, which would indicate the clutter would be spiky with many discrete scatterers. The vertical polarisation is clearly stronger and the vertical lines are much weaker indicating that the reference signal is better reconstructed.

For the third dataset 241, the aircraft had a heading of 27.1° with respect to North and the bistatic geometry is nearly forward scatter. Figure 4.6 shows the range/Doppler images for the vertical and horizontal polarisations and the GIS overlay. They reveal a smaller number of scatterers than datasets 533 and 816 and the reflectivity is slightly stronger for the vertical polarisation. Many of the strong scatterers can be identified as windmills, with the other dominant scattering from other buildings, roads and developed areas. The ambiguity at 22 km is again present with a strong interference component spread along the Doppler bin at 190 Hz. Towards the bottom right of Figure 4.6(iii) is a dashed magenta line that shows the Doppler extent of the receiver. The area between the dashed line and the central region shows the bistatic sea clutter reflection. Unfortunately, the ambiguity noise floor is too high in this dataset, and the sea clutter is not visible.

4.4.2   Model fits

To study the amplitude distributions, a number of data blocks were extracted from both the endo-clutter (clutter+noise) and exo-clutter (noise) regions. These blocks were also

(i) Vertical polarisation

(ii) GIS overlay

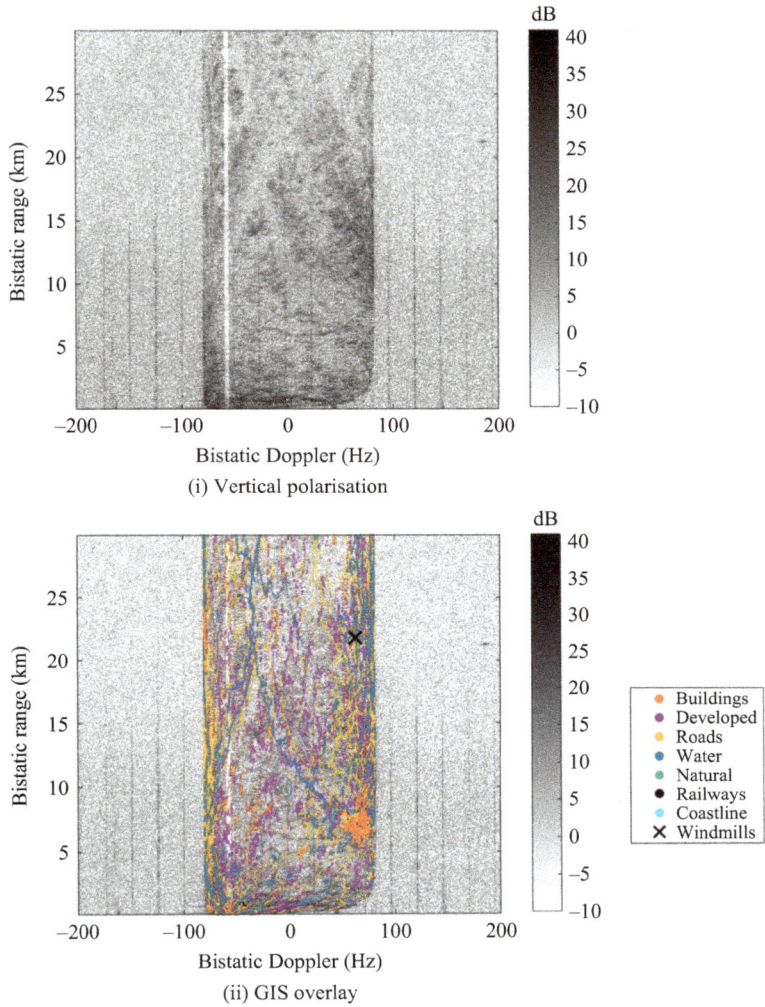

Figure 4.4   Data set 533 range/Doppler image

limited in bistatic range as the CNR changed along range according to the radar range equation. The block size is a tradeoff between having enough samples to measure the statistics and minimising the change in the CNR. Based on this compromise, 50 range bins are chosen which span a total of 1,971 m using a range bin size of 39.4 m. To increase the number of samples, data from four adjacent coherent processing intervals was also used to measure the clutter statistics. The CNR can then be determined by measuring the mean power levels in each region.

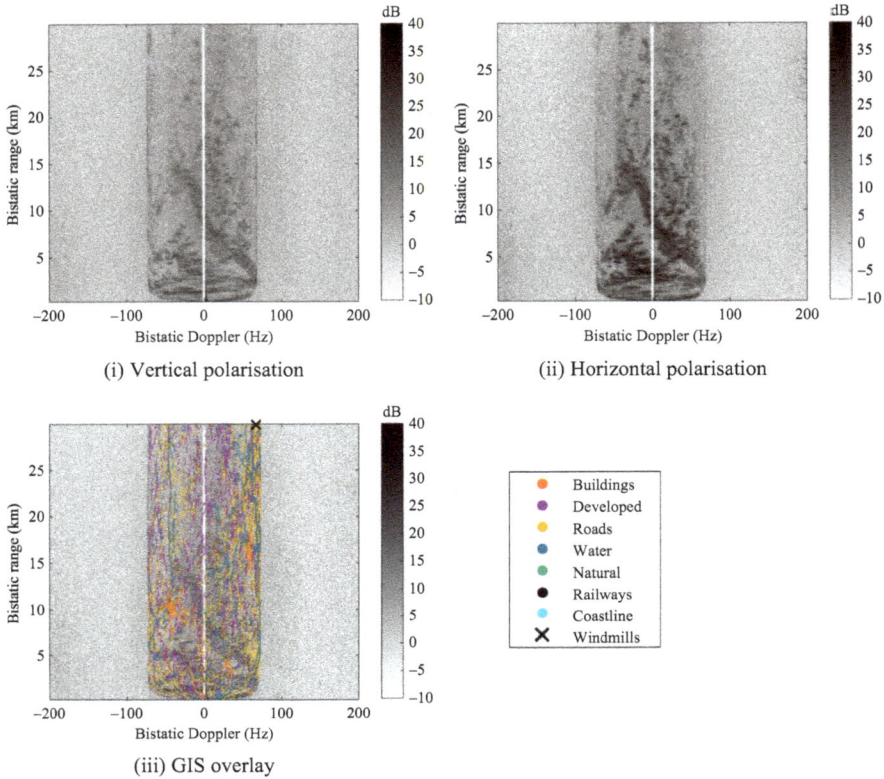

(i) Vertical polarisation

(ii) Horizontal polarisation

(iii) GIS overlay

Figure 4.5   Data set 816 range/Doppler image

To demonstrate the model fits, each dataset has been processed using a 1 s integration time and the first data block (range bins 1–50) has been chosen for analysis. In the following results, the top plot for each pair shows the PDF, while the bottom plot shows the logarithm of the CCDF. Each of the distributions from Section 4.3 has been fitted to the data with the model fit parameters for these examples given in Table 4.2. To highlight a 'good fit' to the data, an ad hoc criteria is defined for model fits with a BD value of less than −20 dB or a TE of less than 1 dB.

The first result in Figure 4.7 considers the dataset 533 vertically polarised data which has a CNR of 8.2 dB and shows the endo-clutter result on the left and the exo-clutter on the right. For the former result, the distributions that show a good fit to the PDF are the log normal, K+N and K+R. However, for the CCDF result, the K+R and the 3MD-distributions show the best match to the data. For the exo-clutter result, the best fits are obtained for the K+R and 3MD-distributions. For dataset 816, results for the vertical and horizontal polarisations are shown in Figures 4.8 and 4.9 with CNRs of 7.2 dB and 11.4 dB respectively. The endo-clutter results are quite similar with

(i) Vertical polarisation

(ii) Horizontal polarisation

(iii) GIS overlay

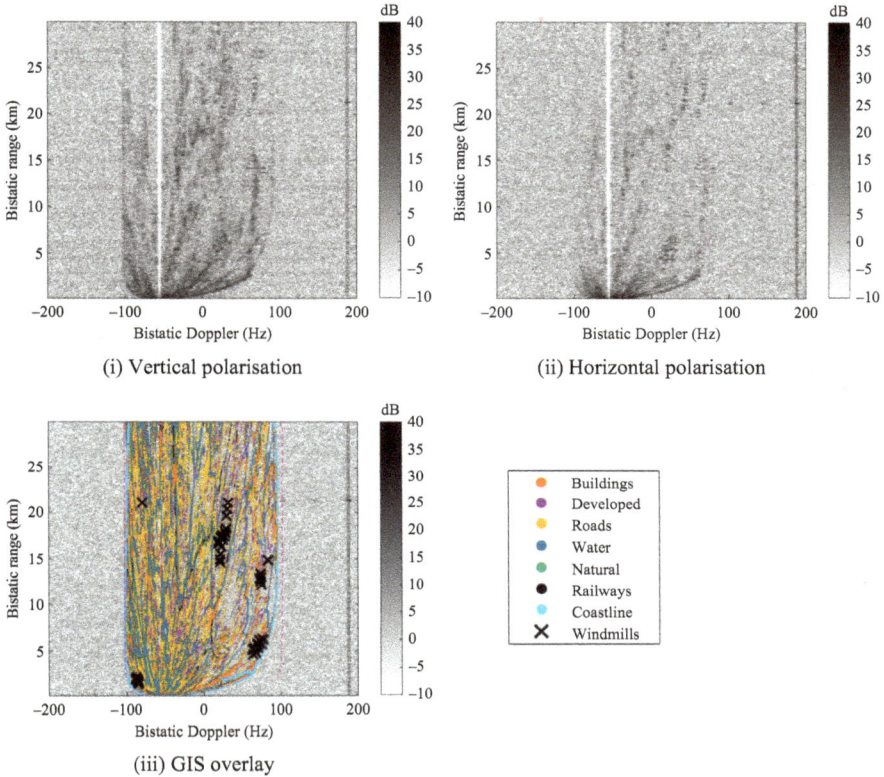

Figure 4.6   Data set 241 range/Doppler image

the LN, K+N and K+R distributions all having a good fit to the PDF, while for the CCDF, the K+N, K+R and 3MD-distributions show the best match. For the exo-clutter region, no distribution matches either polarisation with the K+N, K+R and 3MD each achieving a fit to just one. Figures 4.10 and 4.11 then show the vertical and horizontal polarised results from dataset 241 with CNRs of 10.2 dB and 8.7 dB, respectively. The former figure shows the vertical polarisation where the K+N-distribution fits the endo-clutter PDF, while the LN, K+R and 3MD show a good match to the CCDF. For the exo-clutter region, both the K+N and K+R-distributions show a good fit. With the horizontal polarisation, the results are similar, except now the LN is worse in the endo-clutter region and there are no good matches in the exo-clutter region.

In summary, the best overall PDF model fit for these examples is the K+N distribution, while the K+R and 3MD both consistently match the CCDF and would be suitable for use in target detection. It is also interesting to note that the K-distribution shape parameter for all of the data sets is less than 1, indicating that the clutter is very spiky and the model fits for both the horizontal and vertical polarisations are very similar.

Table 4.2 *Parameter estimates and error measures for the data block examples. Parameters in orange have been declared a 'good' fit. Note that D_{BD} and D_{TE} are both measured in dB and the threshold error is measured when the $\log_{10}(CCDF)$ is -3*

Dataset	Lognormal	K+N	K+R	Pareto+noise	3MD
533 Vert.	$\hat{m}_{LN} = -1.42$, $\hat{\sigma}_{LN} = 1.86$ $D_{BD} = -24.29$ $D_{TE} = 3.70$	$\hat{v}_k = 0.50$, $\hat{b}_k = 0.51$ $D_{BD} = -24.93$ $D_{TE} = 1.35$	$\hat{v}_r = 0.20, \hat{b}_r = 0.25$, $\hat{k}_r = 0.16$ $D_{BD} = -21.84$ $D_{TE} = 0.096$	$\hat{v}_p = 1.59$, $\hat{b}_p = 0.57$ $D_{BD} = -19.54$ $D_{TE} = 1.38$	$\hat{a} = [0.55, 1.43, 3.35]$, $\hat{c} = [0.71, 0.27, 0.021]$ $D_{BD} = -19.50$ $D_{TE} = 0.19$
816 Vert.	$\hat{m}_{LN} = -1.71$, $\hat{\sigma}_{LN} = 1.98$ $D_{BD} = -25.86$ $D_{TE} = 3.95$	$\hat{v}_k = 0.36$, $\hat{b}_k = 0.37$ $D_{BD} = -21.41$ $D_{TE} = 0.71$	$\hat{v}_r = 0.23, \hat{b}_r = 0.25$, $\hat{k}_r = 0.058$ $D_{BD} = -26.02$ $D_{TE} = 0.099$	$\hat{v}_p = 1.45$, $\hat{b}_p = 0.45$ $D_{BD} = -16.75$ $D_{TE} = 1.91$	$\hat{a} = [0.46, 1.14, 2.79]$, $\hat{c} = [0.67, 0.26, 0.066]$ $D_{BD} = -18.25$ $D_{TE} = 0.11$
816 Horiz.	$\hat{m}_{LN} = -1.71$, $\hat{\sigma}_{LN} = 2.00$ $D_{BD} = -25.13$ $D_{TE} = 3.71$	$\hat{v}_k = 0.38$, $\hat{b}_k = 0.39$ $D_{BD} = -20.31$ $D_{TE} = 1.33$	$\hat{v}_r = 0.22, \hat{b}_r = 0.24$, $\hat{k}_r = 0.076$ $D_{BD} = -24.20$ $D_{TE} = 0.25$	$\hat{v}_p = 1.46$, $\hat{b}_p = 0.45$ $D_{BD} = -17.05$ $D_{TE} = 1.42$	$\hat{a} = [0.49, 1.20, 2.95]$, $\hat{c} = [0.70, 0.25, 0.054]$ $D_{BD} = -17.46$ $D_{TE} = 0.32$
241 Vert.	$\hat{m}_{LN} = -2.79$, $\hat{\sigma}_{LN} = 2.29$ $D_{BD} = -19.60$ $D_{TE} = 0.49$	$\hat{v}_k = 0.097$, $\hat{b}_k = 0.099$ $D_{BD} = -24.61$ $D_{TE} = 1.50$	$\hat{v}_r = 0.056, \hat{b}_r = 0.059$, $\hat{k}_r = 0.038$ $D_{BD} = -15.42$ $D_{TE} = 0.30$	$\hat{v}_p = 1.23$, $\hat{b}_p = 0.23$ $D_{BD} = -10.20$ $D_{TE} = 1.27$	$\hat{a} = [0.49, 1.52, 4.70]$, $\hat{c} = [0.87, 0.11, 0.024]$ $D_{BD} = -9.04$ $D_{TE} = 0.79$
241 Horiz.	$\hat{m}_{LN} = -2.86$, $\hat{\sigma}_{LN} = 2.08$ $D_{BD} = -18.84$ $D_{TE} = 5.16$	$\hat{v}_k = 0.057$, $\hat{b}_k = 0.059$ $D_{BD} = -22.00$ $D_{TE} = 1.76$	$\hat{v}_r = 0.038, \hat{b}_r = 0.041$, $\hat{k}_r = 0.024$ $D_{BD} = -19.53$ $D_{TE} = 0.94$	$\hat{v}_p = 1.20$, $\hat{b}_p = 0.20$ $D_{BD} = -10.43$ $D_{TE} = 2.80$	$\hat{a} = [0.46, 5.26]$, $\hat{c} = [0.97, 0.029]$ $D_{BD} = -8.99$ $D_{TE} = 1.13$

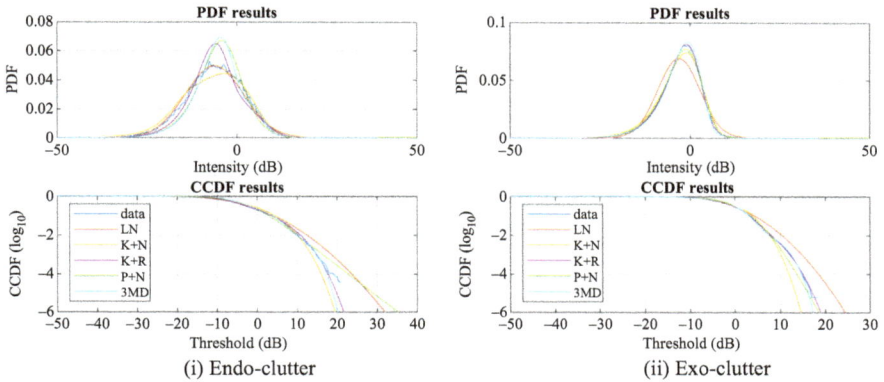

Figure 4.7 Amplitude distributions for dataset 533

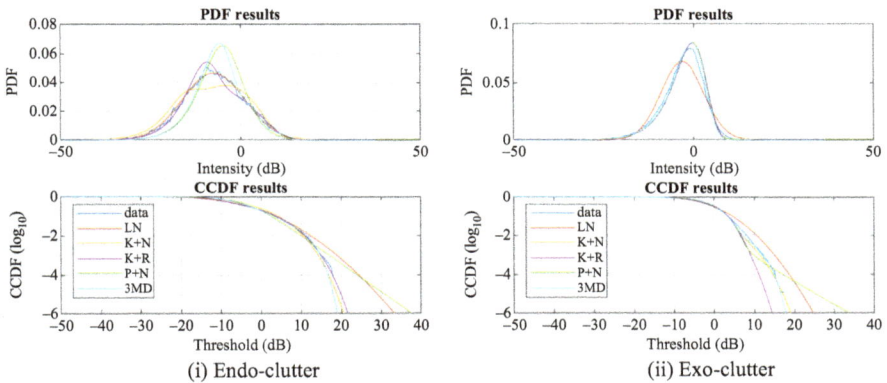

Figure 4.8 Amplitude distributions for dataset 816, vertical polarisation

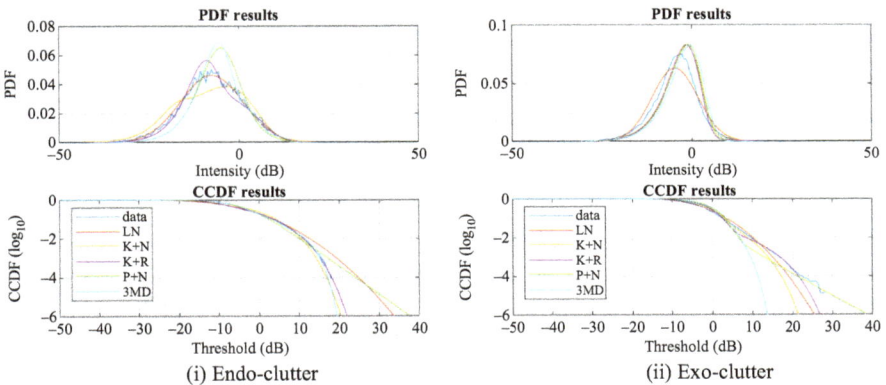

Figure 4.9 Amplitude distributions for dataset 816, horizontal polarisation

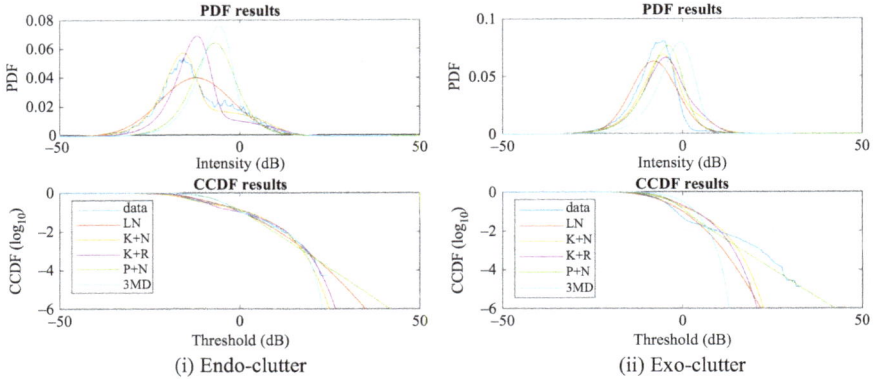

Figure 4.10   Amplitude distributions for dataset 241, vertical polarisation

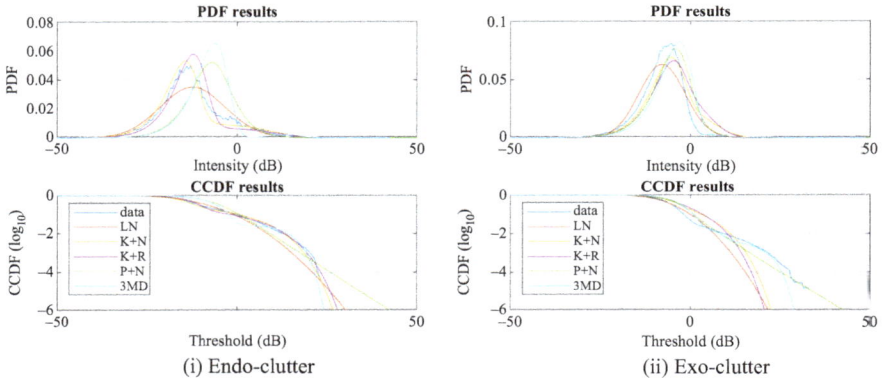

Figure 4.11   Amplitude distributions for dataset 241, horizontal polarisation

4.4.3   Impact of the reference signal quality

Passive radar signal processing requires a reference signal to extract the reflected information. In the previous results, an airborne reference was used to demonstrate performance in a real-world scenario. However, it is also instructive to look at the variation of the clutter statistics if a cleaner reference signal is used. Figure 4.12 shows a comparison of the constellation diagram for dataset 533 using two different reference signals. The first is the MRC signal used previously and the second is from the ground-based receiver. Clearly there is a big difference in the quality of the demodulated signal. Figure 4.13 shows the range/Doppler image using the ground-based reference which is now much cleaner than the result in Figure 4.4(i) with no vertical stripes or ambiguities.

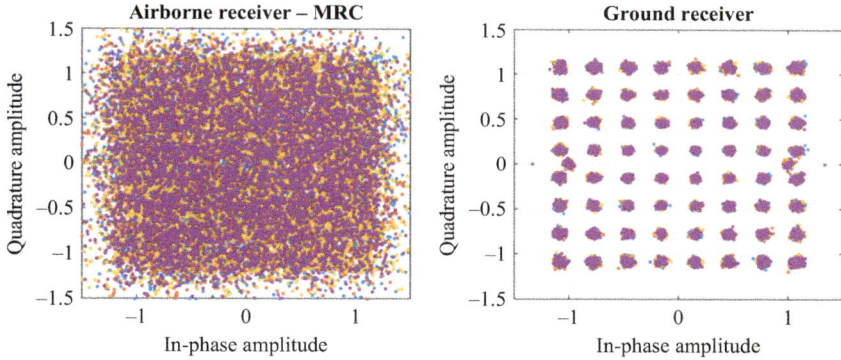

Figure 4.12   Constellation diagram comparing the airborne and ground reference signals

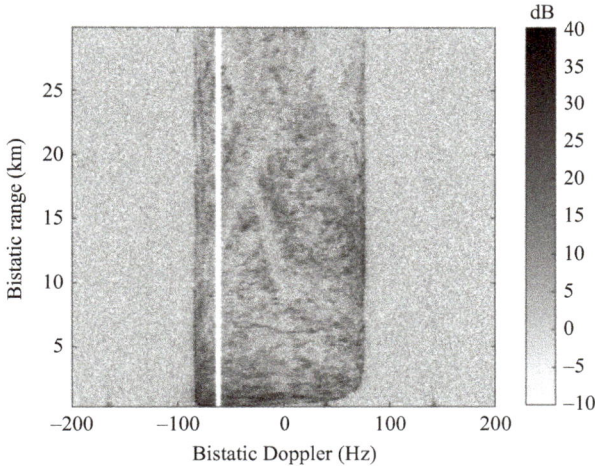

Figure 4.13   Dataset 533 delay/Doppler image using the ground reference

To directly compare the amplitude statistics, Figure 4.14 shows the PDF and CCDF results using both the ground and airborne reference signals. For the former, there is very little difference, with a slightly longer tail observed for the ground-based reference. However, for the exo-clutter region, the PDF for the ground-based reference has shifted to the right, while the tail in the CCDF plot is clearly less pronounced than the airborne reference. This demonstrates that in the exo-clutter region, the mismatched airborne signal is spikier with a lower mean when compared to the ground based reference. Figure 4.15 then shows the model fits to the endo- and exo-clutter regions with the model parameters and error metrics given in Table 4.3. The endo-clutter results are remarkably similar to the airborne reference indicating that

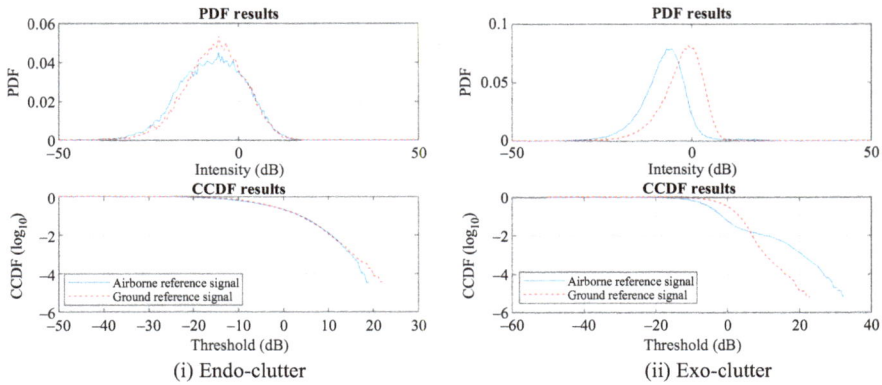

Figure 4.14   *Amplitude distribution for dataset 533 using both the airborne and ground references*

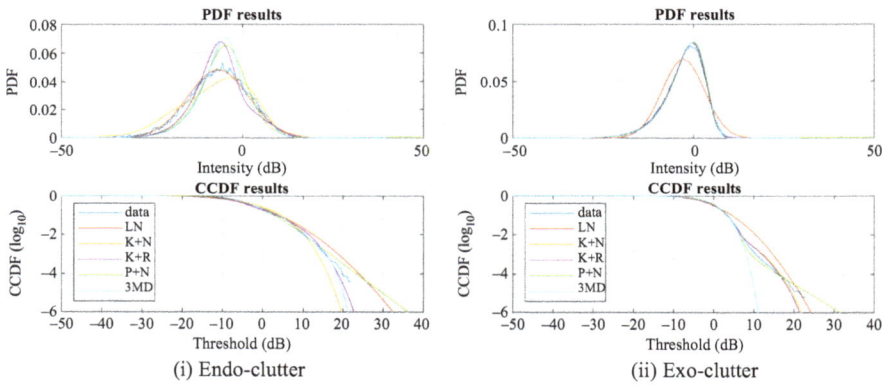

Figure 4.15   *Amplitude distribution for dataset 533 using the ground reference*

the poor quality reference does not greatly impact on the clutter statistics. However, for the exo-clutter region, there is a greater difference and only the K+R-distribution is able to model the data.

4.4.4   *Variation with integration time and bistatic range*

In this section, variation of the integration time and bistatic range is studied with the integration time varying between 0.1 and 1 s, while the bistatic range varies from 2 to 37 km. In each case, the data size is fixed at 50 bistatic range bins to compare with the previous results. For each of the following results, the BD and threshold error are shown with the integration time variation shown on the left and bistatic range

Table 4.3 *Parameter estimates and error measures for the data block example using a ground based reference signal. Parameters in orange have been declared a 'good' fit. Note that D_{BD} and D_{TE} are both measured in dB and the threshold error is measured when the $\log_{10}(CCDF)$ is -3*

Dataset	Lognormal	K+N	K+R	Pareto+noise	3MD
533 Vert.	$\hat{m}_{LN} = -1.54,$	$\hat{\nu}_k = 0.48,$	$\hat{\nu}_r = 0.14, \hat{b}_r = 0.18,$	$\hat{\nu}_p = 1.53,$	$\hat{\mathbf{a}} = [0.56, 1.54, 3.88],$
	$\hat{\sigma}_{LN} = 1.91$	$\hat{b}_k = 0.49$	$\hat{k}_r = 0.19$	$\hat{b}_p = 0.53$	$\hat{\mathbf{c}} = [0.75, 0.23, 0.014]$
	$D_{BD} = -25.38$	$D_{BD} = -20.47$	$D_{BD} = -19.94$	$D_{BD} = -19.12$	$D_{BD} = -18.20$
	$D_{TE} = 3.14$	$D_{TE} = 1.95$	$D_{TE} = 0.17$	$D_{TE} = 0.99$	$D_{TE} = 0.20$

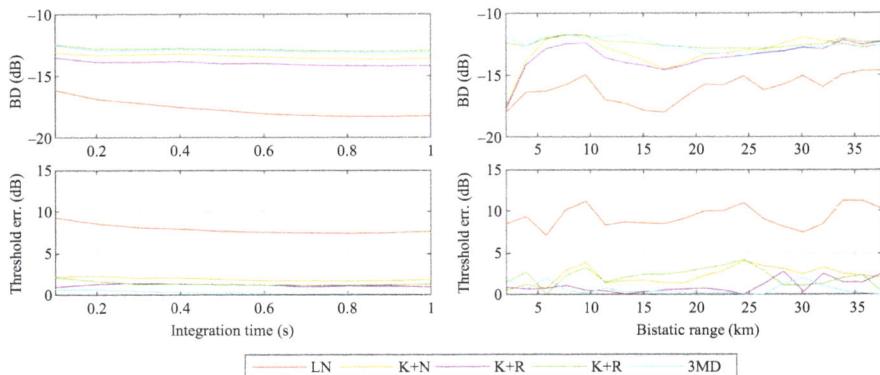

Figure 4.16   Fitting errors for dataset 533, vertical polarisation. Integration time on the left, bistatic range on the right.

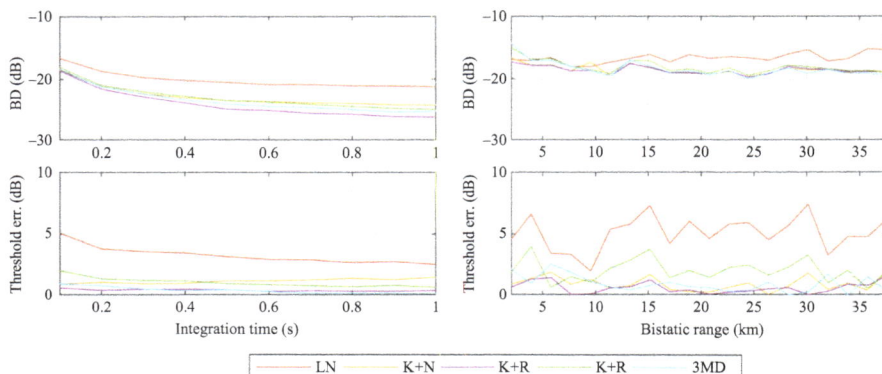

Figure 4.17   Fitting errors for dataset 816, vertical polarisation. Integration time on the left, bistatic range on the right.

variation on the right. Figure 4.16 shows the results for dataset 533 where the LN clearly has the lower BD and the highest absolute threshold error. There is little variation with integration time, while there are some fluctuations along range. For the threshold error, the trend matches the previous results, with the K+R and 3MD-distributions having the lowest error. For dataset 816, Figures 4.17 and 4.18 show the results for the vertical and horizontal polarisations. These results are similar with the K+R showing the best BD with a slight improvement as the integration time increases. The K+N, K+R and 3MD-distributions show low threshold errors across range. The results here agree quite closely with the observations from the previous sections with the best overall PDF model being the K+N distribution, while the K+R and 3MD both consistently match the CCDF. Figures 4.19 and 4.20 then show the vertical and horizontal results for dataset 241, with the K+N showing the lowest BD

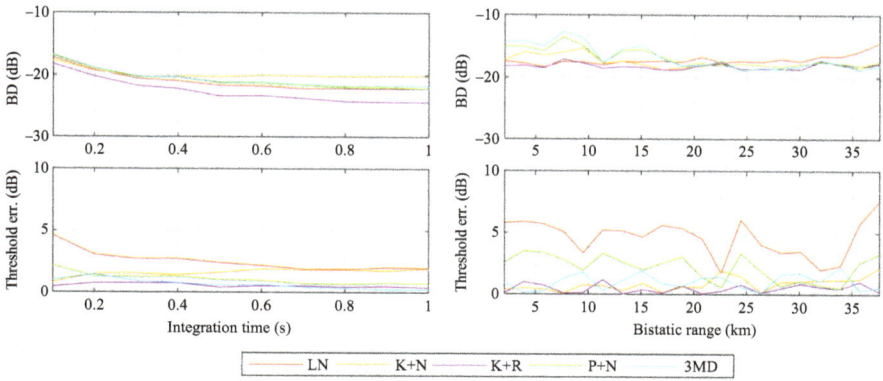

Figure 4.18   Fitting errors for dataset 816, horizontal polarisation. Integration time on the left, bistatic range on the right.

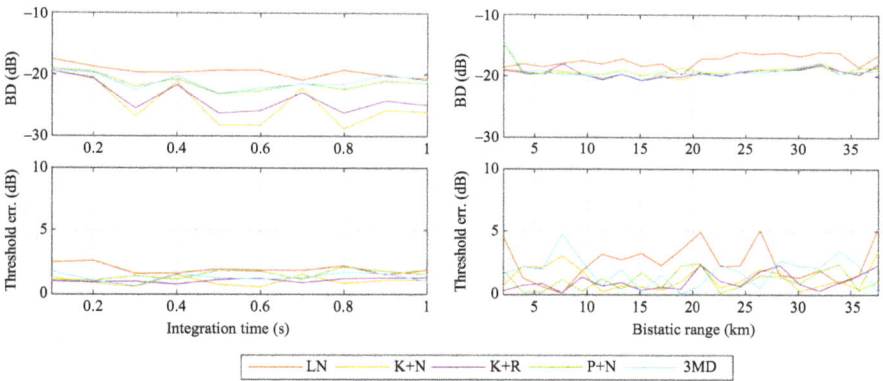

Figure 4.19   Fitting errors for dataset 241, vertical polarisation. Integration time on the left, bistatic range on the right.

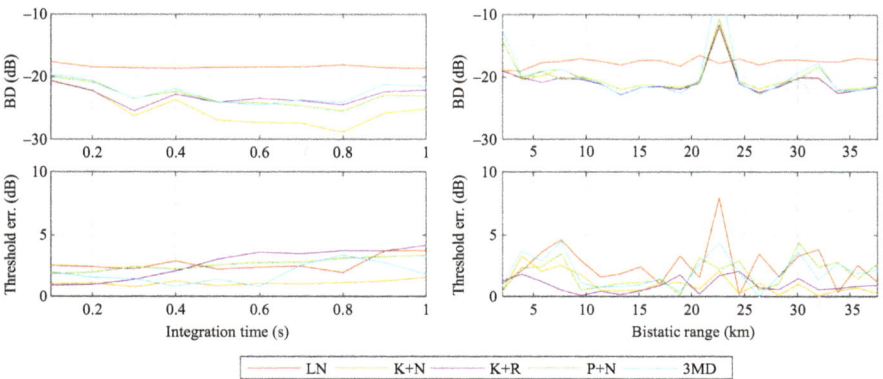

Figure 4.20   Fitting errors for dataset 241, horizontal polarisation. Integration time on the left, bistatic range on the right.

as the integration time varies, while all of the models except the LN have similar BDs across range. In terms of threshold error, the K+N-distribution shows the best results with a number of other distributions also performing well. Also, the ambiguity at 22 km is very strong in the horizontal polarisation causing the models to have a large mismatch with high BD and threshold errors.

4.5 Conclusion

This chapter has studied the amplitude statistics of passive bistatic land clutter collected from an airborne platform. Three datasets were analysed with the scattering for two being predominantly backscatter and the third being forward scatter. The range/Doppler images revealed that the forward scatter datasets had a lower reflectivity with slightly stronger signals in the vertical polarisation. Also, for dataset 241, the strength of the sea clutter was less than the ambiguity noise floor and hence it could not be characterised.

A number of models were considered with the goodness of fit measured for different polarisations and a range of integration times and bistatic ranges. While the lognormal distribution had a good fit to the endo-clutter regions in datasets 533 and 816, the best overall PDF model was the K+N-distribution with the K+R and 3MD both consistently matching the CCDF, making them suitable for use in target detection. It is also interesting to note that the model fits for both the horizontal and vertical polarisations were very similar and the K+N-distribution shape parameter for each of the data sets was less than 1, indicating that the clutter is very spiky. For the exo-clutter region, the statistics were clearly not Gaussian due to interference from the noisy reference signal and the ambiguity due to co-channel interference. The impact of the reference signal quality was measured by comparing the measured statistics from a ground based receiver. This revealed little difference to the airborne reference with the model fits being nearly identical. However, for the exo-clutter region, there was a greater difference and only the K+R-distribution was able to model the data. Also, the ambiguity was present in the horizontal polarisation for dataset 241 which could indicate that the cross-polarised channel is more robust to co-channel interference.

Acknowledgements

The authors wish to thank Diego Cristallini and Philipp Wojaczek from the Fraunhofer FHR and Ashley Summers from DSTG for providing advice and code for the signal processing and GIS extraction.

References

[1] H. D. Griffiths and C. J. Baker. *An Introduction to Passive Radar*. Artech House, 2017.

[2] K. Chetty, K. Woodbridge, H. Guo, and G. E. Smith. Passive bistatic WiMAX radar for maritime surveillance. In *IEEE Radar Conference*, 2010, pp. 188–193.

[3] M. Malanowski, R. Haugen, M. S. Greco, D. W. O'Hagan, R. Pišek, and A. Bernard. Land and sea clutter from FM-based passive bistatic radars. *IET Radar Sonar and Navigation*, 8(2), 2014, pp. 160–166.

[4] J. Raout. Sea target detection using passive DVB-T based radar. In *International Radar Conference*, 2008, pp. 695–700.

[5] F. Pieralice, D. Pastina, F. Santi, and M. Bucciarelli. Multi-transmitter ship target detection technique with GNSS based passive radar. In *International Radar Conference*, 2017, pp. 1–6.

[6] L. Daniel, S. Hristov, X. Lyu, A. G. Stove, M. Cherniakov, and M. Gashinova. Design and validation of a passive radar concept for ship detection using communication satellite signals. *IEEE Transactions on Aerospace and Electronic Systems*, 53(6), 2017, pp. 3115–3134.

[7] J. D. Ouellette and D. J. Dowgiallo. Sea surface scattering of hurricane Maria remnants using bistatic passive radar with S-band satellite illumination. In *IEEE Radar Conference*, 2018, pp. 462–466.

[8] L. Rosenberg, J. D. Ouelette, and D. J. Dowgiallo. Analysis of S-band passive bistatic sea clutter. In *IEEE Radar Conference*, 2019.

[9] B. Billingsley. *Low Angle Radar Land Clutter*. Ed. by The Institution of Engineering and Technology, 2002.

[10] K. D. Ward, R. J. A. Tough, and S. Watts. *Sea Clutter: Scattering, the K-Distribution and Radar Performance*. The Institute of Engineering Technology, 2nd ed., 2013.

[11] N. Willis and H. D. Griffiths. *Advances in Bistatic Radar*. SciTech Publishing, 2007.

[12] W. A. Al-Ashwal. *Measurement and Modelling of Bistatic Sea Clutter*, PhD thesis, Department of Electronic and Electrical Engineering, University College London, 2011.

[13] M. Ritchie, A. Stove, A. Woodbridge, and H. Griffiths. NetRAD: monostatic and bistatic sea clutter texture and Doppler spectra characterization at S-band. *IEEE Transactions on Geoscience and Remote Sensing*, 54(9), 2016, pp. 5533–5543.

[14] L. Rosenberg, J. D. Ouellette, and D. J. Dowgiallo. Passive bistatic sea clutter statistics from spaceborne illuminators. *IEEE Transactions on Aerospace and Electronic Systems*, 56(5), 2020, pp. 7062–7073.

[15] L. Rosenberg. Characterisation of high grazing angle X-band sea-clutter Doppler spectra. *IEEE Transaction on Aerospace and Electronic Systems*, 50(1), 2014, pp. 406–417.

[16] V. Duk, D. Cristallini, P. Wojaczek, and D. W. O'Hagan. Statistical analysis of clutter for passive radar on an airborne platform. In *International Radar Conference*, 2019, pp. 1–6.

[17] M. Glende. PCL-signal-processing for sidelobe reduction in case of periodical illuminator signals. In *International Radar Symposium*, May 2006, pp. 1–4.

[18]   P. Wojaczek, F. Colone, D. Cristallini, and P. Lombardo. Reciprocal filter-based STAP for passive radar on moving platforms. *IEEE Transactions on Aerospace and Electronic Systems*, 55(2), 2019, 967–988.

[19]   P. Wojaczek, D. Cristallini, J. Schell, D. O'Hagan, and A. Summers. Polarimetric antenna diversity for improved reference signal estimation for airborne passive radar. In *IEEE Radar Conference*.

[20]   S. Searle, D. Gustainis, B. Hennessy, and R. Young. Cancelling strong Doppler shifted returns in OFDM based passive radar. In *IEEE Radar Conference*, 2018.

[21]   G. V. Trunk and S. F. George. Detection of targets in non-Gaussian sea clutter. *IEEE Transactions*, AES-6, 1978, 620–628.

[22]   K. D. Ward, C. J. Baker, and S. Watts. Maritime surveillance radar. i. Radar scattering from the ocean surface. In *Radar and Signal Processing, IEE Proceedings F*, vol. 137, 2nd ed., IET, 1990. pp. 51–62.

[23]   D. Blacknell and R. J. A. Tough. Parameter estimation for the K-distribution based on [z log(z)]. *IEE Proceedings of Radar, Sonar and Navigation*, 148(6), 2001, 309–312.

[24]   S. Bocquet. Parameter estimation for Pareto and K distributed clutter with noise. *IET Radar Sonar and Navigation*, 9(1), 2015, 104–113.

[25]   L. Rosenberg, S. Watts, and S. Bocquet. Application of the K+Rayleigh distribution to high grazing angle sea-clutter. In *International Radar Conference*, 2014, pp. 1–6.

[26]   M. A. Sletten. Multipath scattering in ultrawide-band radar sea spikes. *IEEE Transactions on Antennas and Propagation*, 46(1), 1998, pp. 45–56.

[27]   T. Lamont-Smith. Translation to the normal distribution for radar clutter. *IEE Proceedings of Radar, Sonar and Navigation*, 147(1), 2000, 17–22.

[28]   A. Balleri, A. Nehorai, and J. Wang. Maximum likelihood estimation for compound-Gaussian clutter with inverse gamma texture. *IEEE Transactions on Aerospace and Electronic Systems*, 43(2), 2007, pp. 775–779.

[29]   P. Fayard and T. R. Field. Optimal inference of the inverse gamma texture for a compound-Gaussian clutter. In *IEEE International Conference on Acoustics, Speech and Signal Processing*, 2009, pp. 2969–2972.

[30]   M. Farshchian and F. L. Posner. The Pareto distribution for low grazing angle and high resolution X-band sea clutter. In *IEEE Radar Conference*, 2010, pp. 789–793.

[31]   G. V. Weinberg. Assessing Pareto fit to high-resolution high-grazing-angle sea clutter. *IET Electronic Letters*, 47(8), 2011, 516–517.

[32]   L Rosenberg and S. Bocquet. Application of the Pareto plus noise distribution to medium grazing angle sea-clutter. *IEEE Journal of Selected Topics in Applied Earth Observations and Remote Sensing*, 8(1), 2015, pp. 255–261.

[33]   C. H. Gierull and I. C. Sikaneta. Improved SAR vessel detection based on discrete texture. In *European SAR Conference*, 2016, pp. 523–526.

[34]   C. H. Gierull and I. C. Sikaneta. A compound-plus-noise model for improved vessel detection in non-Gaussian SAR imagery. *IEEE Transactions on Geoscience and Remote Sensing*, 99, 2017, pp. 1–10.

[35] S. Angelliaume, L. Rosenberg, and M. Ritchie. Modelling the amplitude distribution of radar sea clutter. *Remote Sensing*, 11, 2019, p. 319.

[36] L. Rosenberg and S. Angelliaume. Characterisation of the tri-modal discrete sea clutter model. In *International Radar Conference*, 2018.

[37] S. Bocquet, L. Rosenberg, and C. H. Gierull. Parameter estimation for the trimodal discrete radar clutter model. *IEEE Transactions of Geoscience and Remote Sensing*, 58(10), 2020, pp. 7062–7073.

[38] D. H. Kil and F. B. Shin. *Pattern Recognition and Prediction with Applications to Signal Processing*. New York, NY: Springer-Verlag New York, Inc., 1998.

[39] A. Bhattacharyya. On a measure of divergence between two statistical populations defined by their probability distributions. *Bulletin of the Calcutta Mathematical Society*, 35, 1943, pp. 99–109.

[40] OpenStreetMap Wiki. Main page—openstreetmap wiki, 2020. https://wiki. openstreetmap.org/w/index.php?title=Main_Page&oldid=2013332.

Chapter 5

Reference signal estimation for an airborne passive radar

Agnès Santori[1], Clément Berthillot[1], Dominique Poullin[2] and Olivier Rabaste[2]

Acronyms

BEM	basis expansion model
BPSK	binary phase shift keying
CE-BEM	complex-exponential BEM
CFAR	constant false alarm rate
CFR	channel frequency response
CIR	channel impulse response
DAB	digital audio broadcasting
DVB-T	digital video broadcast – terrestrial
ESPRIT	estimation of signal parameters via rotational invariance techniques
FFT	fast Fourier transform
GCE-BEM	generalized complex-exponential BEM
ICI	intercarrier interferences
IFFT	inverse fast Fourier transform
ISI	interSymbol interference
KL-BEM	Karhunen–Loeve BEM
LOS	line-of-sight
MDL	minimum description length
MFN	multiple-frequency network
MRC	maximum ratio combining
OFDM	orthogonal frequency division multiplexing
P-BEM	polynomial BEM
PS-BEM	prolate spheroidal BEM
QAM	quadrature amplitude modulation
SFN	single-frequency network
SNR	signal-to-noise ratio
ULA	uniform linear array

Estimating the transmitted signal is a mandatory task in passive radar in order to perform target detection. For DVB-T-based ground passive radar where clutter echoes

[1]Centre de recherche de l'Ecole de l'air, Ecole de l'Air et de l'Espace, France
[2]DEMR, ONERA, Université Paris-Saclay, F-91120 Palaiseau, France

are located at zero-Doppler, the reference signal can be recovered either by focusing the reception array in the transmitter direction, or by decoding the received signal. However, this estimation becomes much more difficult in the case of an airborne receiver since the receiver motion induces clutter multipaths spread in the range-Doppler domain. In this chapter, we thus consider the problem of estimating the reference signal in airborne passive radar applications using DVB-T transmitters. We first briefly recall the principal methods for estimating the reference signal in classical passive applications, i.e. antenna beamforming toward the transmitter direction or OFDM decoding principle for terrestrial configuration. We then present the main features of the aeronautical channel, and several methods for coping with the time-varying channel, notably antenna diversity and basis expansion model (BEM). Finally we present and discuss the performance obtained with these reference signal estimation methods obtained on real airborne signals. It appears that combining antenna diversity together with a BEM representation of the aeronautical channel provides substantial improvement over classical demodulation and these methods considered separately.

5.1 Introduction

Despite the maturity of ground passive systems, developing an airborne passive radar raises a lot of challenges. The first one consists in getting a proper copy of the transmitted signal that is then used as the reference signal along all the data processing. Contrary to the ground motionless environment, the airborne environment is characterized by a large amount of multi-paths, due to the receiver location high in the sky, and due to the transmitter tilt to the ground. This effect is emphasized in configurations where the civilian broadcasting network uses one single frequency to transmit the same signal. The receiver motion induces moreover Doppler shift for all the multi-paths, depending on their angle-of-arrival. Consequently, it appears much more complex to estimate a proper reference signal.

The received signal can be decomposed into several components: the direct path from the transmitter, in case it is actually in line-of-sight (LOS), reflections on ground obstacles that form the clutter, and finally reflections on potential targets. At time t, the received signal can be written:

$$y(t) = y_{DP}(t) + \sum_{i=1}^{N_{CL}} y_{CL,i}(t) + \sum_{i=1}^{N_{TG}} y_{TG,i}(t) + u(t), \tag{5.1}$$

where physical values indexed by DP, CL and TG respectively refer to direct path, clutter and targets. y corresponds to the whole received vector whereas y_{XX} indexed values refer to each received signal component. N_{XX} is the number of paths, u an additive noise.

In order to estimate the reference signal, two major families of solutions exist in the literature. The first one consists in steering an antenna toward the transmitter, either physically or by beam-forming, thus assuming no particular structure of the transmitted signal. The second method aims on the contrary at exploiting the specific structure of the transmitted digital signal by decoding the received signal in order to recover the theoretical transmitted signal. Such an approach is well adapted to digital video broadcast – terrestrial (DVB-T) signals. Error-free decoding would then produce an ideal reference. However, the aeronautical environment introduces heavy constraints that deeply deteriorate the received signal.

Signal processing summary

Figure 5.1 summarizes the passive radar processing steps: first an estimate of the reference signal is obtained by one of the two previously discussed solutions. Then direct path and ground echoes rejection is performed to clean the received signal from any undesired contribution. The classical matched filter is applied to the output of the rejection step in order to maximize the target signal-to-noise ratio (SNR), and finally a detection procedure is performed, for instance according to a constant false alarm rate (CFAR) CFAR test.

Matched filter

Radar target detection is performed via a correlation between the received signal and the transmitted signal $\mathbf{s}(t)$. If τ and ν denote respectively the delay and the Doppler frequency, then the matched filtering correlation can be written:

$$\chi_{(\tau,\nu)}(\mathbf{y},\mathbf{s}) = \int_0^{T_{int}} \mathbf{y}(t)\mathbf{s}^*(t-\tau)e^{-j2\pi\nu t}dt, \tag{5.2}$$

where T_{int} is also called integration time. This processing implies to know the transmitted signal, that needs to be estimated in case of a passive radar. This reference signal will further be denoted as $\mathbf{s}_{ref}$.

The literature [1,2] offers analyses on performance degradation in passive radar processing, for instance impact of a degraded estimation over the clutter rejection. Indeed, an estimated reference signal with estimation errors may lead to multi-path residues after clutter rejection and thus higher sidelobes that may prevent target detection. So, in order to achieve detection with moving platforms, we seek at developing an efficient estimation method to reconstruct a proper reference signal in complex channels.

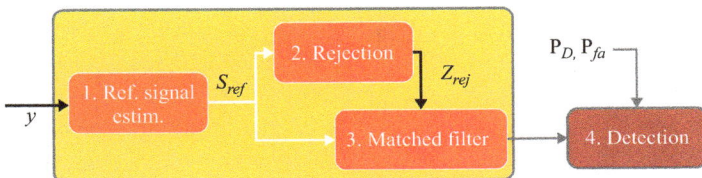

Figure 5.1 *Synoptic of processing steps for passive ground radar*

5.2 Conventional procedures for estimating the reference signal

5.2.1 Signal modeling

Consider a uniform linear array (ULA) with N_{RX} antennas equi-spaced by a distance d. Assuming that the signal is narrow banded (i.e. $c/B_w >> N_{RX}d$), the phase shift between antenna 0 and antenna k is: $\gamma^k = \frac{2\pi kd}{\lambda} \sin(\theta)$, where λ denotes the wavelength and θ the angle-of-arrival. The angular steering vector is defined as:

$$\mathbf{a}(\vartheta) = \left[1, e^{j2\pi\vartheta}, e^{j2\pi 2\vartheta}, \dots, e^{j2\pi(N_{RX}-1)\vartheta}\right]^T,$$

where $\vartheta = \frac{d}{\lambda} \sin(\theta)$. $\mathbf{y}$ refers to the received vector matrix $[\mathbf{y}^0, \mathbf{y}^1, \dots, \mathbf{y}^{N_{RX}-1}]^T$ of size $N_{RX} \times N_{ech}$ where $\mathbf{y}^k$ is a vector of length N_{ech} representing the sampled version of continuous received signal $y^k(t)$. The matched filter output, extended to the spatial domain, becomes:

$$\chi_{(\tau,\nu,\theta)}(\mathbf{y}, s_{ref}) = \sum_{k=0}^{N_{RX}-1} \chi_{(\tau,\nu)}\left(\mathbf{y}^k, s_{ref}\right) e^{-j2\pi \frac{kd}{\lambda}\sin(\theta)}. \tag{5.3}$$

5.2.2 Focused reference signal

Several processing steps require the knowledge of the transmitted signal. Once estimated, it is considered as the reference signal. Several methods for estimating the reference signal can be found in the literature. The first solution consists in directing the antenna toward the transmitter [3,4]. This direction may be either physical (one channel is dedicated to the reference estimation) or digital (thanks to beamforming). The beamformed reference signal is denoted as focalized reference and may be obtained as the eigenvector corresponding to the maximum eigenvalue of the estimated received signal correlation matrix $\mathbf{R}_{yy} = \mathbf{yy}^H/N_{ech}$, referred to as $\mathbf{a}_{FOC}$:

$$s_{ref} = \mathbf{a}_{FOC}^H \mathbf{y}. \tag{5.4}$$

s_{ref} is a vector of length N_{ech}. As it is proposed in [5], the reference signal is assumed to be received only by the mainlobe antenna. It is supposed that clutter and targets come from secondary sidelobes, and so, they are negligible. In the same way, it is assumed that this reference does not contain any multipath, which is generally questionable as shown in [6]. The reference signal thus obtained is composite. Even far less powerful, these multipaths may introduce artefacts or ghost targets while processing the data.

5.2.3 Decoded reference signal

Digital modulations [like DVB-T or digital audio broadcasting (DAB) e.g.] allow a second solution that consists in decoding the received signal in order to reconstruct the transmitted signal from its theoretical model. This reference represents a kind of ideal reference. However, its quality depends on the decoding error rate. Besides it cannot reproduce the transmitter behavior such as non-linearity of transmitting chain, clock jitter, etc. Most digital telecommunication signals used in passive radar are based on orthogonal frequency division multiplexing (OFDM) modulation.

OFDM modulation

The OFDM modulation is built in the frequency domain [7] with equi-spaced subcarriers at frequencies

$$f_k = f_0 + \frac{k}{T_u},$$ (5.5)

where T_u represents the useful OFDM symbol duration and $k = 0, \ldots, N-1$ the subcarrier index. An OFDM symbol n contains N elementary symbols $X_{n,k}$, chosen from a given alphabet. For instance, the alphabet may be binary phase shift keying (BPSK) – see Figure 5.2a – or quadrature amplitude modulation (QAM) – see Figure 5.2b.

The OFDM signal is written as:

$$x(t) = \sum_{n=-\infty}^{+\infty} \sum_{k=0}^{N-1} X_{n,k} \Psi_{n,k}(t),$$ (5.6)

where $\Psi_{n,k}(t) = g_k(t - nT_u)$ and:

$$g_k(t) = \begin{cases} e^{j2\pi f_k t} & 0 \leq t < T_u; \\ 0 & \text{otherwise.} \end{cases}$$ (5.7)

Equation (5.6) means that OFDM symbols are bound to time windows of duration T_u, one after another. Figure 5.3 shows the spectrum of such equi-spaced subcarrier signal. It highlights the fact that the subcarriers of one OFDM symbol are orthogonal to each other. Strictly the orthogonality among subcarriers can be written:

$$\begin{cases} \int_{-\infty}^{+\infty} \Psi_{n,k}(t) \Psi_{n',k'}^*(t) dt = 0 & \text{if } n \neq n' \text{ and } k \neq k'; \\ \int_{-\infty}^{+\infty} ||\Psi_{n,k}(t)||^2 dt = T_u & \text{otherwise.} \end{cases}$$ (5.8)

Figure 5.4 shows the spectrum of one DVB-T OFDM symbol.

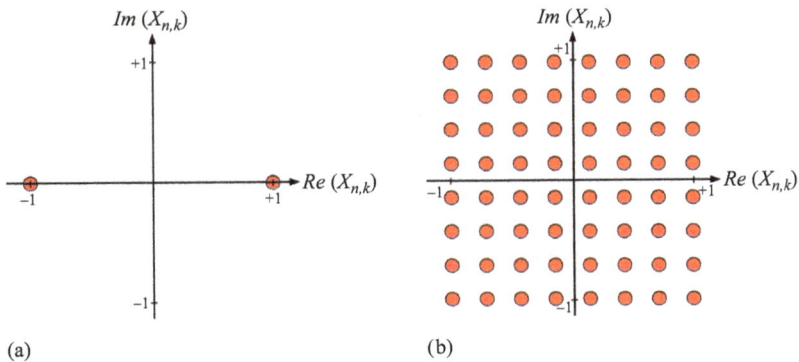

(a)   (b)

Figure 5.2   Alphabet examples used to code elementary symbols. (a) BPSK and (b) QAM64.

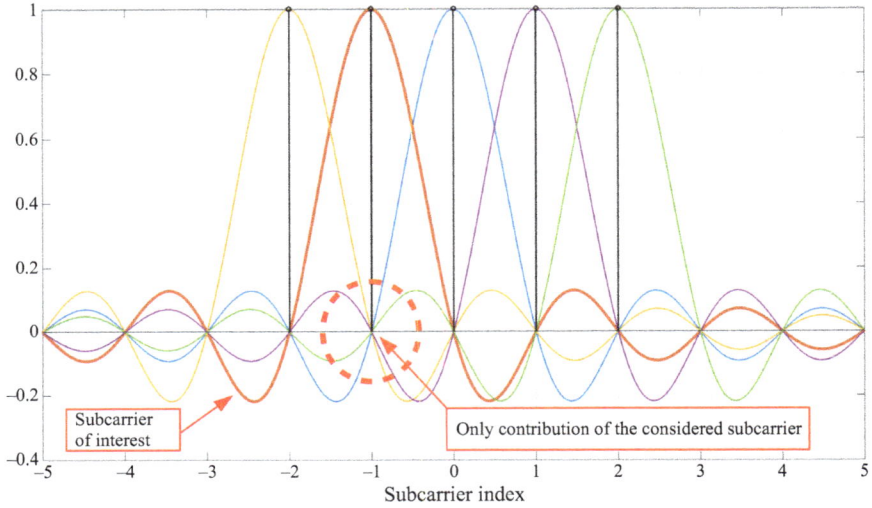

Figure 5.3   *Orthogonality among subcarriers for one OFDM symbol*

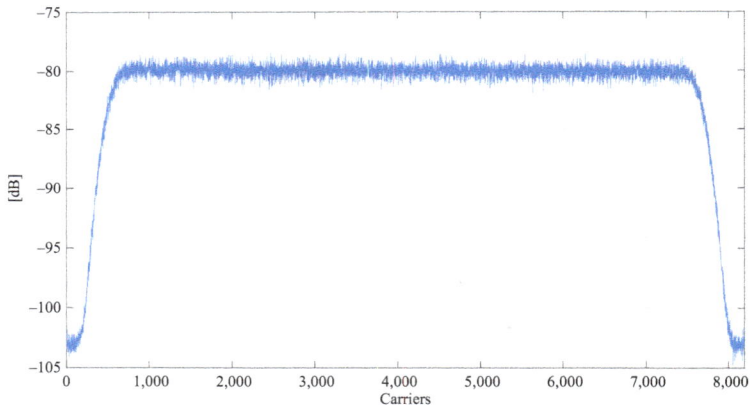

Figure 5.4   *Measured spectrum of one DVB-T OFDM symbol*

In practice, a guard interval is added by copying the G last symbol samples in front of each OFDM symbol so as to face interSymbol interference (ISI). Such a design enables in particular the OFDM modulation to cope with a network composed of several transmitters using the same frequency; such a network is called single-frequency network (SFN). When transmitters use different frequencies, the network is defined as multiple-frequency network (MFN). The whole OFDM symbol duration is then $T_s = \Delta + T_u$, whereas T_u corresponds to the useful part only.

5.2.4 Standard demodulation method

For $nT_s \leq t < nT_s + T_u$ and according to (5.6), the transmitted signal corresponding to the nth OFDM symbol is:

$$x(t) = \sum_{k=0}^{N-1} X_{n,k} \Psi_{n,k}(t). \tag{5.9}$$

It is generating from the information symbol via a simple inverse fast Fourier transform (IFFT) operation. Assuming that the channel response is time invariant over this interval, the received signal may be detailed as:

$$y(t) = \sum_{k=0}^{N-1} H_{n,k} X_{n,k} \Psi_{n,k}(t) + u(t), \tag{5.10}$$

where $t \in [nT_s + \Delta, (n+1)T_s]$, $H_{n,k}$ expresses the complex attenuation of the kth subcarrier and $u(t)$ is additive white gaussian noise. Thanks to the subcarrier orthogonality, the demodulation is operated in the frequency domain after a fast Fourier transform (FFT). We get:

$$Y_{n,k} = H_{n,k} X_{n,k} + U_{n,k}. \tag{5.11}$$

$Y_{n,k}$ thus figures the received signal on subcarrier f_k for the nth OFDM symbol and $U_{n,k}$ is the corresponding noise contribution. Signal demodulation aims at estimating the elementary symbols $X_{n,k}$ from the received samples $Y_{n,k}$. This step is simplified by the transmission of symbols known at the receiver, named pilot symbols. $P_n(k)$ refers to the kth pilot index of the nth OFDM symbol. These symbols allow to estimate the least square channel response. Thanks to the knowledge of $X_{n,P_n(k)}$, the channel frequency response at carrier $f_{P_n(k)}$ can be expressed by:

$$\hat{H}_{n,P_n(k)} = \frac{Y_{n,P_n(k)}}{X_{n,P_n(k)}}. \tag{5.12}$$

The whole channel frequency response $\hat{H}_{n,k}$ is then inferred by interpolation. This method is further referred to as the classic demodulation. The estimated elementary symbols are then:

$$\hat{X}_{n,k} = \frac{Y_{n,k}}{\hat{H}_{n,k}}. \tag{5.13}$$

The discrete estimated nth OFDM symbol, $\hat{\mathbf{X}}_n$, is built in the frequency domain by matching the N elementary symbols to the N subcarriers. The time domain corresponding signal, $\hat{\mathbf{x}}_n$, is inferred by inverse Fourier transform:

$$\hat{\mathbf{x}}_n = \mathbf{F}^H \hat{\mathbf{X}}_n, \tag{5.14}$$

where $.^H$ is the Hermitian transpose and $\mathbf{F}$ refers to the unitary discrete Fourier matrix, that is to say: $\mathbf{F}_{p,q} = 1/\sqrt{N} e^{-j2\pi pq/N}$, with $0 \leq p, q < N$.

Although classic in passive radar, this method is no longer the most suitable when the channel becomes complex, as is the case in moving passive radar.

5.3   Reference signal estimation for an airborne receiver

When the moving platform in passive radar is an aircraft, the aeronautical environment introduces heavy constraints that may deeply deteriorate the received signal.

5.3.1   *The aeronautical channel: a selective frequency and slow-time varying channel*

Aeronautical channel models reasonably assume that the transmitter and the receiver are in LOS. So, to describe the en-route scenario, Ref. [8] proposes to divide the aeronautical channel into two contributions: a LOS, and a more stochastic diffuse backscattered component. A ground reflection component may be associated to this two-ray model [9]. This is all the more relevant that, in the current case, transmitters are generally tilted to the ground. This multipath environment implies that the aeronautical channel is strongly composite.

Moreover the aeronautical environment involves large distances and as a consequence potentially large multipath delays compared to the symbol period. Stronger ground echoes may for instance present a delay corresponding to the propagation time necessary to travel the distance from transmitter to ground and then from ground to receiver, that is about twice the receiver altitude. Thus the channel coherence bandwidth may be far less than the almost 8 MHz bandwidth of DVB-T signal, for example. The aeronautical channel is hence frequency selective, which implies that deep fades may corrupt the channel frequency response (CFR).

In addition it is also characterized by the mobility of the receiver. As a first consequence, the channel impulse response (CIR) may fluctuate over time. Each component of the multipath aeronautical channel may then present a different Doppler frequency. This differential Doppler shift degrades the orthogonality among the subcarriers and may introduce intercarrier interferences (ICI). In other words, the received power does not only come from the desired subcarrier but also from its neighbors. The elementary symbol to be read is therefore corrupted by these interferences (Figure 5.5).

Example 1 (A frequency selective channel). *Assuming that the receiver altitude is limited up to five thousand feet because of low altitude targets and considering that the transmitter is located at the same altitude – see Figure 5.6b: $z_{TX} = z_{RX}$, therefore the main echoes are mostly concentrated during the first 10 µs, that is at most $100 \times T_s$. First it points the fact that the delay spread does not exceed the $1,024 \times T_s$ guard interval. So the ISI is neglected. Moreover it implies, that the channel coherence bandwidth is about 100 kHz, far less than the almost 8 MHz of a DVB-T signal bandwidth. As previously stated, it confirms that the aeronautical channel is frequency selective.*

The temporal channel tap $h_{(n,l)}^{(t)}$ denotes the complex path gain for the nth symbol and path delay lT_s. The received temporal signal for symbol n can be written in a matrix way:

$$\mathbf{y}_n = \mathbf{H}_n^{(t)}\mathbf{x}_n + \mathbf{u}_n, \tag{5.15}$$

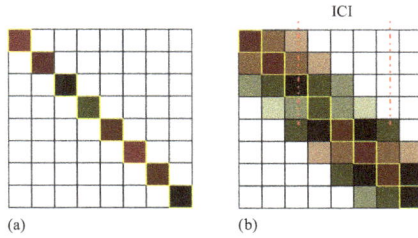

Figure 5.5   Channel matrix, see (5.17). (a) Without ICI and (b) with ICI.

where $\mathbf{u}_n$ is an additive white Gaussian noise and where the lth diagonal of $\mathbf{H}_n^{(t)}$ corresponds to the lth delay $h_{(n,l)}^{(t)}$:

$$\mathbf{H}_n^{(t)} = \begin{bmatrix} h_{(0,0)}^{(t)} & 0 & \cdots & & h_{(0,L-1)}^{(t)} & \cdots & h_{(0,1)}^{(t)} \\ \vdots & & & & & \ddots & \vdots \\ & & \ddots & & & & h_{(L-1,L-1)}^{(t)} \\ h_{(L,L-1)}^{(t)} & \cdots & h_{(L,0)}^{(t)} & \ddots & \ddots & & 0 \\ 0 & & & & & & \vdots \\ \vdots & & \ddots & & & & 0 \\ 0 & \cdots & 0 & & h_{(N-1,L-1)}^{(t)} & \cdots & h_{(N-1,0)}^{(t)} \end{bmatrix}. \tag{5.16}$$

Here the channel matrix presents this specific structure since it is estimated over the L OFDM pilots. In the frequency domain, this can be written:

$$\mathbf{Y}_n = \mathbf{F}\mathbf{y}_n = \mathbf{H}_n\mathbf{X}_n + \mathbf{U}_n, \tag{5.17}$$

where $\mathbf{H}_n = \mathbf{F}\mathbf{H}_n^{(t)}\mathbf{F}^H$ is the channel matrix, and $\mathbf{U}_n$ remains white Gaussian.

In case the channel is time-invariant, $h_{(n,l)}^{(t)} = h_{(m,l)}^{(t)}$ for $l = 0 \ldots L-1$, and all $n, m = 0 \ldots N-1$. So, in that particular case, $\mathbf{H}_n^{(t)}$ is circulant according to (5.16) and $\mathbf{H}_n$ is diagonal. On the contrary, the multipath aeronautical channel introduces interferences from each subcarrier to its neighbors. The channel matrix off-diagonal terms are thus not null anymore. The larger the product $f_{dT} = \Delta f_{Dmax} \times T_u$, the heavier the consequences, where T_u is the useful symbol duration, Δf_{Dmax} is the maximum Doppler shift relatively to the direct path Doppler shift, and f_{dT} is called the normalized Doppler spread – [10, Chapter 1.2.4]. So f_{dT} is the ratio of the maximum Doppler shift and the intercarrier spacing. It measures the loss of orthogonality. At least, only a few neighbor subcarriers can be assumed predominant. $\mathbf{H}_n$ may then be banded, or tailored so as to be considered as such – [11,12]. Figure 5.5 illustrates the channel matrix.

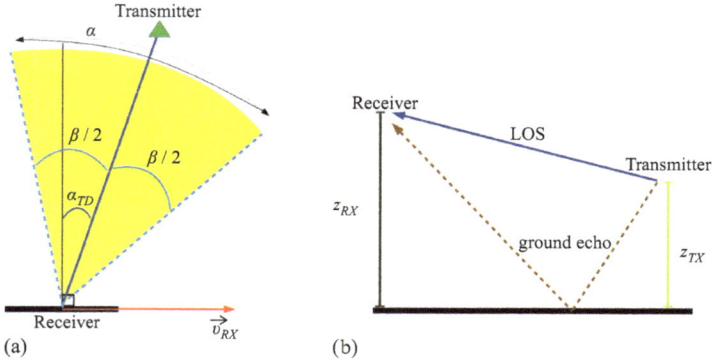

Figure 5.6   Geometrical analyses to reach Doppler (a) and delay (b) estimations

The Doppler shift relatively to the direct path can be written:

$$\Delta f_D(\alpha) = \frac{\|\vec{v}_R\|}{\lambda} \left[\cos\left(\pi/2 - \alpha\right) - \cos\left(\pi/2 - \alpha_{TD}\right)\right], \tag{5.18}$$

where α_{TD} denotes the direct path direction relatively to the receiver normal, $\|\vec{v}_R\|$ the maximum velocity of the receiver and α the angle of the received path from the normal of the receiver. If we consider that main echoes concentrate into a $\beta = 60°$ wide beam – see Figure 5.6, the maximum Doppler shift is then:

$$\Delta f_{Dmax} = \max\left[|\Delta f_D(\alpha_{TD} - \beta/2)|, |\Delta f_D(\alpha_{TD} + \beta/2)|\right]. \tag{5.19}$$

Example 2 (A slow time varying channel). *Assuming a conventional civil aircraft with an embedded receiver, its maximum velocity is about $\|\vec{v}_R\| = 50 \ ms^{-1}$. Let us consider a 600 MHz carrier frequency, and a received path at angle α from the normal of the receiver. Whatever α_{TD}, (5.19) leads to: $\Delta f_{Dmax} \simeq 50$ Hz. Therefore $f_{dT} \simeq 4.5\%$, and thus ICI impact is limited. Indeed, thanks to the slow velocity of the considered aircraft, the channel may be roughly supposed slow time varying.*

5.3.2   Time varying channel estimation method for OFDM signal

As presented in the previous section, OFDM symbols contain some known specific symbols denoted as "pilots" that are interleaved with the real information symbols. These pilot symbols are used to estimate the OFDM channel in classical application when the receiver does not move and thus the propagation channel is not time varying (or slowly time-varying). However, this channel estimation procedure does not take into account time variations or potential ICI. A large panel of solutions have thus been proposed in the literature in order to estimate a time-varying channel in case of OFDM signals. Among these solutions can be found: extension of the interpolation procedure to the time and frequency domains [13,14], derivation of a linear model based either on consecutive symbols, or only one symbol including its guard interval, in order to estimate the channel fluctuations [15], or expansion of such a model to higher order polynomials [16].

The precedent methods do not assume any knowledge on the channel length. Such a knowledge may however be useful to improve the channel estimation, for instance by limiting the number of coefficients to estimate [17]. Some articles in the literature propose solutions to properly estimate this channel length [14], for instance by using the minimum description length (MDL) criteria (see [10]). When the number of channel coefficients is known, the respective delays of the corresponding paths may then be estimated for instance by the estimation of signal parameters via rotational invariance techniques (ESPRIT) algorithm as proposed in [18].

Finally, in the case of time varying channels, some authors [19,20] have proposed basis expansion model (BEM) methods to decompose the CIR on an orthonormal basis composed of a restricted number of time-varying functions that may correctly model the time variations. Many such basis have been proposed in the literature, such as the polynomial BEM (P-BEM), the complex-exponential BEM (CE-BEM), the generalized complex-exponential BEM (GCE-BEM), the Karhunen–Loeve BEM (KL-BEM) [21] or the prolate spheroidal BEM (PS-BEM) [22]. The main advantage of using BEM is that it reduces the number of coefficient to estimate for modeling time-varying channels.

Generally these BEM methods were applied only to model time-variations of the channel over one symbol duration, i.e. they are well adapted to rapidly time-varying channels. However it appears from some real data recorded in the context of airborne passive radar that the recorded signals faced a slow-time varying channel [23]. Of course BEM methods can be extended to slow-time varying by increasing the size of the basis so as to model channel lengths longer than the duration of one single OFDM symbol. This was for instance considered in [24] that showed a performance improvement by using BEM over longer durations. However the drawback of such a method is that it implies consider a much large basis and thus requires to estimate more unknown coefficients. Since the number of CFR samples is limited (it is provided by the number of pilots), it is thus possible to consider channel time variations over only a few consecutive OFDM symbols.

In [23], the authors propose an original solution to deal with a slow-time varying channel. This method considers a large block of consecutive symbols and aims at directly modeling the inter-symbol variations instead of intra-symbol fluctuations, using an inter-symbol BEM model. Since only one coefficient is used per symbol, the number of coefficients to estimate is only linked to the number of consecutive OFDM symbols considered, and thus the complexity is reduced.

Finally, it was shown in [23] that this method combined together with an antenna diversity method exploiting a receiver array composed of several antennas enabled to improve the reference signal estimation. We present in the following this method.

5.3.3 Antenna diversity

As presented in the beginning of this chapter, we assume that the antenna array is composed of N_{RX} sensors. Let $\mathbf{y}^{(a)}$ denote the matrix $\mathbf{y}^{(a)} = \left[\mathbf{y}^{(1)}, \mathbf{y}^{(2)}, \ldots, \mathbf{y}^{(N_{RX})}\right]^T$, where $\mathbf{y}^{(k)}$ is the signal received by antenna k, that may be composed of several successive OFDM symbols. We have already seen that a proper reference signal can be obtained by beamforming in the direction of the direct path. This direction corresponds to the eigenvector $\mathbf{a}_{FOC}$ associated to the maximum eigenvalue of the received signal covariance matrix $\hat{R}_{\mathbf{y}^a \mathbf{y}^a} = \sum \mathbf{y}^{(a)} \mathbf{y}^{(a)^H}$: then the resulting received vector $\mathbf{y}$ is provided by:

$$\mathbf{y} = \mathbf{a}_{FOC}^H \mathbf{y}^{(a)}. \tag{5.20}$$

Interestingly such a method can be linked with the so-called antenna diversity that can help to reduce the impact of channel fluctuations due to multipath fading in communications. Indeed in communications, the probability that each receiver channel faces simultaneously a deep fade lowers as N_{RX} increases – [25, Chapter 13.4]. By weighting properly each receiver channel, it can thus be possible to optimize different output parameters such as the SNR, for instance by considering the maximum ratio combining (MRC) [26,27]. As ICI has been previously assumed limited and, as antenna diversity helps reducing its impact, ICI is neglected in the following. It implies in particular that the CIR may be assumed constant during a symbol period.

5.3.4   BEM channel model

We now present the BEM channel model used in the following. As previously stated, this particular model aims at representing the slow-time varying channel fluctuations over N_a consecutive OFDM symbols. Let us consider here a basis $\mathbf{B}$ composed of $Q + 1$ functions, where the parameter Q can be set depending on the channel to be estimated for the considered application and the modeling accuracy desired. Of course large basis shall provide a more accurate channel model, at the expense of a larger number of coefficients to estimate. Recall that this number of coefficients is limited by the number of available information, provided by the pilots.

Let us denote by $h_{(n,l)}^{(t)}$ the lth channel tap for the nth OFDM symbol, that will be modeled by the BEM coefficients $h_{(q,l)}^{(b)}$.

Denoting by $\mathbf{B}$ the $N_a \times (Q + 1)$ basis matrix, the BEM model enables to write:

$$h_{(n,l)}^{(t)} = \sum_{q=0}^{Q} \mathbf{B}(n,q) h_{(q,l)}^{(b)} + \varepsilon_{n,l}, \tag{5.21}$$

with $\varepsilon_{n,l}$ being the modeling error.

For instance, the coefficients $\mathbf{B}(n,q)$ corresponding to the CE-BEM are defined by [28]:

$$\mathbf{B}(n,q) = \exp\left(j \frac{2\pi n}{N_a} \left(q - \frac{Q}{2}\right)\right). \tag{5.22}$$

Denoting by $\mathbf{h}_q = [h_{(q,0)}^{(b)}, \ldots, h_{(q,L-1)}^{(b)}]^T$ the decomposition coefficients for the qth function, $q = 0 \ldots Q$, it comes that:

$$\mathbf{H}_n^{(t)} = \sum_{q=0}^{Q} \mathbf{B}(n,q) \mathbf{H}^{(q)} + \Xi_n^{(t)}, \tag{5.23}$$

with $\Xi_n^{(t)}$ the modeling error and $\mathbf{H}^{(q)}$ is the circulant matrix defined from $\mathbf{h}_q$. Multiplying by $\mathbf{F}$ on the left and $\mathbf{F}^H$ on the right, we obtain

$$\mathbf{H}_n = \sum_{q=0}^{Q} \mathbf{B}(n,q) \Delta_q + \Xi_n, \tag{5.24}$$

where Ξ_n is the frequency domain modeling error and $\Delta_q = \mathbf{F} \mathbf{H}^{(q)} \mathbf{F}^H$. Note that Δ_q is diagonal since $\mathbf{H}^{(q)}$ is circulant. From this expression of the channel, it comes that

$$\mathbf{Y}_n = \sum_{q=0}^{Q} \mathbf{B}(n,q) \Delta_q \mathbf{X}_n + \mathbf{V}_n, \tag{5.25}$$

where $\mathbf{V}_n$ expresses at once the channel additive noise and the modeling error. Denoting by exponent L the rows corresponding to the pilot symbols, the L corresponding rows can be extracted, so that:

$$\mathbf{Y}_n^{(p)} = \sum_{q=0}^{Q} \mathbf{B}(n,q) \Delta_q^{(p)} \mathbf{X}_n^{(p)} + \mathbf{V}_n^{(p)}. \tag{5.26}$$

From the definition of Δ_q, it can be further shown that $\Delta_q^{(p)}$ can be inferred from the L-sized Fourier transform of $\mathbf{h}_q$.

Computing the ratio $\mathbf{Y}_n^{(p)}(m)/\mathbf{X}_n^{(p)}(m)$ as in classical demodulation provides a sampled estimation of the CFR at the mth pilot, expressed thanks to the fact that Δ_q is diagonal as:

$$\frac{\mathbf{Y}_n^{(p)}(m)}{\mathbf{X}_n^{(p)}(m)} = \begin{bmatrix} \mathbf{B}_{(n,0)} \ldots \mathbf{B}_{(n,Q)} \end{bmatrix} \begin{bmatrix} \Delta_0^{(p)}(m,m) \\ \vdots \\ \Delta_Q^{(p)}(m,m) \end{bmatrix} + \mathbf{V}_n^{(p)}(m). \tag{5.27}$$

Consequently:

$$\mathbf{T} = \begin{bmatrix} \frac{\mathbf{Y}_0^{(p)}(m)}{\mathbf{X}_0^{(p)}(m)} \\ \vdots \\ \frac{\mathbf{Y}_{N_a-1}^{(p)}(m)}{\mathbf{X}_{N_a-1}^{(p)}(m)} \end{bmatrix} = \mathbf{B} \begin{bmatrix} \Delta_0^{(p)}(m,m) \\ \vdots \\ \Delta_Q^{(p)}(m,m) \end{bmatrix} + \begin{bmatrix} \mathbf{V}_0^{(p)}(m) \\ \vdots \\ \mathbf{V}_{N_a-1}^{(p)}(m) \end{bmatrix}. \tag{5.28}$$

where $\mathbf{T}$ is the matrix whose columns are composed of the ratio of the received samples over the transmitted pilots, and $\mathbf{\Delta}^{(p)}$ is the matrix whose mth column is composed of the mth diagonal value of $\Delta_q^{(p)}$ for $q = 0 \ldots Q$. Last Equation (5.28) can be inversed to estimate $\boldsymbol{\Delta}^{(p)}$ from the ratios $\mathbf{T}$, which directly provides the BEM coefficients $\mathbf{H}^{(q)}$ and finally the channel matrix $\mathbf{H}_n$.

As ICI has been assumed negligible, $\mathbf{H}_n$ may be considered as diagonal, for off-diagonal terms may be neglected, simplifying the equalization processing: the estimated OFDM symbol is derived from the ratio of the received vector over the estimated CFR.

5.4   Experimental results

5.4.1   Channel estimation

To analyze the potential ICI, the linear intra-symbol model of the channel time variations as proposed by [15] using the signal received on one of the three channels can be used. Figure 5.7 shows an extraction of the corresponding estimated channel matrix. The zoom view highlights that ICI spreads apart from the diagonal over a few subcarriers only. The channel matrix coefficients decrease from the diagonal to the outside reaches up to 30 dB over these maximum five subcarriers. It confirms the previous assumption that ICI may be neglected.

The BEM-based method has been built with $N_a = 32$ consecutive symbols. The basis $\mathbf{B}$ is provided by the CE-BEM, as defined by (5.22), as it aims at modeling Doppler-induced CIR phase evolution. Q has been empirically set to 32.

Figure 5.8 displays the CFR for three different symbols. There are eight OFDM symbols in between each of the considered symbols. It highlights the deep fades that

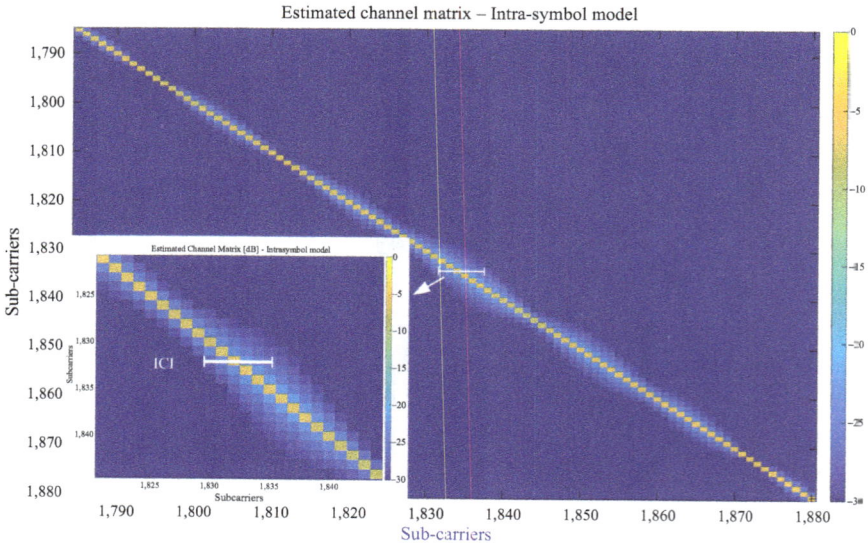

Figure 5.7   Estimated channel matrix – intrasymbol time variation model

Estimated channel frequency response

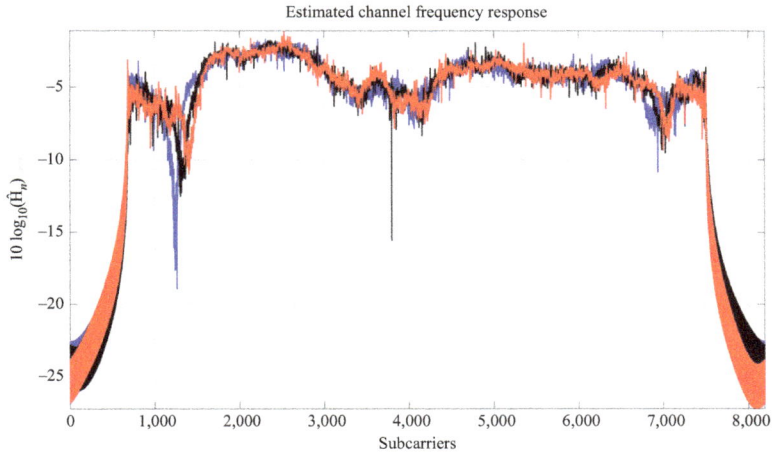

Figure 5.8 *Estimated channel frequency response for three OFDM different symbols – every eight symbols*

characterize the channel we are facing. It can also be noticed that these fades are evolving along time. So the channel estimation has to cope with fading effects and to model their time fluctuations.

5.4.2 Reference signal estimation

In the following, three reference signal estimation methods are compared. The first method corresponds to the classic demodulation: the CFR is inferred from interpolation of its pilot-based LS estimate over one single antenna. The second method, denoted as antenna diversity, exploits the three receiving channels of the reception antenna by combining them as in (5.20) in order to feed the classic demodulation with an improved signal. Finally the third method applies the BEM channel modeling procedure to the output of the antenna diversity processing to improve the channel estimation and thus the reference signal quality. Examples of the final constellations obtained for these three different methods over real data measurements are presented in Figure 5.9. It appears that antenna diversity alone provides improvement over the constellation obtained with the classic demodulation and that the cleanest constellation is obtained for the third method considering jointly the antenna diversity and BEM modeling. The results are more satisfying because the estimated QAM constellation is more homogeneous with the antenna diversity and BEM modeling than in the classic method. Indeed, the edges of the constellation undergo a heavier channel impact, as explained in [29] and as it can be seen in Figure 5.9. Three OFDM symbols are superimposed for each subfigure.

To quantify the ability of the different algorithms to properly retrieve the reference signal, we measure the distance between the estimated point $\hat{\mathbf{X}}_n(k)$ and the closest

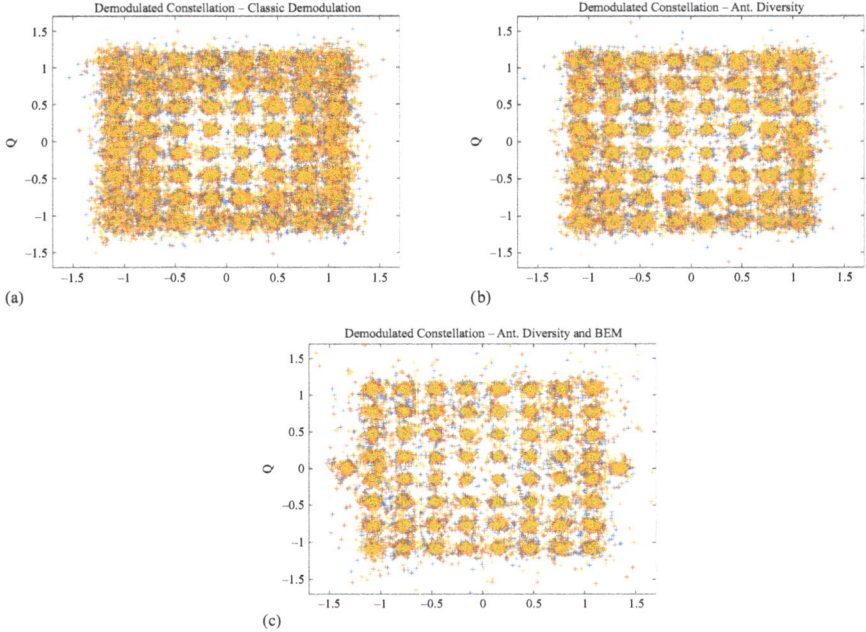

(a)

(b)

(c)

Figure 5.9   Demodulated constellation with three different methods. (a) Classic demodulation; (b) Antenna diversity; and (c) BEM + Antenna diversity.

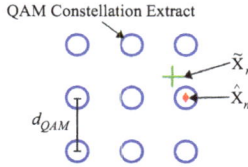

Figure 5.10   Schematic definition of d_{2c} index

constellation point $\hat{\mathbf{X}}_n(k)$ for each subcarrier kth. It is computed relatively to the half-distance (vertical or horizontal) between two constellation points. Let us denote d_{2c} the mean of this parameter over all symbols, presented in Figure 5.10 and defined as

$$d_{2c} = \operatorname*{mean}_{n,k} \left[\frac{\left\| \tilde{\mathbf{X}}_n(k) - \hat{\mathbf{X}}_n(k) \right\|}{d_{QAM}/2} \right] \tag{5.29}$$

Results on real data are presented in Figure 5.11. Again it appears that the antenna diversity alone provides a substantial improvement, but combining it with the BEM modelization enables even better performance.

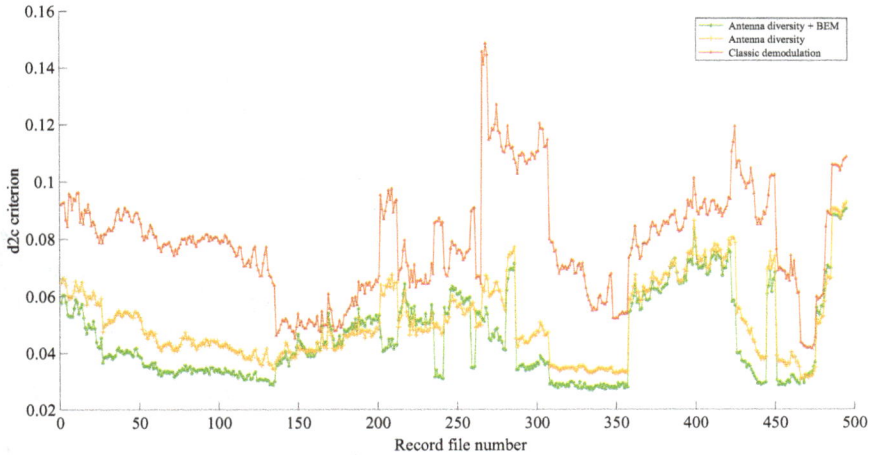

Figure 5.11 *Measured d2c criterion for multiple data records – classic*
demodulation (red), antenna diversity (orange), antenna diversity and
inter-symbol BEM (green)

5.5 Conclusion

Whether it is for a terrestrial or airborne passive radar, the estimation of the reference signal is a key step in the implementation of the detection system. However, the airborne environment induces detrimental effects: first the aeronautical channel involves a large number of multi-paths and second the receiver mobility introduces different Doppler shifts on the different multi-paths depending on their angle-of-arrival. This may introduce large fading in the channel frequency response. Consequently the reference signal reconstruction is made very complex.

The technique chosen strongly depends on the nature of the signal: focusing for an analog signal or decoding for a digital signal. There are many decoding methods, from the most classic to the most original. Reference [23] has shown the importance of knowing the aeronautical channel, particularly in the context of passive radar, in order to choose the most suitable method.

Finally, the reader should keep in mind that quantifying the impact of the reference signal misestimation over the airborne passive radar processing performance is in fact complex, since the reference signal is used both for the rejection step and the matched filter step. Of course, detection performance are also bound to the complexity of the clutter nature and the rejection processing used to remove this clutter contributions.

Acknowledgement

The authors would like to thank the French Air Force for its contribution to the flight tests and Dr. JF Nouvel (ONERA) for sharing his experience for the development and tuning of the RIVERA radar.

References

[1] Cardinali R, Colone F, Lombardo P, *et al.* Multipath cancellation on reference antenna for passive radar which exploits FM transmission. In: *International Conference on Radar Systems*, 2007.

[2] Tang H, Wan X, Hong L, *et al.* Detection improvement by modified modulation error rate of reference signal in passive radar. In: *General Assembly and Scientific Symposium (URSI GASS)*, 2014 XXXIth URSI. IEEE, 2014. p. 1–4.

[3] Colone F, O'Hagan DW, Lombardo P, *et al.* A multistage processing algorithm for disturbance removal and target detection in passive bistatic radar. *IEEE Transactions on Aerospace and Electronic Systems*. 2009;45(2):698–722.

[4] Brown J, Woodbridge K, Griffiths H, *et al.* Passive bistatic radar experiments from an airborne platform. *IEEE Aerospace and Electronic Systems Magazine*. 2012;27(11):50–55.

[5] Colone F, Cardinali R, and Lombardo P. Cancellation of clutter and multipath in passive radar using a sequential approach. In: *2006 IEEE Conference on Radar*. IEEE, 2006. p. 7.

[6] Demissie B. Clutter cancellation in passive radar using GSM broadcast channels. *IET Radar, Sonar & Navigation*. 2014;8(7):787–796.

[7] Lassalle R and Alard M. Principles of modulation and channel coding for digital broadcasting for mobile receivers. *EBU Technical Review*. 1987;224:168–190.

[8] Haas E. Aeronautical channel modeling. *IEEE Transactions on Vehicular Technology*. 2002;51(2):254–264.

[9] Bello PA. Aeronautical channel characterization. *IEEE Transactions on Communications*. 1973;21(5):548–563.

[10] Hijazi H. High speed radio-mobile channel estimation in OFDM Systems [Theses]. Institut National Polytechnique de Grenoble – INPG; 2008. Equipe C2S du Département Images et Signal (DIS). Available from: https://tel.archives-ouvertes.fr/tel-00373806.

[11] Rugini L, Banelli P, and Leus G. Low-complexity banded equalizers for OFDM systems in Doppler spread channels. *EURASIP Journal on Applied Signal Processing*. 2006;2006:1–13.

[12] Zemen T and Mecklenbräuker C. Low-complexity equalization of OFDM in doubly selective channels. *IEEE Transactions on Signal Processing*. 2004;52(4):1002–1011.

[13] Hoher P, Kaiser S, and Robertson P. Two-dimensional pilot-symbol-aided channel estimation by Wiener filtering. *IEEE ICASSP*. 1997;3:1845–1848.

[14] Tomasin S and Butussi M. Analysis of interpolated channel estimation for mobile OFDM systems. *IEEE Transactions on Communications*. 2010;58(5):1578–1588.

[15] Mostofi Y and Cox D. ICI mitigation for pilot-aided OFDM mobile systems. *IEEE Transactions on Wireless Communications*. 2005;4:765–774.

[16] Hijazi H and Ros L. Polynomial estimation of time-varying multipath gains with intercarrier interference mitigation in OFDM systems. *IEEE Transactions on Vehicular Technology*. 2009;58(1):140–151. Available from: https://doi.org/10.1109/TVT.2008.923653.

[17] Berthillot C, Santori A, Rabaste O, *et al.* Improving BEM channel estimation for airborne passive radar reference radar reconstruction. In: *International Radar Symposium (IRS)*. 2015;1:77–82.

[18] Yang B, Letaief KB, Cheng RS, and Cao Z. Channel estimation for OFDM transmission in multipath fading channels based on parametric channel modeling. *IEEE Transactions on Communications*. 2001;49(3):467–479.

[19] Giannakis GB and Tepedelenlioglu C. Basis expansion models and diversity techniques for blind identification and equalization of time-varying channels. *IEEE Proceedings*. 1998;86:1969–1986.

[20] Barhumi I, Leus G, and Moonen M. Time-varying FIR equalization of doubly selective channels. *IEEE Transactions on Wireless Communications*. 2005;4:202–214.

[21] Visitin M. Karhunen–Loeve expansion of a fast Rayleigh fading process. *Electronics Letters*. 1996;32:1712–1713.

[22] Zemen T and Mecklenbräuker C. Time-variant channel estimation using discrete prolate spheroidal sequences. *IEEE Transactions on Signal Processing*. 2005;53(9):3597–3607.

[23] Berthillot C, Santori A, Rabaste O, *et al.* BEM reference signal estimation for an airborne passive radar antenna array. *IEEE Transactions on Aerospace and Electronic Systems*. 2017;53(6):2833–2845.

[24] Tang Z, Leus G, and Banelli P. Pilot-assisted time-varying OFDM channel estimation based on multiple OFDM symbols. In: *IEEE 7th Workshop on Signal Processing Advances in Wireless Communications*; 2006. p. 1–5.

[25] Proakis J and Salehi M. *Digital Communications*, 5th ed. McGraw-Hill Science/Engineering/Math; 2007.

[26] Hutter A, Hammerschmidt J, de Carvalho E, *et al.* Receive diversity for mobile OFDM systems. In: *Wireless Communications and Networking Conference*, vol. 2, 2000. p. 7007–7012.

[27] Ahmed R, Eitel B, and Speidel J. Enhanced maximum ratio combining for mobile DVB-T reception in doubly selective channels. In: *Vehicular Technology Conference (VTC Spring)*, 2015. p. 1–5.

[28] Tang Z, Cannizzaro R, Leus G, *et al.* Pilot-assisted time-varying channel estimation for OFDM systems. *IEEE Transactions on Signal Processing*. 2007;55(5):2226–2238.

[29] Searle S, Davis L, and Palmer J. Signal processing considerations for passive radar with a single receiver. In: *2015 IEEE Conference on International Conference on Acoustics, Speech and Signal Processing (ICASSP)*. IEEE, 2015. p. 5560–5564.

Chapter 6

Passive synthetic aperture radar with DVB-T transmissions

Michail Antoniou[1] and George Atkinson[1]

6.1 Introduction

Passive synthetic aperture radar (SAR) is a special case of bistatic SAR, which relies on illuminators of opportunity to provide radar images of a target area. The essential requirement for applying SAR techniques in such a scenario is that at least the transmitter or the receiver exhibits motion relative to the scene to be imaged.

Illuminators of opportunity are defined here as transmitting sources not originally designed for radar applications, which includes communication and broadcasting systems. Transmitters such as these have several relative merits compared to active imaging radars. Key among them is silent operation – by separating the transmitter from the receiver, and not transmitting radar signals, it is possible to conceal both the location of the receiver and the fact that an area is under surveillance. Numerous such systems are also designed for global, persistent coverage, thus enabling passive SAR to operate persistently, either on its own or as a gap filler. Furthermore, the reuse of already existing and allocated parts of the spectrum contributes to more efficient EM spectrum management, especially at lower parts of the spectrum (e.g. UHF) where frequency licensing is an issue, and the reduction of EM pollution. On the other hand, as these transmitters were not created for radar purposes, the key radar performance parameters, such as the power budget and resolution, should be identified to best defined the passive SAR application space.

A number of different illuminators have been considered in the past for passive SAR. Primarily, those have focused on spaceborne systems. In such a configuration, satellite motion can be used for aperture synthesis and thus it is sufficient to deploy fixed receivers on the ground. One of the first passive SAR systems relied on an almost geostationary TV satellite and a fixed receiver to provide area imaging [1]. Around the same time, considerations were made on the design of a passive SAR using a receiver in a geosynchronous orbit and transmissions from a geostationary, digital audio broadcasting (DAB) satellite [2]. Global navigation satellite systems (GNSS), such as the global positioning system (GPS), Glonass, Galileo and Beidou

[1]University of Birmingham, UK

have been considered at length [3–5], to the degree where advanced imaging modes including multistatics [6–9], coherent change detection [10], differential interferometry [11] and tomography [12] have been experimentally demonstrated. An overview of the fundamental principles behind GNSS-based SAR can be found in [5]. Recently, the use of Inmarsat satellite transmitters and a moving, ground-based receiver was demonstrated using Doppler beam sharpening too [13].

On the other hand, passive SAR with terrestrial illuminators of opportunity was less considered until recently. This has been in contrast to passive coherent location (PCL), where transmitters such as DAB and terrestrial digital video broadcasting (DVB-T) traditionally dominate. In 2015–2016, the first papers showcasing the feasibility of DVB-T SAR imaging with experimental airborne campaigns emerged from the Swedish Defense Research Agency (FOI) and the Warsaw University of Technology [14,15]. Airborne demonstrations followed by Fraunhofer FHR [16] and the University of Birmingham [17]. It is noted here that placing the receiver on an aircraft may be the configuration followed here, but it is not the only one. Indeed, experimental results using receivers on moving ground vehicles have been demonstrated [18,19], while spaceborne receivers have also been considered at the theoretical level [20].

The aim of the chapter is to provide an introduction to the topic of passive SAR using DVB-T transmitters, in which an airborne receiver is assumed. The chapter provides an overview of the essential SAR parameters that define system performance, a brief description of a suitable image formation algorithm, as well as signal processing techniques to account for common DVB-T SAR artefacts, namely the presence of the direct DVB-T signal in imagery as well as defocusing due to uncompensated aircraft motion errors during flight. The development of an appropriate technology demonstrator is presented. Finally, experimental images from an extended airborne campaign are presented and discussed.

The chapter is broken down as follows: Section 6.2 derives the DVB-T SAR power budget, sensitivity, and spatial resolution. Section 6.3 summarises image formation, including direct signal suppression and autofocus techniques. Section 6.4 provides an overview of the airborne demonstrator and the experimental setup, and discusses experimental results obtained. Finally, Section 6.5 concludes the chapter and discusses potential future research strands for DVB-T SAR.

6.2 Power budget and resolution

To better understand the essential capabilities of a DVB-T SAR imaging system, it is useful to derive the available power budget and the available spatial resolution. The direct signal power at the output of the radar's receiving antenna can be used to gauge limits on the stand-off distance between the transmitter and the receiver. The SAR sensitivity is a primary metric of image quality and, similarly, can also be used to establish limits on target area size. Finally, the calculation of the spatial resolution and its dependence on the acquisition geometry allows an indication of potential application areas for this type of system.

A simplified geometry of the system is shown in Figure 6.1. A DVB-T broadcasting station acts as the transmitter (T_x), with the receiver (R_x) being onboard an aircraft. The receiver moves with a velocity, V, and its instantaneous distance from the T_x during flight is the baseline, R_B. The synthetic aperture length over a time, T_d is given by VT_d.

It is assumed that the receiver records the direct signal from the transmitter, with an instantaneous time delay proportional to R_B, plus the reflected signal from a target, with an instantaneous time delay proportional to the transmitter-to-target (R_T), and receiver-to-target (R_R), distances.

Calculations are based on the parameters of a transmitter that was used in experimental work described in this chapter. This is the Sutton Coldfield broadcasting station in the UK (Lat: 52.600556°, Long: −1.833889°). The location of the station and its coverage are shown in Figure 6.2. The station has a single mast with a height of 270.5 m above the ground. The transmitter has a maximum output power of 250 kW for FM radio, and a 200 kW maximum output power for digital DVB-T signals. The DVB-T channels transmitted from the station are listed in Table 6.1, including their carrier frequency, effective isotropic radiated power (EIRP), and polarisation. Work reported in this chapter assumes operation with a single DVB-T channel, channel 43 (650 MHz, with an EIRP of 200 kW).

A hypothetical system can then be assumed for the calculations that follow, whose parameters are shown in Table 6.2. An airborne receiver is assumed, with a separate antenna pointed towards the transmitter.

We should highlight at this point that the precise derivation of direct signal power is a complex study that requires many considerations that are not taken into account here, including transmit antenna patterns and the propagation channel itself. Hence, calculation results should be viewed as the theoretical upper limits which are difficult, if possible at all, to reach. At the same time, the calculations show that the margins that ensure long distance operation are indeed very wide. Direct signal strength is important to measure in a passive system as this signal is required to maintain coherence between the transmitter and the receiver. Taking into account the height of the transmitter and the receiver, it is not unreasonable to assume a free-space propagation

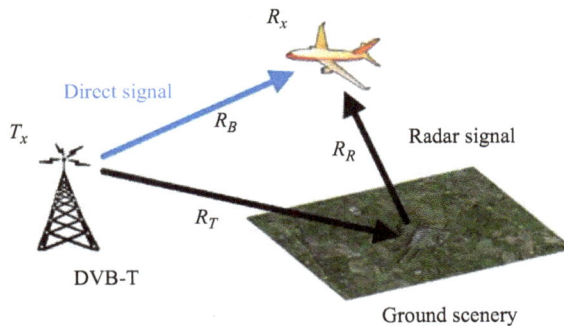

Figure 6.1 DVB-T SAR geometry

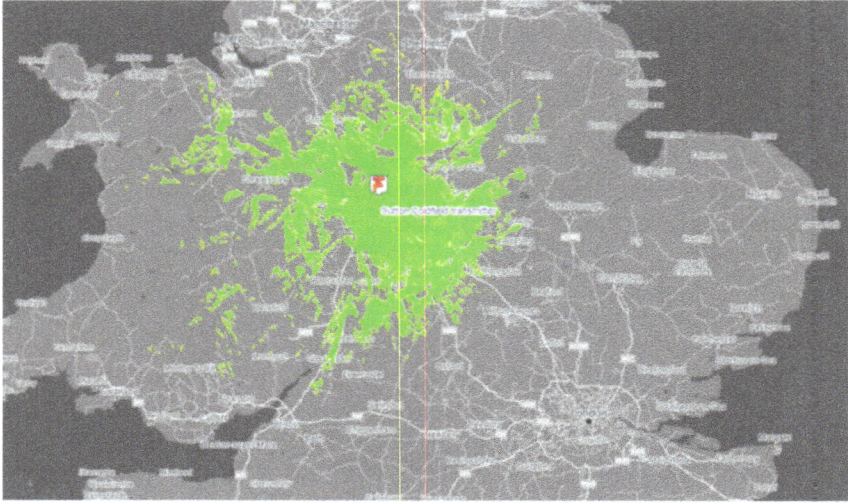

Figure 6.2   Sutton Coldfield transmitter location (red marker) and coverage (green coloured area) [21]

Table 6.1   Sutton Coldfield transmitted DVB-T signals

Channel no.	Carrier frequency	EIRP	Polarisation	System
33	570 MHz	89.2 kW	H −3.5 dB	DVB-T2
35	586 MHz	86 kW	H −3.7 dB	DVB-T2
39+	618.2 MHz	200 kW	H max	DVB-T
40+	626.2 MHz	200 kW	H max	DVB-T2
42	642 MHz	200 kW	H max	DVB-T
43	650 MHz	200 kW	H max	DVB-T
45	666 MHz	200 kW	H max	DVB-T
46	674 MHz	200 kW	H max	DVB-T
51	714 MHz	10 kW	H −13 dB	DVB-T

loss. Under this condition, the direct signal-to-noise ratio (SNR) at the output of the receiving antenna onboard the aircraft is given by:

$$\text{SNR}_\text{D} = \frac{P_T G_R \lambda^2}{(4\pi)^2 R_B^2 k T_0 B N_f L} \tag{6.1}$$

where k is Boltzmann's constant and T_0 is the receiver operating temperature, assumed to be 300 K.

Based on the parameters of Table 6.2, the variation of SNR_D with transmitter–receiver distance is shown in Figure 6.3. If a minimum SNR of 20 dB is required for the direct signal to maintain overall system coherence, prior to any other possible

Table 6.2 Calculation parameters

Parameter	Symbol	Value	Unit
EIRP	P_T	200	kW
Carrier frequency	f_c	650	MHz
Radar wavelength	λ	0.46	m
Channel bandwidth	B	7.61	MHz
Receiver antenna gain	G_r	10	dB
Receiver noise figure	N_f	5	dB
Receiver velocity	v	100	m/s
System losses	L	6	dB

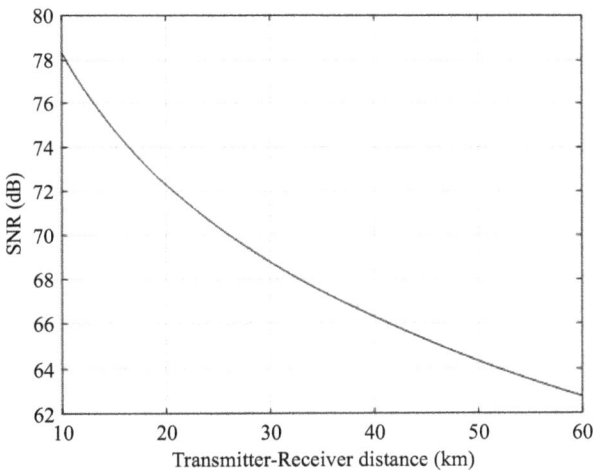

Figure 6.3   SNR of the direct signal vs. transmitter–receiver distance

signal processing gains obtained during the signal synchronisation process, the figure shows potential for large stand-offs between the transmitter and the receiver, even if additional losses due to factors not considered in (6.1) are taken into account.

The spatial resolution in range and azimuth is calculated next. In bistatic SAR, both the resolution and the direction it is measured upon should be specified, since the range and azimuth directions are generally not orthogonal in bistatic SAR. Those have been derived in detail in [22].

In range, the resolution along the bisector of the bistatic angle is given by:

$$\Delta R = \frac{c}{2B \cos \frac{\beta}{2}} \tag{6.2}$$

where c is the speed of light and β is the bistatic angle, defined as the angle between the transmitter and receiver lines of sight (LOS) to a target, when the receiver is at the midpoint of its synthetic aperture. If the range resolution is to be measured along the receiver LOS, ΔR should be projected to this direction, i.e. the cos () factor in (6.2) is squared. It is noted that the bistatic angle depends on the target as well as the receiver co-ordinates. Therefore, it is possible for extended target areas to be imaged at different bistatic angles.

For the parameters in Table 6.2, the range resolution is plotted as a function of the bistatic angle in Figure 6.4. The finest achievable resolution is when $\beta = 0°$, i.e. the quasi-monostatic configuration. For DVB-T SAR using a single DVB-T channel for imaging, this amounts to nearly 20 m. Beyond this point, the range resolution degrades until $\beta = 180°$, where the range resolution is altogether lost. This is the so-called forward scatter configuration. At $\beta = 140°$, the resolution degrades by a factor of 3 compared to its finest possible value.

On the other hand, azimuth resolution is in the direction of the effective velocity vector of the receiver relative to the target and is given by [22]:

$$\Delta A = \frac{\lambda}{2\omega_E T_d} \tag{6.3}$$

where ω_E is the effective angular speed of the receiver and T_d is the dwell time on target. Azimuth resolution is again dependent on the relative bistatic geometry, the receiver-to-target range, and the receiver speed, all of which combined determine ω_E. Note that ΔA in DVB-T SAR does not depend on transmitter parameters as the latter is fixed on the ground and hence does not contribute to the synthetic aperture formation. For a quasi-monostatic configuration, a target at a distance of 10 km from the receiver, a dwell time on target of 10 s, and all other parameters as in Table 6.2, ΔA is about 2.3 m. The key bottleneck in achieving fine resolution DVB-T SAR imagery is thus the range resolution, which is limited by the transmit signal bandwidth.

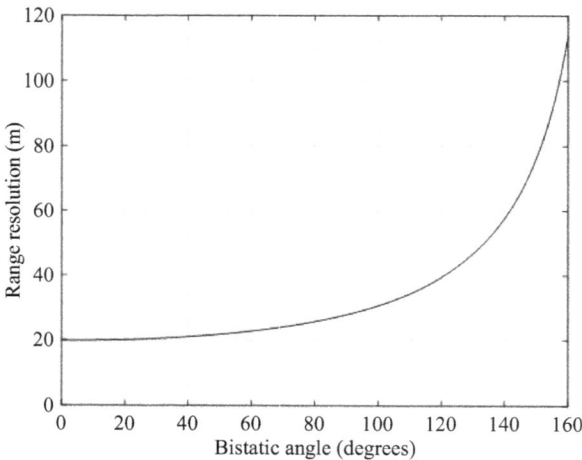

Figure 6.4   Range resolution vs. bistatic angle

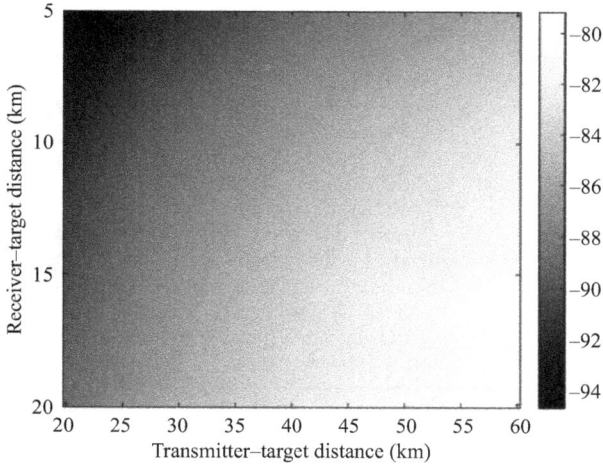

Figure 6.5 NESZ as a function of transmitter–target and receiver–target distances

A key metric in any SAR system is its sensitivity. This is usually expressed via the noise equivalent sigma zero (NESZ) parameter (6.4). The NESZ is the equivalent back-scatter coefficient that thermal noise would possess if it were processed into a SAR image. Since thermal noise will always be contained in a radar image, an equivalent noise floor is introduced that ultimately determines image contrast.

For a monostatic SAR, a step-by-step derivation of the NESZ can be found in [23]. Following the same reasoning, the NESZ for a bistatic SAR can be derived as [20]:

$$\text{NESZ} = \frac{(4\pi)^3 R_T^2 R_r kTVN_f}{P_T G_R \lambda^3 \Delta RL} \tag{6.4}$$

For the parameters of Table 6.2, the NESZ can be plotted in dB as a function of the transmitter–target and receiver–target distances, R_T and R_R, respectively, as reported in Figure 6.5. As expected, the sensitivity is the highest at the shortest R_T and R_R, and lowest when they are at their longest. Regardless, in ideal conditions, it can be seen that a DVB-T SAR may have a modest spatial resolution but can produce highly sensitive imagery at long stand-offs.

6.3 Image formation

The ideal DVB-T signal can be written as [24]:

$$g(t) = \sum_{m=0}^{\infty} \sum_{l=0}^{L-1} \sum_{k=0}^{K-1} c_{mlk} e^{j2\pi \frac{k}{T_U}(t - T_U - lT_S - mLT_S)} \cdot w\left(t - lT_S - mLT_s\right) \tag{6.5}$$

The parameters in (6.5) are listed in Table 6.3.

Table 6.3   DVB-T signal parameters

Symbol	Definition
k	Carrier number
l	OFDM symbol number
m	OFDM frame number
K	Number of transmitter carriers
T_S	Symbol duration
T_U	Useful symbol duration
Δ	Complex quadrature amplitude modulation (QAM) symbol for carrier k, OFDM symbol l, and frame m
$w(.)$	Rectangular window

After quadrature demodulation, the direct and reflected signals can be written as functions of slow-time, u, as:

$$d(t,u) = g\left[t - \frac{R_B(u)}{c}\right]\exp\left[-j\frac{2\pi}{\lambda}R_B(u)\right] \tag{6.6}$$

$$s(t,u) = g\left[t - \frac{R_T(u)+R_R(u)}{c}\right]\exp\left\{-j\frac{2\pi}{\lambda}[R_T(u)+R_R(u)]\right\} \tag{6.7}$$

where c is the speed of light. In (6.6) and (6.7), $u \in \left[\frac{-T_d}{2}, \frac{T_d}{2}\right]$ and $t \in [0, T_s]$. The latter implies that as DVB-T symbols are transmitted continuously, an effective pulse repetition interval (PRI) can be chosen that is equal to the total DVB-T symbol duration. Of course, this selection is arbitrary and flexible – ultimately the major criterion in selecting it depends on having a pulse repetition frequency (PRF) that adequately samples the SAR signal in the slow-time dimension.

The main purpose of the direct signal is to maintain the coherence required for imaging. This is due to the fact that the transmitter and the receiver have independent clocks, leading to relative clock slippage and oscillator drift if left unchecked. The direct and reflected signals contain common transmitter and receiver artefacts, and therefore using the direct signal as the reference function for subsequent signal processing operations forms a simple basis for removing them.

In passive radar systems in general, it is common practice to use separate RF channels (albeit with a common clock and local oscillator) with highly directional antennas pointed towards the transmitter and the target area to record the direct and reflected signals, respectively. Those are referred to as the heterodyne and radar channels hereafter.

While using high-gain antennas for the heterodyne and radar channels is desirable for an airborne receiver, it is not always possible but also not strictly necessary. First, even using high-gain antennas, the direct signal can normally be observed at the surveillance channel due to its high EIRP. Second, at UHF, to achieve a high antenna gain, the dimensions of the antenna are relatively large, while physical space within the aircraft could be limited and any external installation could require subsequent flight certification. To fit within a small aircraft cabin, or even smaller airborne platforms, such as drones, receiving antennas should be physically smaller. In that case, their mainlobe-to-backlobe ratios are substantially reduced (could be as low as -3 dB for patch antennas), to the degree that the direct signal could be recorded from the backlobe of the antenna with almost as high strength as it could have been recorded from its mainlobe, on an average.

It is thus possible to use a single, low-gain antenna within an airborne DVB-T SAR system to record both the direct and reflected signals and form coherent SAR imagery. In the following sections, signal processing techniques that take advantage of this modality and are able to address some of its caveats are shown.

6.3.1 Back-projection algorithm (BPA)

A suitable algorithm for bistatic SAR systems is the BPA. Although computationally intensive, it is a straight-forward algorithm capable of focusing bistatic SAR data collected from arbitrary geometries. A brief overview of a conventional bistatic BPA is provided here. For faster BPA implementations, the reader is prompted to [25]. To aid understanding of the BPA algorithm, it is assumed that the direct and reflected signals are captured separately and are perfectly isolated from each other, while the reflected signal originates from a single point scatterer with (ground range, cross-range) co-ordinates of (x_i, y_j) within the target area. In that case, a block diagram of the algorithm is shown in Figure 6.6.

The first step in the algorithm is range compression, implemented via a matched filtering process in the fast-time direction. The reference signal used in the matched filter is $d(t, u)$. The matched filtering operation can be performed efficiently in the frequency domain, by multiplying the Fourier transform (FT) of $s(t, u)$ in the fast-time

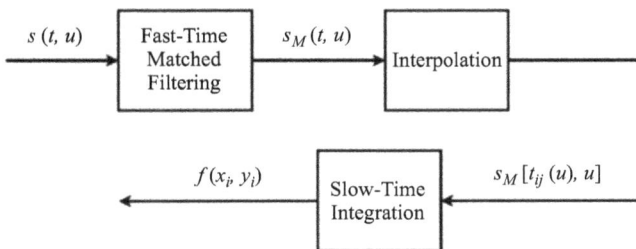

Figure 6.6 BPA block diagram (adapted from [26])

direction to that of the complex conjugate of $d(t, u)$, followed by an inverse FT. The output of the matched filter can be written as:

$$s_M(t, u) = G\left[t - \frac{R_T(u) + R_R(u) - R_B(u)}{c}\right]$$

$$\times \exp\left\{-j\frac{2\pi}{\lambda}\left[R_T(u) + R_R(u) - R_B(u)\right]\right\} \tag{6.8}$$

where $G(.)$ is the auto-correlation function of the transmitted signal $g(.)$.

For a given slow-time position, u, the range history of a given target, s_M, is projected into the spatial domain to determine the return from a reflector at position (x_i, y_j), giving back-projection its name. This process requires an interpolation step in the fast-time direction to extract the precise range history of the target, $s_M(t_{ij}, u)$, where t_{ij} is the instantaneous time delay of the target with co-ordinates (x_i, y_j) at slow-time u. The reconstruction of the target function, or the point spread function (PSF), is finally given by:

$$f(x_i, y_i) = \int_u s_M\left(t_{ij}, u\right) \Phi_{ij}(u) du \tag{6.9}$$

where $\Phi_{ij} = \exp\{j\frac{2\pi}{\lambda}\left[R_T(u) + R_R(u) - R_B(u)\right]\}$ is the phase history of a target at (x_i, y_j).

The short overview of the BPA shows that image formation requires knowledge of the transmitter and the receiver's locations to perform the back-projection operation. The transmitter locations are generally straight-forward to find or measure, while the receiver's instantaneous location is normally measured by onboard self-sensing systems such as INS/GPS devices. Of course, the accuracy to which receiver locations can be measured is finite, and for airborne systems there are perturbations from nominal flight trajectories which necessitate autofocus techniques.

Algorithmically, the only difference between using two separate channels to record the direct and reflected signals and using a single channel to record both lies in the range compression mechanism of the algorithm. In the former case, matched filtering is performed using the direct signal in the heterodyne channel, thus the output of the matched filter is the cross-correlation function between the heterodyne and radar channels (Figure 6.7a). In the latter case, the auto-correlation function of the radar channel is computed instead (Figure 6.7b).

6.3.2   *Effects of DVB-T pilot signals*

Having covered the essentials of BPA, we can begin to consider the case where both the direct and reflected signals are collected by a single antenna, and investigate possible artefacts in DVB-T SAR imaging. The ones we specifically consider here are any artefacts introduced by DVB-T pilot signals, methods of direct signal removal, and

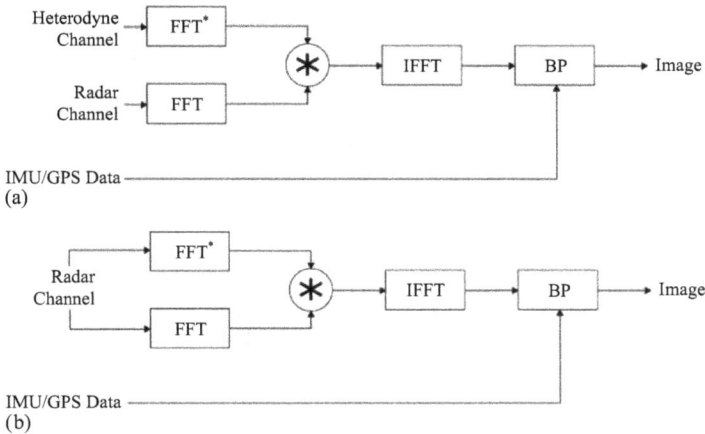

Figure 6.7   Correlation schemes for range compression: (a) cross-correlation scheme and (b) auto-correlation scheme [17]

methods for correcting motion errors due to receiver trajectory deviations. In this section, we investigate the former.

In conventional passive coherent location (PCL) systems relying on DVB-T, target ambiguities in range and Doppler arise due to the periodicity in DVB-T pilot signals. This problem has been known since the early days of DVB-T for passive radar, e.g. [27], and the necessity of effective ways to suppress it has been highlighted. Substantial efforts from the scientific community have culminated in a number of important works proposing a range of solutions (e.g. [28–32]) that can include reconstruction of the transmitted signal without its pilots using DVB-T demodulation techniques, even at high Doppler frequencies that could compromise OFDM orthogonality in DVB-T signals [33]. Such sophisticated but complex approaches could be extended to SAR, but first the severity of the problem in SAR should be investigated.

The signal recorded by the receiving antenna, ignoring amplitude terms, is given by:

$$r(t, u) = s(t, u) + d(t, u) \tag{6.10}$$

Rather than doing range compression in the way described in the previous section, we instead compute the auto-correlation function of (6.10) [34]. In doing so, at the output of the range compression operation, for a single point target, there are four correlation functions arising of which only one is desired. Those are the auto-correlation functions of the direct and reflected signals, the cross-correlation function between the direct and the reflected signal (desired), i.e. (6.5), and vice versa. Each of these are additional terms in (6.5), and each have a relative time delay and phase. However, they are not indicative of a target with a phase history $\Phi_{ij}(u)$ and hence they are defocused by the BPA.

A similar effect can be shown for DVB-T pilots. To align this analysis more closely to PCL, we write the cross-correlation between the direct and reflected signals for a single moment in slow-time, u, as:

$$G(\tau) = \int_{-\infty}^{\infty} g(t - \tau_0)g^*(t - \tau_1 - \tau)dt \qquad (6.11)$$

where $\tau_0 = \frac{R_B}{c}$ and $\tau_1 = \frac{R_T + R_R}{c}$. Writing $t - \tau_0 = t$ for simplicity, (6.11) becomes:

$$G(\tau) = \int_{-\infty}^{\infty} g(t)g^*(t - \Delta t - \tau)dt \qquad (6.12)$$

where $\Delta t = \frac{R_T + R_R - R_B}{c}$. We can now compare (6.12) to the ambiguity function for PCL [22], written as:

$$AF(\tau, v) = \int_{-\infty}^{\infty} d(t)s^*(t - \tau)\exp(-j2\pi vt)dt \qquad (6.13)$$

where v is the target velocity. It can be derived that $G(\tau) = AF(\tau - \Delta t, 0)$. Under this condition, the analyses of the ambiguity function for PCL state that the location of range-compressed scattered pilots will be at [28]:

$$\tau = \Delta t \pm \left\{0, \frac{k}{12}T_u, T_s\right\}, \quad k = 1, 2, ..., 12 \qquad (6.14)$$

Equation (6.14) shows that the location of ambiguous peaks in range is separated by $\frac{k}{12}T_u$. Thus, we may expect the range history of the same target to repeat itself every $\frac{k}{12}T_u$ s. However, the BPA acts as a spatial filter, rather than a temporal one. For $T_u = 896\mu s$ [24], and a quasi-monostatic data acquisition geometry the first ambiguity occurs at $c\Delta t + 22.4\,km$, but its range and phase variation are that of a target at $c\Delta t$. It will thus be severely defocused by the BPA, and ambiguities further out will be suppressed even further.

As an example, we consider the matched filter outputs of bistatic azimuth signals (Φ_{ij} in (6.9)) obtained under a quasi-monostatic configuration. The transmitter is located 20 km behind the receiver and its cross-range co-ordinate coincides with the midpoint of the receiver's synthetic aperture. The receiver moves along the cross-range axis at a constant speed of 100 m/s for a dwell time of 16 s, and a target is 10 km away from the receiver. The operating frequency is 650 MHz. The auto-correlation function of Φ_{ij} in this case is shown in Figure 6.8 as a solid line. Next, we assume the same signal, appearing 22.4 km further as a pilot would, but processed by BPA as though it were a target at that range. The output of the matched filter is shown as a dashed line. The results show a suppression of approximately 25 dB, which seems effective and well within the limits of a practical SAR sensitivity. This result does not take into account that the range, as well as the phase history, will be different too, and that affects the interpolation step of the BPA leading to further suppression to what is shown here.

Therefore, despite that removing or suppressing pilots prior to image formation is still optimal, their effects do appear diminished by the image formation algorithm operation itself.

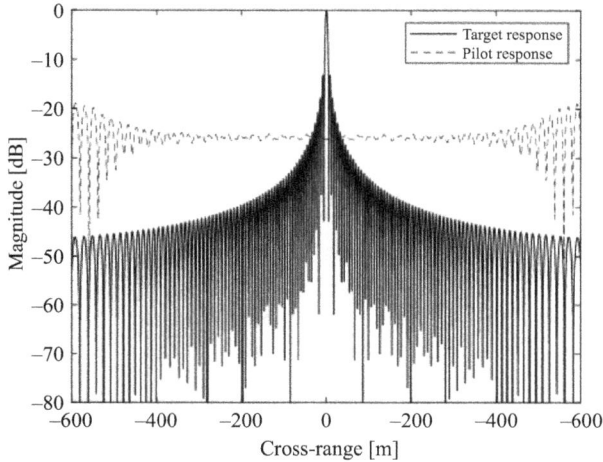

Figure 6.8   Azimuth compression with matched and mismatched filtering

6.3.3   Direct signal suppression and autofocus

6.3.3.1   Direct signal suppression

The next two sets of problems to consider are those of the direct signal present in the radar channel as well as dealing with unknown receiver motion errors due to flight irregularities.

In a previous section, calculations showed the SNR of the direct signal at the output of a receiving antenna. Regardless of whether the direct signal is recorded via separate receiver channels and regardless of their antenna gains, its signal strength is usually high enough to be visible in radar imagery. Further, as this signal strength is generally higher than that of targets in an imaged scene, even sidelobes emanating from a defocused direct signal in a passive radar image can be inadvertently pronounced.

This problem is not new to passive radar and is not new to passive SAR either. For example, in passive SAR using navigation satellites the direct signal is clearly visible over a range of bistatic geometries [8,9], and some of the first DVB-T SAR images, e.g. [34], show similar behaviours too.

In PCL, a common approach to remove direct signal artefacts relies on using CLEAN techniques such as those reported in [35,36]. The general principle behind CLEAN-like techniques is to de-convolve a radar output (e.g. a range-Doppler map or a radar image), and then re-convolve it without the direct signal. To do so requires prior knowledge of the direct signal structure, either by knowing the exact transmit waveform a priori, which is not possible in many passive radar environments unless a radar transmitter of opportunity is used [37], or by measuring its characteristics with high fidelity. When using a dedicated heterodyne channel, with highly directional antennas pointed towards the transmitter, the latter method has been used with success for PCL and hence it could be extended to passive SAR. However, applying the same approaches to the case where a single low-gain antenna is used to capture both the direct and reflected signals, as is the case here, was not always found to be effective.

The fundamental issue found experimentally was that in a single receiving channel the direct and reflected signals co-exist, thus compromising the estimation of the direct signal parameters. Therefore, a modified CLEAN technique should be derived to obtain the reference direct signal needed for CLEAN to operate. A possible approach is a modification of the algorithm in [37] to overcome these problems, and its block diagram is shown in Figure 6.9. The method mainly includes three parts: feature extraction (red dashed line), matrix correlation coefficient calculation (blue dashed line), and image subtraction (purple dashed line).

The direct signal suppression block is applied after the range compression step in the BPA (Figure 6.6), but before the interpolation step. As such, the inputs to the block are range-compressed data from a given receiver position along its synthetic aperture length. It is reminded here that the PRI used is equal to one DVB-T symbol, i.e. T_s.

The algorithm operates on the assumption that after range compression, and over a limited slow-time, the phase of the direct signal remains stable but the phase of complex echoes from the entire target area does not. This effect may thus be accentuated by averaging the range-compressed signal over a number of receiver positions (PRI's and, in this case, range-compressed DVB-T symbols). This can be implemented by applying a progressive mean on the data and forms the averaged compressed radar signal in Figure 6.9. The ith averaged range line, where i is an index denoting the current receiver position is extracted, and the location of its maximum intensity, Jr_{max}, is found. This is sample $s_{cd}(i, Jd_{max})$ within the averaged data. The

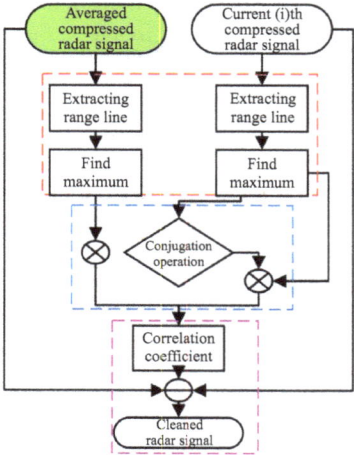

Figure 6.9   Modified CLEAN algorithm for direct signal suppression [38]

same operations take place for the current, ith range line, without any averaging, similarly yielding $s_{cr}(i, Jr_{max})$.

The matrix correlation coefficient is [38]:

$$C_{mcc} = \frac{\sum_{i=1}^{N} s_{cr}(i, Jr_{max}) s_{cd}^*(i, Jd_{max})}{\sum_{i=1}^{N} s_{cr}(i, Jr_{max}) s_{cr}^*(Jr_{max})} \tag{6.15}$$

where N is the number of receiver azimuth positions used in the averaging. The CLEANed data is then the ith range line minus the averaged data weighted by the correlation coefficient, i.e.:

$$s_{cc}(i,j) = s_{cr}(i,j) - C_{mcc} s_{cd}(i,j) \tag{6.16}$$

where i and j are the azimuth and range indices, respectively. Following this operation, s_{cc} could be used to follow through the rest of the BPA, i.e. interpolation and slow-time integration.

6.3.3.2 Autofocus

Coherent radar imaging via SAR requires precise positional knowledge of the platform carrying the SAR instrument. This is normally measured by INS/GPS systems installed on the platform and used as an input to the image formation algorithm. Unfortunately, the accuracy of this measurement is not always sufficient to fully focus an image, and this problem is exacerbated in airborne platforms which experience flight irregularities due to turbulence.

This problem is traditionally dealt with using SAR autofocus algorithms. There are numerous such algorithms, with their own relative merits and drawbacks in terms of complexity and accuracy. For monostatic SAR, there is a plethora of such algorithms – [39] and [40] are excellent references containing details on a number of seminal autofocus algorithms and their implementation.

The algorithm adopted for DVB-T SAR here is based on mapdrift autofocus (MDA) [39]. The algorithm is chosen because of its relative simplicity and its ability to sufficiently correct second-order (quadratic) phase errors, which are the major source of image defocus in SAR, and because we are dealing with a SAR system operating at a relatively lower frequency.

There are a few other practical reasons for this choice. Of course, more powerful algorithms like phase gradient autofocus (PGA) [41] can potentially be adopted. However, its application is more complex and this is not only because of its computational cost. In a bistatic SAR, the spatial resolution directions for range and azimuth are no longer orthogonal [42] and this could cause further complications in estimating azimuth errors from SAR imagery. The MDA could be more robust to such effects as it operates mostly in the data domain, rather than the image domain.

A block diagram of the MDA routine for passive SAR using BPA is shown in Figure 6.10.

Similar to MDA for monostatic SAR, the algorithm works by splitting the entire synthetic aperture into two sub-apertures and forming two separate images of the same target area with BPA. If direct signal suppression is used, the MDA block would be used after this operation. According to the MDA principle of operation, if the

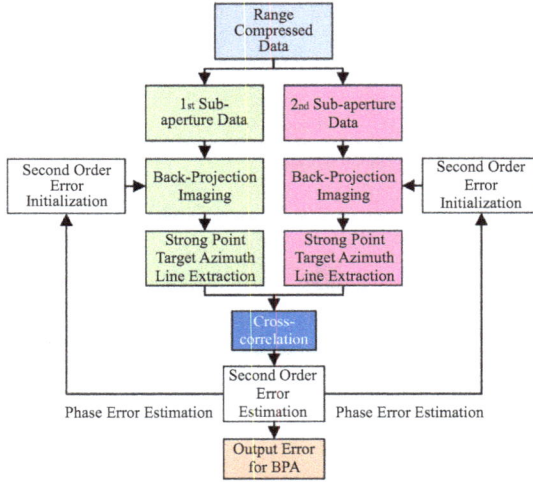

Figure 6.10   MDA block diagram [38]

velocity of the receiving platform is precisely measured, the two sub-aperture images will be identical. If not, the images will appear shifted in cross-range relative to each other, and this shift is directly proportional to the velocity error.

Velocity errors manifest as quadratic phase errors in the data, and those can be extracted by cross-correlating the sub-aperture images. The process is iterative, yielding progressively finer estimates of the error until a condition is met. It should be stressed here that the MDA, like many other autofocus algorithms, is conventionally used in conjunction with image formation algorithms operating in frequency domains, e.g. the range-Doppler algorithm, where the mapping relationship of the estimated phase errors between the data and image domain is more straightforward. However, the BPA focuses data in azimuth in the time domain. The MDA is still possible to apply, but to get a sufficiently accurate first estimate of the phase errors may require more iterations than conventionally required.

A quadratic phase error due to imprecise velocity measurement can be modelled as $\Phi(u) = \exp\left(-j\pi b u^2\right)$, where b, measured in Hz/s, is the rate of frequency change induced by the velocity error and is the parameter to be estimated. By measuring the number of samples, the sub-aperture images are shifted by, N_{drift}, and b can be estimated as:

$$b = 2N_{drift}\left(\frac{PRF}{N_a}\right)^2 \tag{6.17}$$

where N_a is the total number of receiver positions along the synthetic aperture. The algorithms for both the direct signal suppression and image formation are verified experimentally and their performance is assessed in the following section.

6.4   Experimental campaign and results

6.4.1   Experimental system

A passive SAR receiver was built and flown in a series of bistatic campaigns, whose major components are shown in Figure 6.11a. The aircraft used was a Cessna 172N Skyhawk (Figure 6.11b). The receiver was based around the National Instruments (NI) Universal Software Radio Peripheral (USRP) 2950R. The USRP is a software-defined radio device controlled by a laptop and was programmed to capture a single DVB-T channel with a carrier frequency of 650 MHz.

   The receiver had the option to record data via a heterodyne and a radar channel, or a radar channel only. The antennas and front-ends used were custom-built. To fit within the aircraft cabin, patch antennas that could be internally attached to the side windows of the aircraft were preferred. The design of the antennas was based on [43]. Each antenna (Figure 6.12a) had overall dimensions of 32×30 cm approximately, with approximate measured gains and horizontal beamwidths of 10 dBi and 50°, respectively.

   The front-end comprised of a low-noise amplifier (LNA) and a band-pass filter (BPF). Both the LNAs and BPFs were off the shelf components from mini-circuits and were mounted directly to the antenna mounting supports. The LNA was a ZRL-700+, which has a typical gain of 30 dB and a typical noise figure of 2 dB for the DVB-T frequencies received by the patch antenna. The BPF was an RBP-650+, with a pass-band between 624 and 680 MHz. A schematic of the front-end is shown in Figure 6.12b.

   A spatial FOG GPS/INS [44] was used to record receiver locations during flight. Its horizontal and vertical accuracies were 0.8 m and 1.5 m, respectively, while positional data were stored to the control laptop at a rate of 20 Hz. The positional accuracies could be improved by two orders of magnitude if base stations around the target areas used were available to enable real-time kinematic (RTK) outputs, however such base stations were not available in the measurement locations. The USRP and the spatial FOG used an external 10 MHz clock signal provided by the same GPS-disciplined oscillator (GPSDO), so that radar and positional data were synchronised in time.

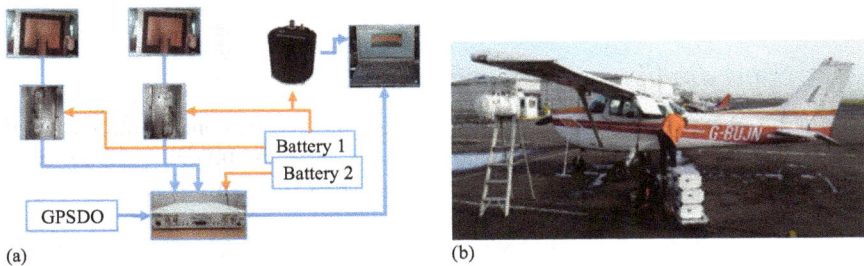

Figure 6.11   *(a) Passive SAR receiver architecture and (b) aircraft used in experimental campaign [17]*

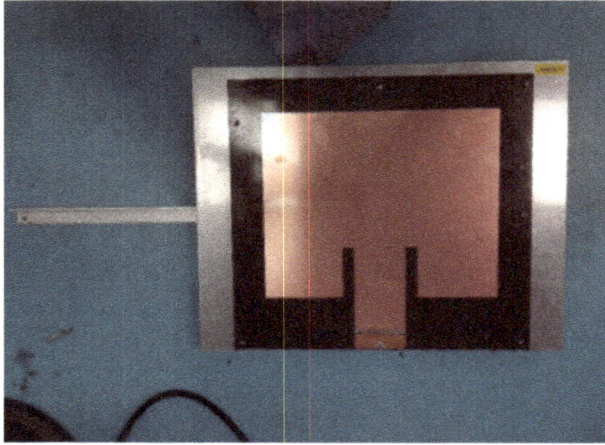

(a)

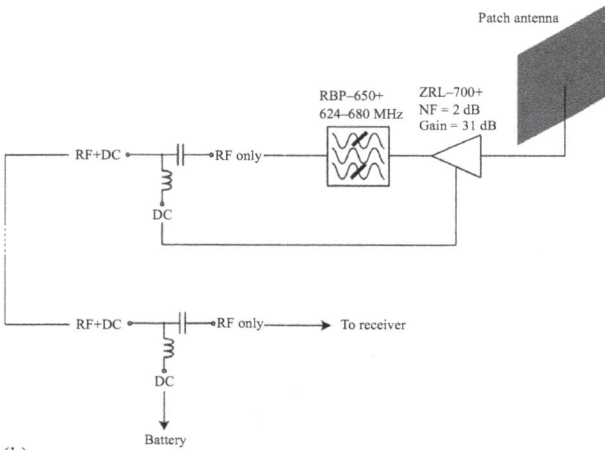

(b)

Figure 6.12   (a) Receiving antenna and (b) front-end schematic

6.4.2   Short dwell time, quasi-monostatic campaign

The first set of experiments were in a nearly quasi-monostatic configuration to verify the system and begin to explore some of its characteristics. Images were generated with a short dwell time on target, over which the use of autofocus techniques may not be required.

The target area was Bruntingthorpe Airfield at a (latitude, longitude) of (52.4906, −1.1314) degrees. The area surrounding the airfield contains a number of farms and villages with multiple one- and two-story buildings. As well as the typical potential imaging targets expected from an airfield, such as the runway, parked aircraft, and

Figure 6.13   Google Earth image of the Bruntingthorpe target area with the flight path of the aircraft shown (black line)[17]

large buildings and aircraft hangars, the airfield itself is also completely surrounded by trees. The maximum size of the target area imaged was 10 km by 4 km in range and cross-range.

In a nearly quasi-monostatic configuration, the location of the receiver was approximately 45 km away from the transmitter (Figure 6.13). The centre of the target area was about 4.75 km from the receiver's trajectory. On the day of the experiment no major turbulence was experienced during flight. 20 dB attenuators were added to the antenna front-ends to prevent the receiver being driven to saturation due to the high direct signal strength measured in flight.

The experimental parameters used in this set of experiments are shown in Table 6.4.

An example image obtained from the collected data, superimposed on a satellite photograph of the target area, is shown in Figure 6.14. The intensity in the image is plotted in dB, with 0 dB corresponding to the highest echo intensity within it. The dynamic range is artificially clipped to 40 dB. The x-coordinate in the image corresponds to distance from the receiver, while the y-axis is along the nominal receiver trajectory. As a first step, the image was generated by using the direct signal recorded at the heterodyne channel as the reference signal for range compression, followed by the BPA described in Section 6.3.1 and Figure 6.7a.

Inspection of the image shows good coincidence between the locations of high echo intensity in the image and areas in the target area that could yield it. This can readily be identified by tracing the outline of the airfield in Figure 6.14, but also by comparing parts of the target area and the image more closely. As an example, a side-by-side comparison between the image and the target area is shown in Figure 6.15. This is over a span of 2 km in range and cross-range, respectively. It can be seen that

Table 6.4   Experimental parameters

Parameter	Value
Carrier frequency	650 MHz
Bandwidth	7.6 MHz
Transmitter EIRP	200 kW
Transmitter height	264 m
Transmitter distance	45.1 km
Transmitter grazing angle	$\sim 0.5°$
Receiver antenna horizontal beamwidth	50°
Bistatic angle	$\sim 4°$
Dwell time	8s
Average receiver ground speed	240 km/h
Aircraft heading	12°
Aircraft altitude above mean sea level	600 m

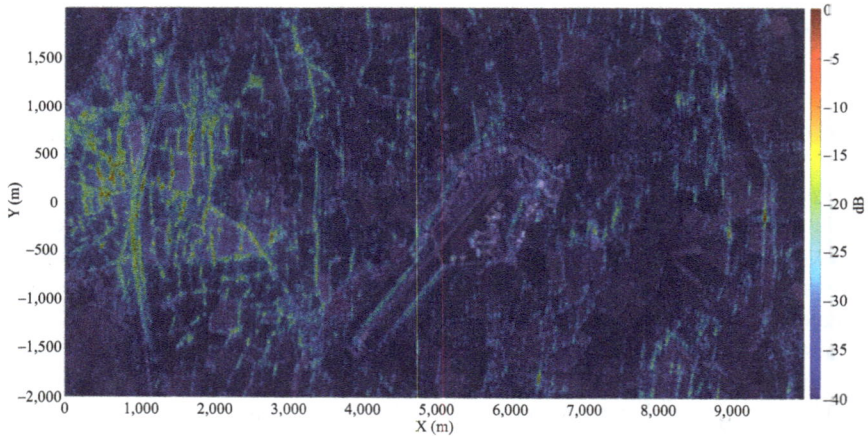

Figure 6.14   DVB-T SAR image of the Bruntingthorpe target area superimposed on satellite photograph of target area

the location and orientation of trees and hedgerows outlining field boundaries, as well as buildings, are clearly identified in the image. Reflections from terrain are difficult to discern, however at the low grazing angles of the transmitter and the receiver (e.g. at 45 km distance the transmitter is near-grazing, see Table 6.4) and the dynamic range of 40 dB used this could be expected. No artefacts like ambiguities due to pilots were observed in any of the images generated.

Images obtained by different range compression schemes are compared next. Figure 6.16a shows the image obtained using the auto-correlation function of the radar channel data, as in Section 6.3.1 and Figure 6.7b. The image of Figure 6.14, formed using the same part of the data set, is shown for comparison in Figure 6.16b,

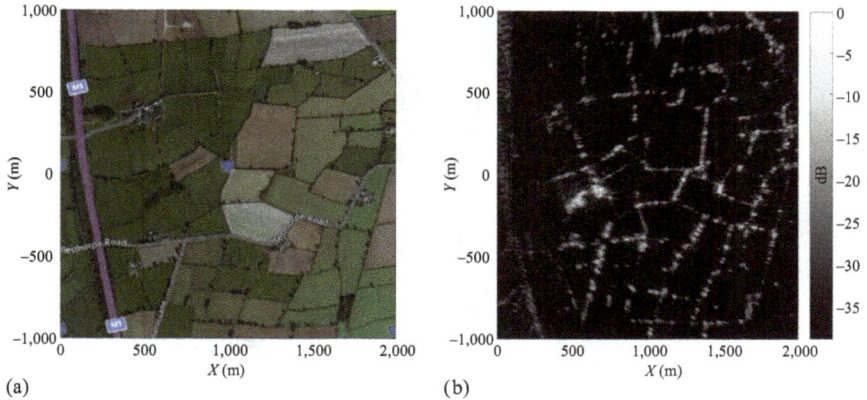

Figure 6.15   Close-ups of (a) target area and (b) DVB-T SAR image [17]

without the target area in the background. The two techniques can be seen to yield nearly identical results, with no visible artefacts in using auto-correlation to perform range compression. Following the same reasoning as in Section 6.3.2, this too can be expected. Using auto-correlation, multiple ambiguous peaks can appear after range compression (4 per target), however, BPA operation is sufficient to suppress them to an adequate degree. In doing so, and similar to Figure 6.8, ambiguous peaks are substantially defocused. The drawback in this approach is that as the image now contains energy from mismatched targets in azimuth the overall background level in the image can be expected to rise, thus reducing image contrast, however this effect was not directly visible in the reported experimental case. In the two images of Figure 6.16, the mean background level seems higher using auto-correlation compared to using cross-correlation, however the reason for this is different. Using auto-correlation, the direct signal enters the radar channel antenna through its backlobe, while in the heterodyne channel, it is received through the mainlobe of an identical antenna. The antennas used had a measure mainlobe-to-backlobe ratio of approximately 3 dB, hence we can expect for an image generated using the heterodyne channel to have a dynamic range that is 3 dB better. Indeed, the mean background level of the image using auto-correlation can be measured as -42.1 dB, while in Figure 6.16b it was measured as -45.5 dB, i.e. about 3.5 dB difference.

In both images, the presence of direct signal artefacts can be seen as the diagonal lines starting from the origin of the image and spanning the entire scene in the range dimension.

Focal quality and the spatial resolution can be assessed next. In an ideal scenario, the functionality of the image formation algorithm can be confirmed by imaging a point-like target and analysing the resulting point spread function (PSF) within a SAR image. This is usually achieved in monostatic SAR by placing a calibrated target, such as a corner reflector, within the target scene at a known location. Despite such calibrated targets were not available in our target area, they are fundamentally

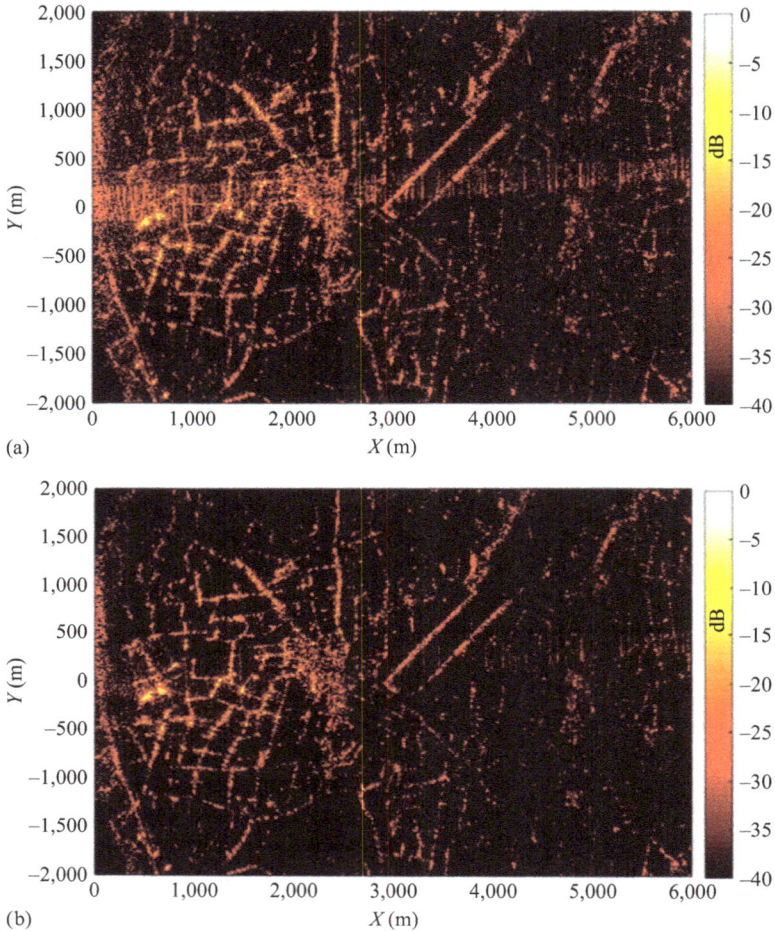

Figure 6.16   Images obtained using (a) auto-correlation of the radar channel and (b) cross-correlation between the heterodyne and radar channels for range compression

possible to install [45] but there are caveats. First, the dimensions of a corner reflector for a radar operating at UHF could be formidable, especially if a long stand-off is desired. Perhaps more importantly, traditional calibrated targets such as trihedral corner reflectors are designed to return high strength echoes in the direction of signal transmission, which compromises their suitability in a general bistatic scattering environment where the locations of signal transmission and echo reception can be substantially different.

In this instance, we can instead concentrate on targets of opportunity already in the target area, whose compressed echoes can be seen as point-like within an image,

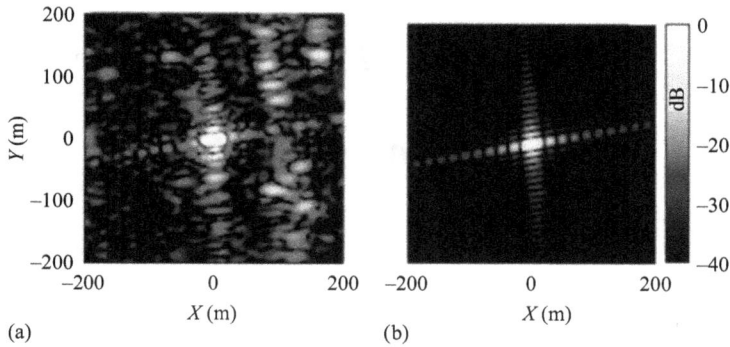

Figure 6.17 *(a) PSF obtained from real target and (b) theoretically expected PSF at the same location [17]*

and which are isolated from those of other targets. One such return was found at a range of approximately 9 km from the receiver and close to $Y = 0$ m and is visible as a bright spot in Figure 6.16a. This return corresponds to a single wind turbine, at a (latitude, longitude) of (52.486487, −1.066874) degrees. Of course, physically speaking, a wind turbine is far from a point target – however, at a distance of 9 km and with a transmit signal bandwidth of nearly 8 MHz, the size of the resolution cell is wide enough to encompass the entire structure. A close-up of the image in the vicinity of the wind turbine is shown in Figure 6.17a, where (0, 0) is the location of the wind turbine and 0 dB corresponds to the highest echo intensity within the presented image segment. It can be seen that the response from this target resembles a PSF.

This return can now be compared to the theoretically expected PSF. This can be computed from [42], where the form of the bistatic PSF was derived. It is important to highlight that the PSF was calculated at the same location as the echo from the wind turbine. The bistatic PSF, both in terms of resolution cell size and orientation, is a function of the bistatic angle and the effective velocity vector of the receiver relative to the target, and for the size of the target area those can vary considerably across the image. The calculated PSF is shown in Figure 6.17b.

A comparison of the theoretical and experimental results shows high similarity between them. To further validate the image formation approach, cross-sections of the theoretical and experimental results were taken along the range (X) and cross-range (Y). Those are shown in Figure 6.18. Both range and cross-range cross-sections are in good coincidence. In both cases, the mainlobes of the responses are nearly identical, other than a slight broadening seen in the experimental results. Even though the wind turbine is not the only target in Figure 6.17a (other echoes can be seen in the ±200 range and cross-range intervals around it), the sidelobe structure in both the range and cross-range profiles are again similar to the theoretical expectation. The first sidelobes in the experimental cross-range profile are slightly raised and imbalanced (the first sidelobe to the right is at −10 dB, while the one on the left is at −12 dB), which is indicative of uncompensated motion errors.

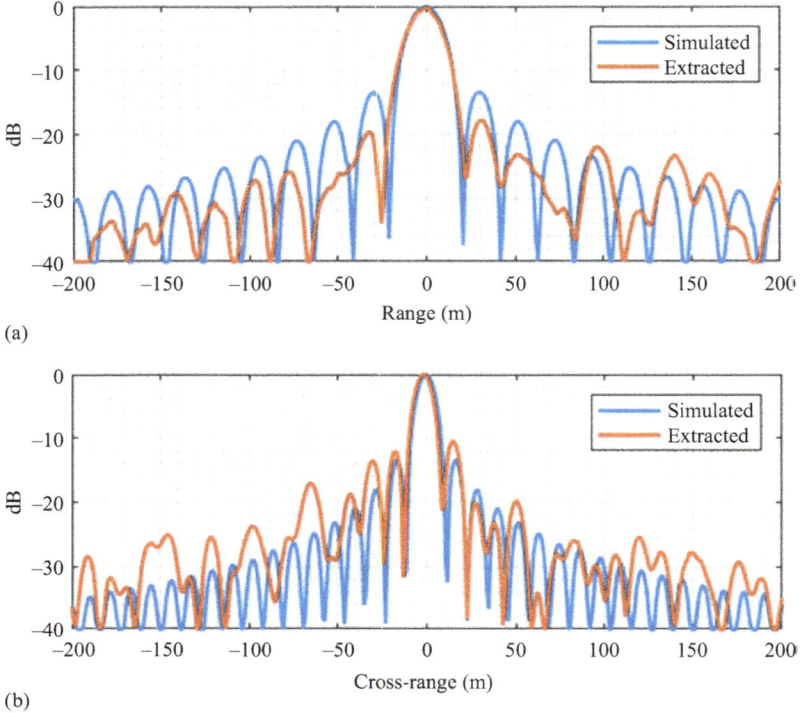

(a)

(b)

Figure 6.18 *(a) Range and (b) cross-range cross-sections of theoretical and experimental PSFs [17]*

The spatial resolution can also be measured from the cross-sections of Figure 6.18 and compared to theoretical expectations. The peak-to-null width of the range response is approximately 20 m, which is consistent with the theoretical range resolution of 19.75 m for a bandwidth of 7.61 MHz and a bistatic angle of 4° (6.2). Similarly, the 3-dB width of the cross-range profile is approximately 8 m. Again, this is close to the resolution predicted using (6.3).

6.4.2.1 Long dwell time, quasi-monostatic results
Having confirmed the essential elements of the DVB-T SAR system using short dwells, the next step is to assess the validity of the direct signal suppression and autofocus algorithms. To do that, data collected from the same target area, again under a similar and nearly quasi-monostatic configuration to the one described in the previous section (Figure 6.13), were processed over longer dwell times on target. The experimental parameters were the same as those listed in Table 6.4. The major difference is that the dwell time is now 16 s rather than 8 s, while the average aircraft speed was roughly 67 m/s rather than 64 m/s.

Table 6.5   Cross-range resolution

PSF	Resolution before autofocus	Resolution after autofocus	Predicted value
...	7.4 m	3.8 m	3.6 m
...	6.7 m	3.4 m	3.3
...	14 m	5.5 m	3.7

The image obtained under these conditions using radar channel auto-correlation is shown in Figure 6.19a. The direct signal sidelobes can be readily observed and so can image defocus, especially by inspection of the wind turbine echo at approximately 9 km distance from the receiver. Figure 6.19b shows the image obtained after application of the CLEAN algorithm and MDA. Each sub-aperture comprised 9,240 range lines, subsequently formed into sub-images which were then cross-correlated to estimate the phase error coefficient b (Section 6.3.3.2) as in Figure 6.10. The iteration of the motion error estimation was set to stop when the change in the estimated error coefficient between the current and the previous estimate was 0.1% or less.

Application of both techniques improves image quality. The direct signal is effectively removed in Figure 6.19b. A closer inspection does not reveal any legitimate targets being removed by the CLEAN operation either, although to state this with certainty in a complex target area with limited ground truth is difficult. The dynamic range of the image after the two techniques has been raised by approximately 6 dB.

Next, the performance of the autofocus algorithm is assessed using in-scene, point-like compressed echoes. Three such targets can be identified in the target area, located at (9,183, −1,367) m, (8,675, −1,463) m, and (8,803, 64.82) m. Their extracted responses from the images in Figure 6.19 are shown in Figure 6.20. Their cross-sections along the cross-range direction are shown in Figure 6.21. Inspection of Figures 6.20 and 6.21 shows that after MDA target cross-range resolution improves, while in Figure 6.21 it can be observed that sidelobes in target responses decrease. The measured cross-range resolutions before and after autofocus for each target, as well as the theoretical expectations at their locations, are listed in Table 6.5. The cross-range resolution improves by more than a factor of 2 in some cases and approaches theoretical limits. In terms of image magnitude, the improvement is more modest. The overall image contrast is measured to have improved after autofocus by approximately 0.1 dB, while the peak intensity of the three point targets was measured to have increased by 0.2 dB, 1.4 dB and 0.3 dB, respectively.

6.4.3   Bistatic experiments

DVB-T SAR is a bistatic system and hence it inherently possesses a spatial diversity that may be used as a degree of freedom. It is thus important to understand the relative merits and limitations of DVB-T SAR imagery as the bistatic angle varies. This is a

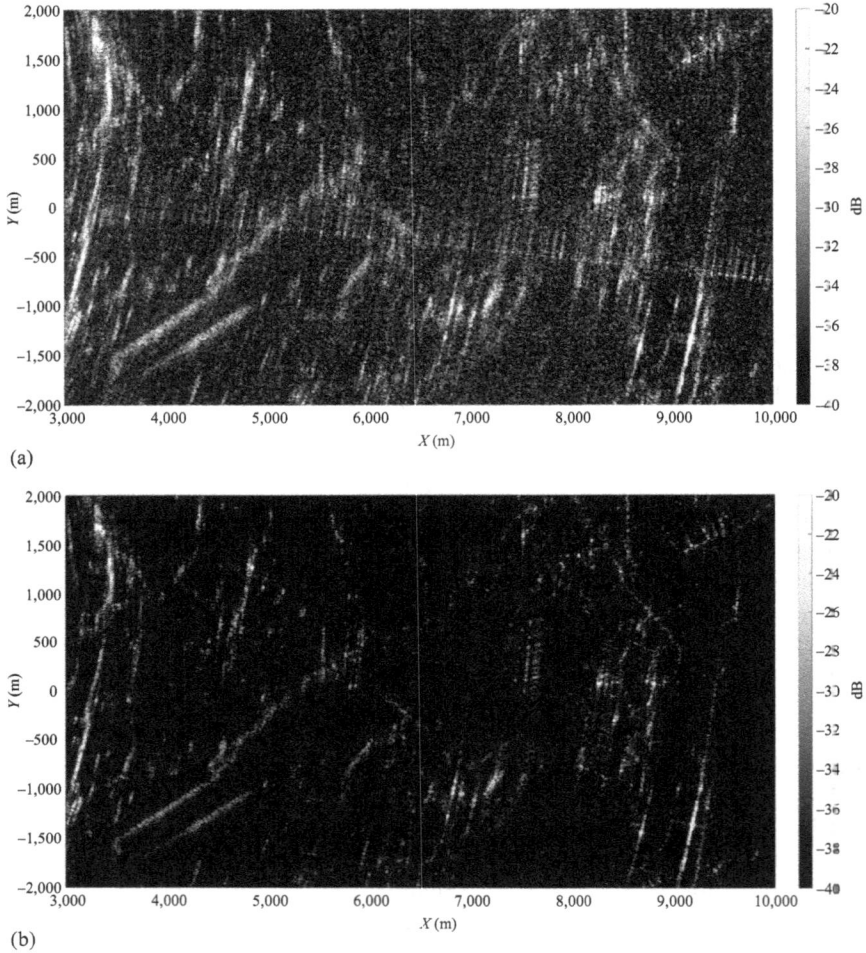

Figure 6.19   DVB-T SAR imagery (a) before and (b) after direct signal suppression and autofocus [19]

complex topic as the system is both bistatic and low-frequency and requires intense modelling and experimental work which, as this type of system is relatively new, is yet to be consolidated.

What can be provided here is some first experimental results from an airborne bistatic campaign to begin aiding our understanding of how the bistatic geometry affects DVB-T SAR imagery, and which may be used as a stepping stone for further theoretical and experimental works in the future.

The campaign was conducted at Bruntingthorpe, with the same airborne demonstrator described in a previous section. The flight path took the aircraft around the

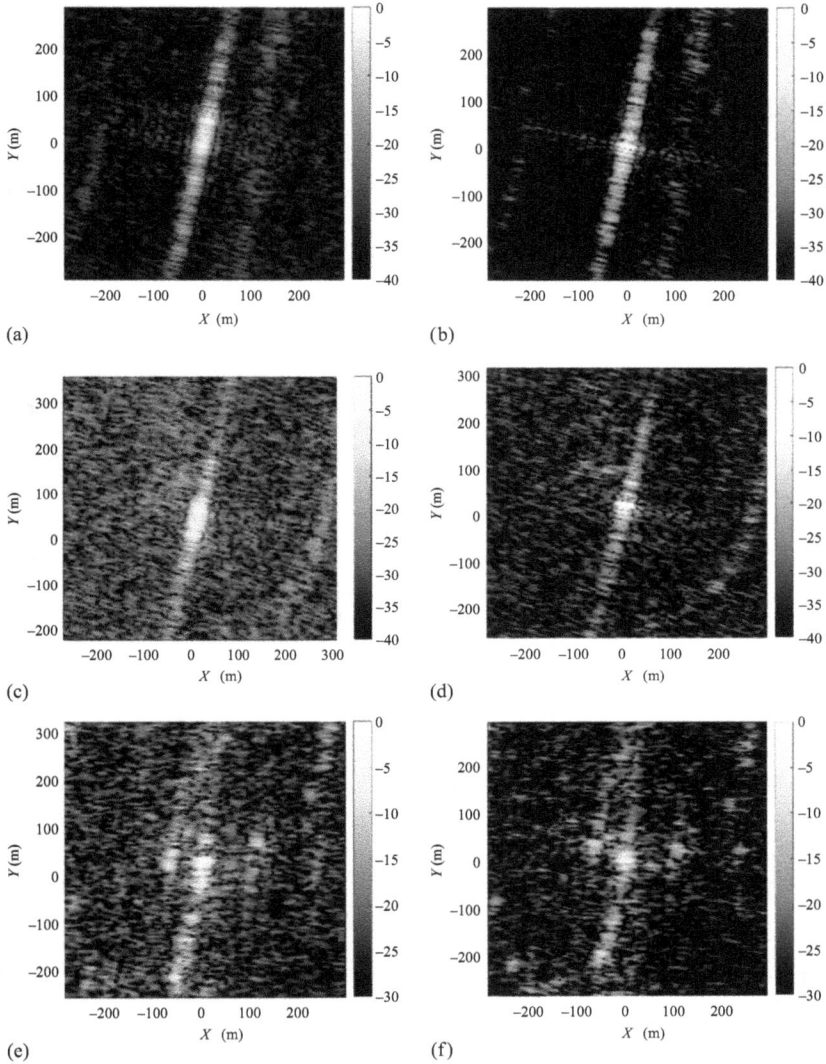

Figure 6.20 *Extracted PSF of three point-like targets for the 16 s aperture dwell time. (a, c, e,) Before and (b, d, f) after direct signal suppression and autofocus [38].*

target scene along six straight line segments, which formed a half-hexagon trajectory from an aerial view (Figure 6.22). Five of the straight line segments were used to capture bistatic radar measurements, while the longest segment, which was in a forward scatter configuration, was not used due to expected loss in range resolution. Flight trajectory 3 corresponds to a nearly quasi-monostatic configuration. Thus, the

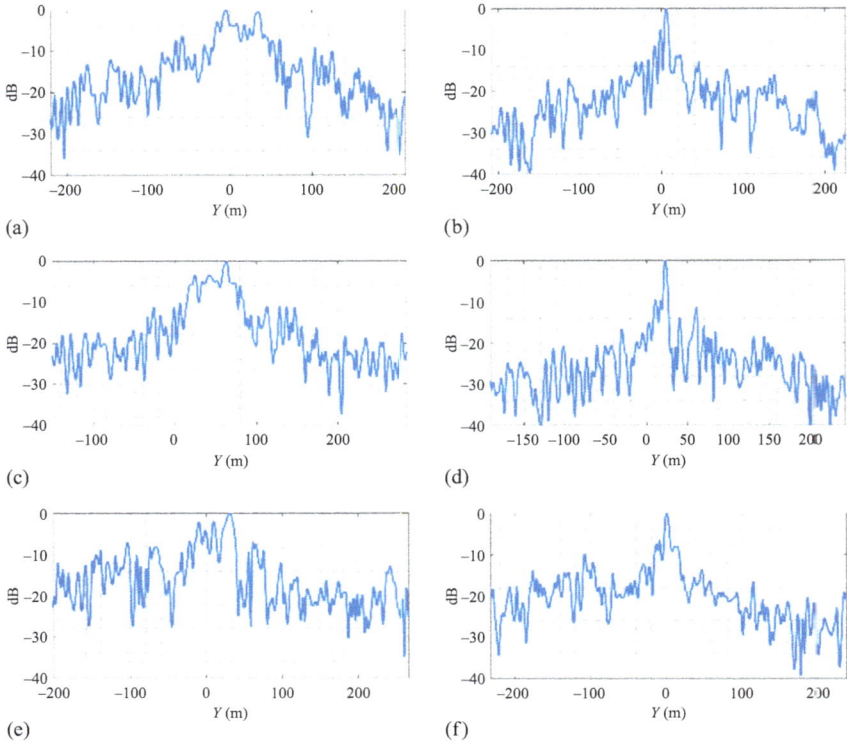

Figure 6.21   Cross-range profiles of three point target PSFs. (a, c, e,) Before and (b, d, f) after direct signal suppression and autofocus (adapted from [38]).

orientation of trajectories 2 and 4 relative to the transmitter are nearly 45°, while those of 1 and 5 are about 90°.

The experimental parameters are similar to those listed in Table 6.4, except the dwell time on target, which was set to 4 s. This was done so that a larger number of images were generated at smaller bistatic angle increments, and also to reduce any effects of uncompensated motion errors on images provided, thus aiding comparative analyses between them. It also relieved the computational burden of forming a series of images using autofocus/direct signal suppression. The aircraft altitude was increased to 1.1 km above mean sea level too. The bistatic angles and the receiver grazing angles (defined in Figure 6.22b) varied across the span of the target area for a single aperture. The total spread in bistatic angle was measured to be from 4 to 140 degrees for all apertures. A total of 53 images were formed across this span, although the distribution of these measurements was not uniform in terms of bistatic angle. Images were individually inspected for defocus before being retained for further analyses,

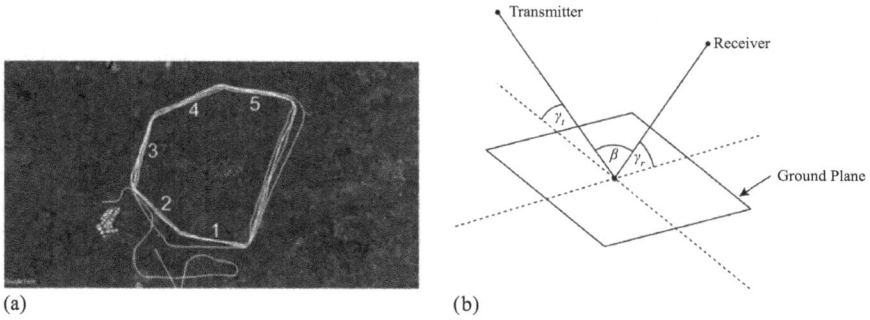

(a) (b)

Figure 6.22 *(a) GPS recorded flight path for airborne measurements and (b) bistatic angle, transmitter and receiver grazing angle definitions*

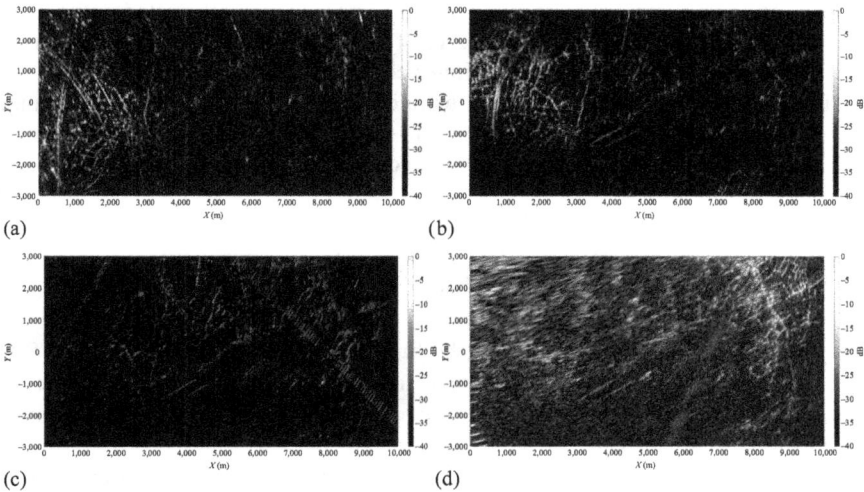

(a) (b)

(c) (d)

Figure 6.23 *Example images from flight trajectories (a) 2, (b) 3 (quasi-monostatic), (c) 4, and (d) 5*

by observing PSFs obtained from known point-like reflectors within the scene. All measurements were conducted on the same day and within a 2-h interval.

Example images are shown in Figure 6.23 for flight trajectories 2–5 (shown in Figure 6.22a), respectively. The direction of the direct signal sidelobes can be used to gauge the orientation of the aircraft relative to the target area. All images have been normalised to the same value and have the same dynamic range of 40 dB.

It can be seen that as the orientation of the aircraft changes, images obtained can have strong similarities but also striking differences. Different parts of the same target area can appear pronounced in some images, while suppressed in others, and this can partly be explained through the imaging geometry. Take, for example, the upper left corner in each image (X between 0 and 2 km, Y between 0 and 3 km). In Figure 6.23c, high strength echoes are practically non-existent and that is because this area was outside the receive antenna beam. Figure 6.23b is a quasi-monostatic image, obtained by the aircraft flying along the Y-axis. In this case, the bistatic angle is less than 10° and at close range the same area in the image is well pronounced. Conversely, at longer distances echo strength decreases by comparison. In Figure 6.23d, however, the opposite effect happens. Here, the aircraft flies parallel to the X-axis. The upper left corner in the image is imaged at bistatic angles of at least 130°, resulting in substantial loss of resolution. On the other hand, the intensity at further ranges is substantially improved in the image ($X > 7.5$ km).

Similar observations can be made for other parts in the target area. Thus, while from the four images the quasi-monostatic one appears to be the most informative on its own overall, it is not unreasonable to believe there are multi-static image combination schemes that could enhance image information space more than any single image can. Simple such schemes have been experimentally tested with GNSS-based SAR [9], where there is an abundance of spatial diversity at least on the transmit side, and those or more complex techniques could be applied here too.

It is important to note that effects illustrated above are linked to the bistatic angle not only in terms of resolution but also in terms of echo intensity due to bistatic scattering. This analysis is rather complex and lengthy and is thus outside the scope of this chapter.

6.5   Conclusions and outlook

This chapter provided an overview of the fundamental elements of a DVB-T SAR system. The power budget, sensitivity and resolution have been calculated. These calculations indicate potential for long distance operation and a modest spatial resolution when using a single DVB-T channel for imaging.

An image formation processor has been presented. The algorithm is based on BPA, so while it is computationally expensive it is able to form images for arbitrary bistatic data acquisition geometries. For airborne receivers, where physical space could be limited, it could be sufficient to use a single receiving channel to collect both the direct and reflected signals for imaging. DVB-T pilot signals seem to have little effect on imaging. An analysis on effects of DVB-T pilot signals on imaging shows their effects do not seem to be pronounced and that is due to the BPA operation as a spatial, rather than temporal filter. Algorithms to compensate for direct signal artefacts and for uncompensated aircraft motion errors have been derived.

An experimental system using mostly off-the-shelf components has been described. Software-defined radios are relatively straightforward for this purpose. An experimental campaign designed to verify system components and to begin further exploration of DVB-T SAR has been presented and experimental results have been discussed, in quasi-monostatic and multiple bistatic measurements.

Overall, through independent efforts from numerous scientific groups, the feasibility of DVB-T SAR has been solidified. The natural next step is to probe this technology further to fully understand and extract its potential, and the following are the authors' own thoughts.

One possible route to explore is DVB-T SAR phenomenology. A passive system like this is both bistatic and low-frequency, while at long stand-offs the transmitter is likely to be at near-grazing angle. As such, it is important to understand possible indirect propagation effects. On the one hand, this can set limits on the operational distance due to the local landscape, but on the other hand, it can investigate potential for beyond-the-hill vision, where higher frequency systems face restrictions. A glimpse of such capability has been shown in [19], where a receiver was on a ground moving vehicle and the resulting imagery produced echoes behind a hill blocking visual LOS. Additionally, UHF bands are known to have an adequate degree of penetration through foliage, while adding bistatics has the potential to increase signal-to-clutter ratio (SCR) compared to monostatic, low-frequency measurements (e.g. [46]). Demonstration of capabilities such as those could add operational value for DVB-T SAR.

Another strand could be to consider advanced SAR signal processing techniques, and the possibilities here are many. To name a few, one example is to investigate methods of spatial resolution improvement and that could be done in a number of ways. DVB-T stations broadcast on multiple RF channels (Table 6.1). Ways of combining their bandwidths, as has been shown for DVB-T inverse SAR [47], could be a way of improving range resolution. Considering multistatic approaches to resolution improvement, through coherent [8] or non-coherent [7,9] methods could be another way of enhancing image information space.

Acknowledgements

The authors thank Prof. Mikhail Cherniakov, Dr Alp Sayin, and Dr Yue Fang for their invaluable contributions to the work presented in this chapter. The authors also thank Almat Aviation Ltd for their support in organising and conducting our airborne trials.

References

[1] Cazzani L, Colesanti C, Leva D, *et al.* A ground based parasitic SAR experiment. In: *IEEE 1999 International Geoscience and Remote Sensing Symposium.* IGARSS'99 (Cat. no. 99CH36293). vol. 3, 1999. p. 1525–1527.

[2] Prati C, Rocca F, Giancola D, *et al.* Passive geosynchronous SAR system reusing backscattered digital audio broadcasting signals. *IEEE Transactions on Geoscience and Remote Sensing.* 1998;36(6):1973–1976.

[3] Cherniakov M, Saini R, Zuo R, *et al.* Space surface bistatic SAR with spaceborne non-cooperative transmitters. In: *European Radar Conference,* 2005. EURAD 2005, 2005. p. 9–12.

[4] Antoniou M, Saini R, and Cherniakov M. Results of a space-surface bistatic SAR image formation algorithm. *IEEE Transactions on Geoscience and Remote Sensing.* 2007;45(11):3359–3371.

[5]   Antoniou M and Cherniakov M. GNSS-based bistatic SAR: a signal processing view. *EURASIP Journal on Advances in Signal Processing.* 2013;2013(1):98. Available from: https://asp-eurasipjournals.springeropen. com/articles/10.1186/1687-6180-2013-98.

[6]   Santi F, Antoniou M, and Pastina D. Point spread function analysis for GNSS-based multistatic SAR. *IEEE Geoscience and Remote Sensing Letters.* 2015;12(2):304–308.

[7]   Santi F, Bucciarelli M, Pastina D, *et al.* Spatial resolution improvement in GNSS-b SAR using multistatic acquisitions and feature extraction. *IEEE Transactions on Geoscience and Remote Sensing.* 2016;54(10): 6217–6231.

[8]   Nithirochananont U, Antoniou M, and Cherniakov M. Passive coherent multistatic SAR using spaceborne illuminators. *IET Radar, Sonar & Navigation.* 2020;14(4):628–636. Available from: https://onlinelibrary.wiley.com/ doi/10.1049/iet-rsn.2019.0425.

[9]   Nithirochananont U, Antoniou M, and Cherniakov M. Passive multistatic SAR – experimental results. *IET Radar, Sonar & Navigation.* 2019;13(2):222–228. Available from: https://onlinelibrary.wiley.com/ doi/10.1049/iet-rsn.2018.5226.

[10] Liu F, Antoniou M, Zeng Z, *et al.* Coherent change detection using passive GNSS-based BSAR: experimental proof of concept. *IEEE Transactions on Geoscience and Remote Sensing.* 2013;51(8):4544–4555.

[11] Hu C, Li Y, Dong X, *et al.* Three-dimensional deformation retrieval in geosynchronous SAR by multiple-aperture interferometry processing: theory and performance analysis. *IEEE Transactions on Geoscience and Remote Sensing.* 2017;55(11):6150–6169.

[12] Hu C, Zhang B, Dong X, *et al.* Geosynchronous SAR tomography: theory and first experimental verification using Beidou IGSO satellite. *IEEE Transactions on Geoscience and Remote Sensing.* 2019;57(9):6591–6607.

[13] Rowlatt JP, Hristov S, Daniel L, *et al.* Doppler beam sharpening in passive bistatic radar with spaceborne illuminators of opportunity. In: *2018 19th International Radar Symposium (IRS),* 2018. p. 1–10.

[14] Ulander LMH, Frölind PO, Gustavsson A, *et al.* VHF/UHF bistatic and passive SAR ground imaging. In: *2015 IEEE Radar Conference (RadarCon),* 2015. p. 0669–0673. ISSN: 2375–5318.

[15] Gromek D, Kulpa K, and Samczyński P. Experimental results of passive SAR imaging using DVB-T illuminators of opportunity. *IEEE Geoscience and Remote Sensing Letters.* 2016;13(8):1124–1128.

[16] Walterscheid I, Wojaczek P, Cristallini D, *et al.* Challenges and first results of an airborne passive SAR experiment using a DVB-T transmitter. In: *EUSAR 2018; 12th European Conference on Synthetic Aperture Radar,* 2018. p. 1–4.

[17] Atkinson G, Sayin A, Stove A, *et al.* Passive SAR satellite (PASSAT) system: airborne demonstrator and first results. *IET Radar, Sonar & Navigation.*

2019;13(2):236–242. Available from: https://onlinelibrary.wiley.com/doi/10.1049/iet-rsn.2018.5225.

[18] Gromek D, Krysik P, Kulpa K, *et al.* Ground-based mobile passive imagery based on a DVB-T signal of opportunity. In: *2014 International Radar Conference*, 2014. p. 1–4.

[19] Atkinson GM, Sayin A, Cherniakov M, *et al.* Passive SAR satellite system (PASSAT): ground trials. In: *2018 International Conference on Radar (RADAR)*, 2018. p. 1–6.

[20] Antoniou M, Stove AG, Sayin A, *et al.* Passive SAR satellite constellation for near-persistent earth observation: prospects and issues. *IEEE Aerospace and Electronic Systems Magazine.* 2018;33(12):4–15.

[21] Sutton Coldfield (Birmingham, England) Full Freeview transmitter. Available from: https://ukfree.tv/transmitters/tv/Sutton_Coldfield.

[22] Cherniakov M. *Bistatic Radar: Emerging Technology.* Chichester, England: John Wiley & Sons, 2008.

[23] Richards MA, Scheer J, Holm WA, *et al.*, editors. *Principles of Modern Radar.* Raleigh, NC: SciTech Pub, 2010.

[24] Digital Video Broadcasting (DVB); Framing Structure, Channel Coding and Modulation for Digital Terrestrial Television. European Broadcasting Union, 1997. Available from: https://pdfs.semanticscholar.org/cde9/8a0be0311505aee20afe13c1578819a4ca0f.pdf.

[25] Ulander LMH, Froelind PO, Gustavsson A, *et al.* Fast factorized backprojection for bistatic SAR processing. In: *8th European Conference on Synthetic Aperture Radar*, 2010. p. 1–4.

[26] Soumekh M. *Synthetic Aperture Radar Signal Processing with MATLAB Algorithms.* New York: John Wiley, 1999.

[27] Saini R and Cherniakov M. DTV signal ambiguity function analysis for radar application. *IEE Proceedings – Radar, Sonar and Navigation.* 2005;152(3):133. Available from: https://digital-library.theiet.org/content/journals/10.1049/ip-rsn_20045067.

[28] Colone F, Langellotti D, and Lombardo P. DVB-T signal ambiguity function control for passive radars. *IEEE Transactions on Aerospace and Electronic Systems.* 2014;50(1):329–347.

[29] Palmer JE, Harms HA, Searle SJ, *et al.* DVB-T passive radar signal processing. *IEEE Transactions on Signal Processing.* 2013;61(8):2116–2126.

[30] Searle S, Palmer J, Davis L, *et al.* Evaluation of the ambiguity function for passive radar with OFDM transmissions. In: *2014 IEEE Radar Conference*, 2014. p. 1040–1045.

[31] Kuschel H, Ummenhofer M, O'Hagan D, *et al.* On the resolution performance of passive radar using DVB-T illuminations. In: *11th International Radar Symposium*, 2010. p. 1–4.

[32] Poullin D. Passive detection using digital broadcasters (DAB, DVB) with COFDM modulation. *IEE Proceedings – Radar, Sonar and Navigation.* 2005;152(3):143. Available from: https://digital-library.theiet.org/content/journals/10.1049/ip-rsn_20045017.

[33]   Berthillot C, Santori A, Rabaste O, *et al*. BEM reference signal estimation for an airborne passive radar antenna array. *IEEE Transactions on Aerospace and Electronic Systems*. 2017;53(6):2833–2845.

[34]   Ulander LMH, Frölind PO, Gustavsson A, *et al*. Airborne passive SAR imaging based on DVB-T signals. In: *2017 IEEE International Geoscience and Remote Sensing Symposium (IGARSS)*. 2017. p. 2408–2411.

[35]   Kulpa K. The CLEAN type algorithms for radar signal processing. In: *2008 Microwaves, Radar and Remote Sensing Symposium*, 2008. p. 152–157.

[36]   Garry JL, Baker CJ, and Smith GE. Evaluation of direct signal suppression for passive radar. *IEEE Transactions on Geoscience and Remote Sensing*. 2017;55(7):3786–3799.

[37]   Kulpa K, Samczynski P, Malanowski M, *et al*. The use of CLEAN processing for passive SAR image creation. In: *2013 IEEE Radar Conference (RadarCon13)*, 2013. p. 1–6.

[38]   Fang Y, Atkinson G, Sayin A, *et al*. Improved passive SAR imaging with DVB-T transmissions. *IEEE Transactions on Geoscience and Remote Sensing*. 2020;58(7):5066–5076.

[39]   Carrara WG, Goodman RS, and Majewski RM. Spotlight synthetic aperture radar: signal processing algorithms. In: *The Artech House Remote Sensing Library*. Boston, MA: Artech House, 1995.

[40]   Oliver C and Quegan S. Understanding synthetic aperture radar images. In: *Artech House Remote Sensing Library*. Boston, MA: Artech House, 1993.

[41]   Jakowatz CV, editor. *Spotlight-Mode Synthetic Aperture Radar: A Signal Processing Approach*. Boston, MA: Kluwer Academic Publishers, 1996.

[42]   Zeng T, Cherniakov M, and Long T. Generalized approach to resolution analysis in BSAR. *IEEE Transactions on Aerospace and Electronic Systems*. 2005;41(2):461–474.

[43]   Rosado-Sanz J, Jarabo-Amores MP, Mata-Moya D, *et al*. Design of a broadband patch antenna for a DVB-T based passive radar antenna array. In: *2017 IEEE 17th International Conference on Ubiquitous Wireless Broadband (ICUWB)*, 2017. p. 1–5.

[44]   Spatial FOG – Advanced Navigation. Available from: http://www.advancednavigation.com.au/product/spatial-fog.

[45]   Frölind PO, Gustavsson A, Haglund A, *et al*. Analysis of a ground target deployment in an airborne passive SAR experiment. In: *2017 IEEE Radar Conference (RadarConf)*, 2017. p. 273–278.

[46]   Ulander LMH, Barmettler A, Flood B, *et al*. Signal-to-clutter ratio enhancement in bistatic very high frequency (VHF)-band SAR images of truck vehicles in forested and urban terrain. *IET Radar, Sonar & Navigation*. 2010;4(3):438. Available from: https://digital-library.theiet.org/content/journals/10.1049/iet-rsn.2009.0039.

[47]   Olivadese D, Giusti E, Petri D, *et al*. Passive ISAR with DVB-T signals. *IEEE Transactions on Geoscience and Remote Sensing*. 2013;51(8):4508–4517.

Chapter 7
Passive radar for GMTI

*Philipp Markiton[1], Fabiola Colone[2] and
Pierfrancesco Lombardo[2]*

Abstract

The tasks of clutter suppression and ground moving target indication are a typical
and also very challenging problem for airborne radar systems for both defence and
civilian applications. Nowadays, research aims at applying these tasks for passive
radar systems on moving platforms. This chapter addresses the problem of clutter
cancellation and slowly moving target detection in orthogonal-frequency-division-
multiplex-based passive radar systems mounted on moving platforms. Conventional
signal processing approaches which exploit multiple receiving channels can be inef-
fective for the considered application due to the impossibility to control the employed
waveform of opportunity. Therefore, a processing scheme, which aims at ground moving target indication needs to address this issue. The space-time adaptive processing
scheme, proposed in this Chapter, exploits the benefits of the reciprocal filtering
strategy applied at a range compression stage together with a flexible displaced phase
center antenna approach. The effectiveness of the proposed scheme is demonstrated
via application to a simulated dataset as well as on experimental data collected by a
multichannel passive radar on an airborne moving platform.

7.1 Introduction

Passive radar [or Passive Coherent Location (PCL)] for stationary applications has
reached a stage of maturity in recent years, such that it can be deployed for both
military and also civilian applications, e.g. for improved air and flight security.

Passive radar on moving platforms for applications such as Ground Moving
Target Indication (GMTI) and Synthetic Aperture Radar (SAR) are a driver of current
research. The required transmitter (TX) does not need to be maintained by the radar
operators, instead common communication and broadcast transmitters are hijacked

[1] Fraunhofer Institute FHR, Germany
[2] University of Rome 'La Sapienza', Italy

for radar operations. This fact of operating covertly make PCL systems on moving (airborne) platforms especially interesting for military applications.

As the required receiving hardware is typically lightweight and easy to be built and maintained, passive radar is favorable to be mounted on small airborne platforms, such as unmanned aerial vehicles or ultra-light aircrafts. This provides the operator to monitor large areas covertly, for example for border or coast-lines control, thus contributing to reconnaissance operations and to situational awareness.

Over the past years, digital broadcast and communication transmitters (typically transmitting using the Orthogonal Frequency-Division Multiplexing (OFDM) scheme) were the preferred choice as Illuminator of Opportunity (IO), as the signal bandwidth remains constant due to its independency of the signal content, thus providing constant range resolution. Furthermore, the digital transmission format makes an almost ideal recovery of the transmitted signal possible: A dedicated reference channel pointing to the transmitter to collect a clean reference signal is not necessary. Instead the reference signal can be recovered using the received signals from the surveillance channels. A dedicated reference channel pointing to the transmitter to collect a clean transmitted signal is therefore not necessary. Consequently the total number of required receiving channels is reduced, which relaxes the hardware requirements.

With particular reference to GMTI applications, research described in [1–6] has focused on both theoretical analyses and experimental validation.

Whilst bistatic STAP has been addressed in [7–9], accounting for the noncooperative waveform in addition to the geometric complexities poses an additional set of challenges for the signal processing. On the other hand, there are still open questions on the effects induced by the receiver (RX) motion, for instance, on the direct signal reconstruction if digital illuminators of opportunity are used [10–12].

In this chapter, we focus on the problem of clutter cancellation and the detection of slowly moving targets for passive radar systems mounted on moving platforms exploiting OFDM-based waveforms. We show that the main limitation in clutter cancellation is due to the time-varying characteristics of the employed waveforms of opportunity. Additionally, we aim to provide a solution that can handle the time-varying characteristics. Specifically, we are interested in the application of two different filters used in the range compression stage and their influence on a subsequent clutter filtering stage. One of the filters for the range compression stage used in our analysis is the reciprocal filter, which will be compared to the widely used matched filter. We demonstrate that the reciprocal filter is beneficial compared to the matched filter in terms of removing the temporal variability of the system impulse response, so that an ideal clutter cancellation using spatio-temporal approaches can be obtained. In addition, we present a strategy, which alleviates the constraint of the equivalent pulse-repetition frequency at the DPCA stage, such that the proposed method of reciprocal filtering is effective also if the considered continuous waveform is processed on a batch-wise basis.

7.1.1 Outline

This chapter is organized as follows: The keypoints of our findings are listed in Section 7.1.2, which is followed by the Notation. The signal model is developed in

Section 7.2. In Sections 7.3 and 7.4, we analyze the application of the matched filter and of the reciprocal filter respectively, in the range compression stage. In Section 7.5, we apply the data against real data from an airborne measurement campaign. And finally in Section 7.6, we summarize our findings and the chapter.

7.1.2 *Key points*

Important concepts developed throughout the chapter are summarized here:

- A theoretical analysis of the impact of the widely used matched filter for range compression on clutter suppression.
- A theoretical analysis of the advantage using the reciprocal filter for range compression in lieu of the matched filter and subsequent clutter suppression.
- A comparison between matched filter and reciprocal filter by theoretical results and simulated data.
- Verification by evaluation of experimental real data.

7.1.3 *Notation*

A_q	complex amplitude of echo from qth scatterer
c_0	propagation velocity (speed of light) (m/s)
c_m	mth constellation symbol of corresponding constellation map
d	interelement spacing (m)
f_{Dq}	bistatic Doppler frequency of qth point-like scatterer (Hz)
$g_n[l]$	impulse response of range compression filter for nth batch
$h_n[l]$	range compression filter for nth batch
h_P	altitude of the receiver platform (m)
k, k'	multiplicative constant to keep thermal noise level at output of filtering process constant
L	number of samples in one batch
l_{τ_q}	bistatic propagation delay of qth scatterer in discrete notation
M_C	order of constellation of employed modulation scheme
N_R	number of bistatic range cells
N	number of batches in one coherent processing interval
$P_C^{(in)}$	clutter power at system input
$P_C^{(out)}[l, m]$	clutter power at delay-Doppler cell (l,m)
R_{Bq}	bistatic range of qth point-like scatterer (m)
$r(t), r[l]$	signal received at surveillance channel in continuous and discrete time notation
$r_C(t), r_C[l]$	clutter echoes received at surveillance channel in continuous and discrete time notation
$r_G(t), r_G[l]$	target echoes received at surveillance channel in continuous and discrete time notation

$r_N(t)$, $r_N[l]$	thermal noise contribution at surveillance channel in continuous and discrete time notation
$r_0(t)$, $r_0[l]$	equivalent baseband signal for single point-like scatterer in continuous and discrete time notation
$\bar{r}[l, n]$	signal received at surveillance channel in 2-dimensional matrix notation
$s(t)$	equivalent baseband signal emitted by transmitter in continuous time notation
$s_n[l]$	nth batch of equivalent baseband signal emitted by transmitter in discrete time notation
T	temporal duration of one batch (s)
T_{OFDM}	temporal duration of one OFDM symbol (s)
T_D	DPCA time delay (s)
T_K	integer multiple of duration of one batch for application of flexible DPCA in slow-time domain (s)
ΔT	fine time delay for application of flexible DPCA in Doppler frequency domain (s)
v_P	velocity of the receiver platform (m/s)
v_{Bq}	bistatic velocity of qth point-like scatterer (m/s)
$x[l, n]$	range compressed signal in 2-dimensional matrix notation
$y[l, n]$	2-dimensional signal after clutter filtering in matrix notation
$z[l, n]$	2-dimensional signal after DFT in matrix notation
α	angle between platform velocity vector and Rx-scatterer LOS in radians
λ	signal carrier wavelength (m)
τ_q	bistatic propagation delay of qth scatterer (s)
μ	multiplicative constant dependent on employed modulation scheme
σ_s^2	variance of illumination waveform

7.1.4 Acronyms

A list of the used acronyms:

AF	ambiguity function
ARX	airborne receiver
AWGN	additive white Gaussian noise
CNR	clutter-and-noise ratio
CPI	coherent processing interval
CW	continuous wave
DFT	discrete Fourier transform
DPCA	displaced phase center antenna
DSI	direct signal interference
DVB-T	digital-video-broadcast – terrestrial

ERP	effective radiated power
GMTI	ground moving target indication
ICM	internal clutter motion
IDFT	inverse discrete Fourier transform
IMU	inertial measurement unit
IO	illuminator of opportunity
IQ	in-phase and quadrature
LA	leading antenna
MF	matched filter
MRC	maximal ratio combining
OFDM	orthogonal frequency-division multiplexing
PCL	passive coherent location
PRF	pulse repetition frequency
PRI	pulse repetition interval
QAM	quadrature amplitude modulation
QPSK	quadrature phase-shift keying
RX	receiver
RpF	reciprocal filter
SAR	synthetic aperture radar
SCNR	signal-to-clutter-and-noise ratio
SET	sensors & electronics technology
SNR	signal-to-noise ratio
SRX	stationary receiver
STAP	space–time adaptive processing
TA	trailing antenna
TX	transmitter
ULA	ultra light aircraft
USRP	universal software radio peripheral

7.2   Signal model

We consider a passive radar mounted on a moving platform. The passive radar exploits a ground-based TX as an illuminator of opportunity, as shown in Figure 7.1. The platform carrying the receiving system is assumed to move with constant velocity v_P along a direction aligned with the y-axis at height h_P. The ground is assumed to be flat for simplicity. The receiving system is equipped with two receiving channels, where each receiving channel is connected to one of two identical antenna elements. The antenna elements are mounted in pure side-looking configuration and displaced in the along-track direction by $d = \lambda/2$. The antenna positioned first in flight direction is called the Leading Antenna (LA) and the antenna positioned second in flight direction is called Trailing Antenna (TA).

   In ideal conditions, the TA occupies the same spatial position of the LA after $T_D = d/v_P$ seconds. While the RX is moving along the defined trajectory, it receives

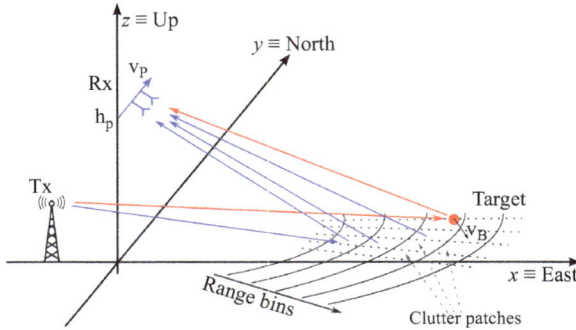

Figure 7.1 System geometry

echoes $r_C^{(\gamma)}(t)$ from stationary objects, i.e. clutter, and echoes $r_G^{(\gamma)}(t)$ from moving objects, i.e. targets, at both antenna elements:

$$r^{(\gamma)}(t) = r_C^{(\gamma)}(t) + r_G^{(\gamma)}(t) + r_N^{(\gamma)}(t) \tag{7.1}$$

$r_N^{(\gamma)}$ defines Additive White Gaussian Noise (AWGN) at both receiving channels $\gamma := [LA, TA]$; statistically independent between both receiving channels.

The equivalent baseband signal scattered by a single point-like scatterer for short observation times and simultaneously collected at the two Rx channels can be written as a function of time t as

$$r_0^{(LA)}(t) = A_0 s(t - \tau_0) \exp\{j2\pi f_{D0}t\}$$

$$r_0^{(TA)}(t) = A_0 s(t - \tau_0) \exp\{j2\pi f_{D0}t\} \exp\{-j2\pi \frac{d}{\lambda} \cos\alpha_0\} \tag{7.2}$$

The bistatic Doppler frequency f_{D0} can be expressed as the sum of two contributions: $f_{D0} = \frac{v_P}{\lambda} \cos\alpha_0 - \frac{v_{B0}}{\lambda}$. The first contribution is related to the platform motion and the Rx-scatterer geometry and applies even to stationary scatterers. The second term is due to the scatterer's intrinsic motion which depends on the bistatic velocity $v_{B0} = v_{Tx-s_0} + v_{Rx-s_0}$ defined as the sum of the target's velocity projection along the target-Tx direction $Tx-s_0$ and the target-Rx direction $Rx-s_0$ when both the Tx and the Rx are assumed to be stationary (therefore, $v_{B0} = 0$ for stationary scatterers).

Figure 7.2 shows the basic processing architecture employed in OFDM-based PCL systems. Here we evaluate the bistatic range-Doppler map by applying a batching approach [13,14]. This strategy operates first by synchronizing on the OFDM signal, and afterwards by:

1. Dividing the received signals into short consecutive batches.
2. Performing a range compression on a batch-by-batch basis, where the remodulated version of the reference signal is used [15,16].
3. Performing clutter suppression by application of Displaced Phase Center Antenna (DPCA).

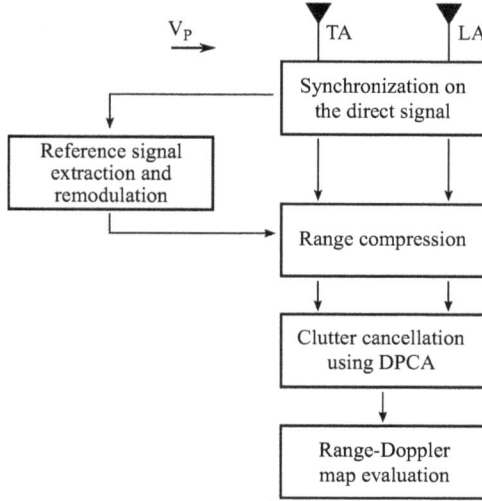

Figure 7.2   Flowchart of the processing

4. Applying a Discrete Fourier Transform (DFT) across consecutive batches to evaluate the range-Doppler map.

This approach adopts the classical slow-time/fast-time framework of a pulsed radar operating at a given Pulse Repetition Interval (PRI), by applying the stop-and-go approximation for the received pulses. This batching strategy is of high practical interest especially for passive radar exploiting wide bandwidth signals (e.g., DVB-T transmissions [17]). For this kind of signals, it might be unfeasible to implement a real-time evaluation of the optimum range-Doppler map, which is based on the evaluation of the cross-ambiguity function between the reference and the surveillance signals. Furthermore, it is quite natural to process the digital transmissions on a batch-wise basis, as OFDM signals are based on consecutive symbols with a similar structure within a frame. Finally, it is essential when a signal reconstruction stage is present. Such a stage exploits the signal arrangement given by the communication signal standards in order to successfully demodulate and reconstruct the reference signal. For the following analysis, we rewrite the transmitted signal $s(t)$ as a sequence of fragments $s_n(t)$ of duration T, i.e. $s(t) = \sum_n s_n(t - nT)$ in order to take such a batching strategy into account. T is deliberately selected to be $T = T_{\mathrm{OFDM}}$. The discrete version of the signals in (7.2) can be approximated as:

$$r_0^{(\mathrm{LA})}[l] = A_0 \sum_n s_n[l - l_{\tau_0}] \exp\{j2\pi f_{D0} nT\}$$

$$r_0^{(\mathrm{TA})}[l] = A_0 \sum_n s_n[l - l_{\tau_0}] \exp\{j2\pi f_{D0} nT\} \exp\{-j2\pi \frac{d}{\lambda} \cos \alpha_0\} \qquad (7.3)$$

The discrete version of target contributions in (7.1) can be readily written using (7.3): $r_G^\gamma[l] = r_0^\gamma[l]$, where $v_{B0} \neq 0$ holds for the bistatic velocity of a moving target.

Clutter contributions can be written as superposition of echoes from stationary scatterers:

$$r_C^{LA}[l] = \sum_{q=1}^{N_R} \int_{\Phi_q} A_q(\alpha) \sum_n s_n[l - nL - l_{\tau_q}] \exp\left(j2\pi \frac{v_P}{\lambda} \cos\alpha\, nT\right) d\alpha$$

$$r_C^{TA}[l] = \sum_{q=1}^{N_R} \int_{\Phi_q} A_q(\alpha) \sum_n s_n[l - nL - l_{\tau_q}] \exp\left(j2\pi \frac{v_P}{\lambda} \cos\alpha\, nT\right) \qquad (7.4)$$

$$\cdot \exp\left(-j2\pi \frac{d}{\lambda} \cos\alpha\right) d\alpha$$

The stationary scatterers appear at different bistatic range cells R_{B_q} for $q = 1, \ldots, N_R$ and corresponding delays $\tau_q = R_{B_q}/c_0$. A scatterer appears under the angle α within the angular sector Φ_q with amplitude $A_q(\alpha)$. The amplitudes are modeled as Gaussian random variables with zero mean and variance $\sigma_{C_q}^2(\alpha) = \mathbb{E}\left\{|A_q(\alpha)|^2\right\}$. In (7.4), Internal Clutter Motion (ICM) is neglected by the assumption that the amplitudes $Aq(\alpha)$ are constant within a Coherent Processing Interval (CPI). Furthermore, amplitudes related with stationary scatterers from different bistatic range cells and different angles are statistically independent: $\mathbb{E}\left\{A_q(\alpha)A_p^*(\alpha')\right\} = 0$ for $q \neq p$, $\alpha \neq \alpha'$.

The signal model in (7.1) might also include the Direct Signal Interference (DSI) from the TX by using (7.4). However, in the following theoretical development and analysis it is assumed that the DSI has been suppressed using a dedicated filtering stage [13,18–23].

The baseband signals from (7.3) are arranged in a fast-time/slow-time data matrix $\bar{r}^{(\gamma)}$ for the range compression as:

$$x^{(\gamma)}[l, n] = \bar{r}^{(\gamma)}[l, n] * h_n[l], \quad n = 0, \ldots, N - 1 \qquad (7.5)$$

The nth column of $\bar{r}^\gamma[l, n]$ contains the nth batch of the received signal.* h_n defines the filter, that is matched to the batch n.

The next step of the processing is the clutter cancellation via a dedicated clutter cancellation stage (see Figure 7.2). The DPCA technique is the most straightforward approach to perform clutter cancellation using multiple antennas mounted on a moving platform [7,24]. It aims at collecting the radar echoes from two displaced receive antenna elements at the time that their two-way phase centers occupy the same spatial position, while they are moving along the (assumed locally rectilinear) platform trajectory. To this purpose, it is assumed that the Pulse Repetition Frequency (PRF) is matched to the platform velocity v_P and the distance $d/2$ between the two-way antenna phase centers, i.e. PRF $= K \cdot 2v_P/d$. Under this condition, the two-way phase center of the TA occupies the same location of the two-way phase center as the LA after

---

*The operator '$*$' defines the convolution operation and operates on single batches.

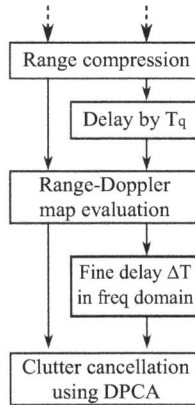

Figure 7.3   Adapted flowchart of the processing shown in Figure 7.2

an integer number K of PRIs. The clutter cancellation is then easily performed by subtracting the (temporally displaced) samples of the radar echoes received by the two antennas at the same spatial position. The echoes backscattered by moving targets have a different phase in the two acquisitions and are therefore preserved, while the echoes corresponding to stationary objects are filtered out. In principle moving targets can then be detected even in a strong clutter scenario.

After range compression, a DFT is performed across the slow-time dimension, namely across index n in (7.5), to evaluate the range-Doppler map for the clutter filtered data.

For passive radar systems employing a stationary TX and a batch-wise processing, the condition above can be modified as $PRF = K \cdot v_P/d$, since the distance between the two-way phase centers coincides with the distance d between the receiving antennas.

Obviously, the strict DPCA condition of phase-center alignment of both antenna elements after an integer multiple of the duration of one PRI is difficult to meet in real data acquisitions. In this case a two-stage co-registration of the sampled data is required [25,26], which can be achieved using the approach called 'flexible DPCA' (or short flex-DPCA), as described in [27]:

$$T_K = \left\lfloor \frac{d}{Tv_p} \right\rfloor T \tag{7.6}$$

$$\Delta T = T_D - T_K, \ \Delta T \in [0, T] \tag{7.7}$$

It applies first an integer delay T_K of the sampled data of the LA in slow-time domain, and afterwards a fine time delay compensation in Doppler frequency domain by applying a linear phase law $2\pi f \Delta T$. It must be noted that for this technique to be applicable, the platform velocity must be accurately measured during the data acquisition, for example by using an Inertial Measurement Unit (IMU).

Furthermore, the sequence of the flowchart of the processing shown in Figure 7.2 must be adapted as shown in Figure 7.3: after range compression the co-registration is first performed in slow-time domain using (7.6) followed by a DFT across batches to create the range-Doppler map for the data sets from both antenna elements. Afterwards, the fine co-registration is applied using (7.7). Different from the processing shown in Figure 7.2, the clutter subtraction with DPCA is the last stage in Figure 7.3 and applied by subtraction of both range-Doppler maps.

As the flexible DPCA technique adds a stage of temporal adaptivity, it can also be referred to as space–time adaptive processing (STAP).

7.3   Range compression using matched filter

It is well known that clutter returns appear spread in the Doppler domain due to the platform motion. It is therefore likely that they mask echoes from slowly moving targets appearing in endo-clutter region. As the Ambiguity Function (AF) of the employed waveforms of opportunity shows a generally high sidelobe level both in range and in Doppler, this effect is more severe in PCL systems than compared to active radar systems with carefully designed waveforms. Basically, clutter echoes originating from a given range cell are likely to affect the detection of targets appearing at other cells even in the presence of a large range-Doppler separation due to their sidelobes.

To better understand the effects of this phenomenon, a conventional signal processing based on matched filtering at the range compression stage followed by a clutter filtering stage is considered, where one batch corresponds to a single OFDM symbol and consists of L samples.

To simplify the derivation and analysis in the following, it is assumed that the DPCA condition with an integer multiple of the duration of one PRI holds. Therefore, the processing is carried out as described in Figure 7.2.

Also note that in the following analysis and simulations the transmitted and received waveforms are simulated Digital-Video-Broadcast-Terrestrial (DVB-T) signals. However, the conclusions drawn from the results can be readily extended to other types of OFDM-based waveforms.

7.3.1   Single point-like scatterer

Stationary point-like scatterer
First, a single point-like scatterer is considered. The output of the range compression stage in (7.5) is then:

$$x^{(LA)}[l, n] = A_0 g_n^{(MF)}[l - l_{\tau_0}] \exp\left(j2\pi \frac{v_P}{\lambda} \cos\alpha_0 nT\right) \exp\left(-j2\pi \frac{v_{B0}}{\lambda} nT\right)$$

$$x^{(TA)}[l, n] = A_0 g_n^{(MF)}[l - l_{\tau_0}] \exp\left(j2\pi \frac{v_P}{\lambda} \cos\alpha_0 nT\right) \exp\left(-j2\pi \frac{v_{B0}}{\lambda} nT\right) \qquad (7.8)$$

$$\cdot \exp\left(-j2\pi \frac{d}{\lambda} \cos\alpha_0\right)$$

where

$$g_n^{(\mathrm{MF})}[l] = s_n[l - nL_S] * h_n^{(\mathrm{MF})}[l] = \kappa \mathscr{F}^{-1}\{|S_n[m]|^2\} \qquad (7.9)$$

is the output of the filter matched to the nth batch $s_n[l - nL_S]$, $S_n[m]$ being the DFT of $s_n[l]$, i.e. $S_n[m] = \mathscr{F}\{s_n[l]\}$ and $\mathscr{F}^{-1}$ defining the Inverse Discrete Fourier Transform (IDFT). κ is a multiplicative constant, which is used in order to achieve unit gain to thermal noise in the signal processing chain.

The payload content of the employed waveform varies from symbol to symbol, therefore the impulse response $g_n[l]$ after matched filtering for a particular symbol $s_n[l]$ is different from the output response $g_p[l]$ for a symbol $s_p[l]$, if $n \neq p$:

$$
\begin{aligned}
s_n[l] \neq s_p[l], \text{ for } n \neq p &\Rightarrow \quad h_n[l] \neq h_p[l] \\
&\Rightarrow \quad g_n[l] \neq g_p[l]
\end{aligned}
\qquad (7.10)
$$

In Figure 7.4, the impulse response of one matched filtered DVB-T symbol is shown. One can clearly see strong peaks approximately 20 dB below the main peak, which appear due to the deterministic components (pilots) in the waveform and usually show periodic repetitions across consecutive OFDM symbols within a frame. Furthermore one notices a quite constant background level of approximately 40 dB below the main peak. This noise floor is due to correlation of the non-deterministic payload data transmitted on the data sub-carriers and varies from impulse response to impulse response. The power of the impulse response was aligned relative to the main peak, which is independent of n if equal energy fragments are used.

In the following analysis, the deterministic components are neglected. Instead the signal is modeled as a random waveform with zero-mean and variance $\sigma_s^2 = \mathbb{E}\{|s[l]|^2\}$.

Specifically a Continuous Wave (CW) transmission with constant power during the observation time is considered. The signal fragments $s_n[l]$, $n = 0, \ldots, N_S - 1$ are statistically independent.

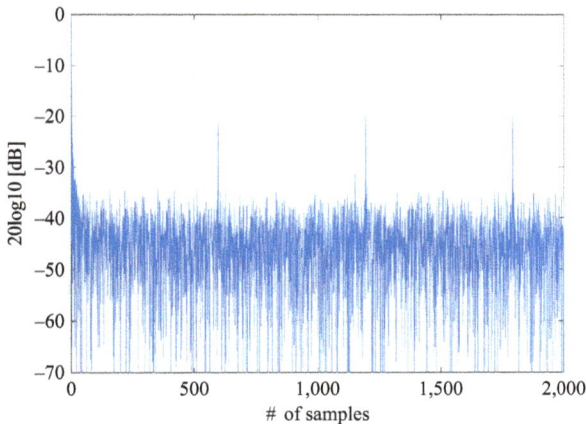

Figure 7.4   Impulse response of one matched filtered DVB-T symbol

Under these assumptions, the samples of the autocorrelation $g_n^{(MF)}[l] = \mathbb{E}\{s[n]s^*[n-l]\}$ can be treated as random variables, statistically independent across consecutive batches, with mean value and variance given by:

$$\mathbb{E}\{g_n^{(MF)}[l]\} = \mathbb{E}\left\{g_0^{(MF)}[l]\right\} = kL\sigma_s^2\delta[l] \ \forall \ n = 0, \ldots, N-1 \tag{7.11}$$

$$\mathrm{Var}\{g_n^{(MF)}\} = \mathrm{Var}\left\{g_0^{(MF)}[l]\right\} = \kappa^2 L\sigma^4(\mu - 1) \ \forall \ n = 0, \ldots, N-1 \tag{7.12}$$

where $\mu = \frac{1}{M_C} \sum_{m=0}^{M_C-1} |c_m|^4$ is dependent on the adopted modulation scheme, and $M_C = \{4, 16, 64\}$ for Quadrature phase-shift keying (QPSK), 16-Quadrature amplitude modulation (QAM), and 64-QAM. The corresponding constellation is defined so that it has a unitary average power, $\frac{1}{M_C} \sum_{m=0}^{M_C-1} |c_m|^2 = 1$, and $\{c_m\}_{m=0,\ldots,M_C-1}$ being the constellation symbols of the corresponding constellation map.

In [28], a more general result was developed at the output of the range-Doppler map evaluation. The focus of (7.11) and (7.12) is on the range compression stage outputs that represent the input of the clutter cancellation stage under investigation based on DPCA. However, the results reported in (7.11) and (7.12) can be easily derived from the corresponding equations of [28] by neglecting the effect of the Doppler processing stage.

Apparently, for the QPSK case, the variables $g_n^{(MF)}[l]$ become deterministic as all the transmitted symbols have equal energy and the power spectral density of any signal fragment is therefore deterministic and continuously flat. This result might slightly differ from the actual output, as the presence of pilot carriers, which are transmitted at boosted power level, was neglected in the equations above. The discussion in the following will be mostly referred to the widely exploited 16-QAM and 64-QAM schemes.

After co-registration of the data and the DPCA-based clutter cancellation stage, the scatterer contribution becomes:

$$
\begin{aligned}
y_0[l, n] =& x_0^{(TA)}[l, n] - x_0^{(LA)}[l, n - K] \\
=& A_0 \left(g_n^{(MF)}[l - l_{\tau 0}] - \exp\left(j2\pi \frac{v_{B0}}{\lambda}KT\right) g_{n-K}^{(MF)}[l - l_{\tau 0}] \right) \\
& \cdot \exp\left(j2\pi \frac{v_P}{\lambda}\cos\alpha_0(n - K)T\right) \exp\left(-j2\pi \frac{v_{B0}}{\lambda}nT\right)
\end{aligned}
\tag{7.13}
$$

In the case $v_{B0} \neq 0$ (i.e. for a moving target), its contribution is not totally filtered out due to its phase shift of $2\pi KTv_{B0}/\lambda$, as expected from the DPCA application.

However, unlike to a conventional active radar, which exploits periodic time-invariant waveforms, the contribution from a stationary scatterer provides a non-zero output although a constant amplitude A_0 at temporally displaced observations was assumed. This is because the shape of the impulse response changes for each n (see (7.10)).

To achieve a range-Doppler map, the output of the DPCA-based stage is then integrated by a DFT:

$$
\begin{aligned}
z_0[l,m] &= \sum_{n=0}^{N-1} y_0[l,n] \exp\left(-j2\pi \frac{m}{N} n\right) \\
&= A_0 \exp\left(-j2\pi \frac{v_P}{\lambda} \cos\alpha_0 KT\right) \sum_{n=0}^{N-1} \exp\left(j2\pi \left(f_{D_0}T - \frac{m}{N}\right)n\right) \\
&\quad \cdot \left(g_n^{(MF)}[l - l_{\tau 0}] - \exp\left(-j2\pi \frac{v_{B0}}{\lambda} KT\right) g_{n-K}^{(MF)}[l - l_{\tau 0}]\right)
\end{aligned} \tag{7.14}
$$

where $f_{D_0} = \frac{1}{\lambda}(v_P \cos\alpha_0 - v_{B0})$. Equation (7.14) represents the output contribution of a point-like scatterer belonging to the delay-Doppler bin $[l_{\tau 0}, f_{D_0}NT]$ observed at the delay-Doppler bin $[l,m]$. That means, as the echo of a stationary point-like scatterer ($v_{B0} = 0$) is not perfectly canceled in (7.13), its contribution will be spread all over the range-Doppler map, thus leading to an increased floor moving targets have to compete with.

Based on the assumption of a random noise-like waveform, the mean value of the output in (7.14) can be written as:

$$
\begin{aligned}
\mathbb{E}\{z_0[l,m]\} &= A_0 \exp\left(-j2\pi \frac{v_P}{\lambda} \cos\alpha_0 KT\right) \exp\left(j\pi \frac{v_{B0}}{\lambda} KT\right) \kappa L\sigma^2 \delta[l - l_{\tau 0}] \\
&\quad \cdot \left(-2j \sin\left(\pi \frac{v_{B0}}{\lambda} KT\right)\right) \exp\left(j2\pi \left(f_{D_0} - \frac{m}{NT}\right)(N-1)T\right) \\
&\quad \cdot \mathrm{dsinc}\left[\pi \left(f_{D_0} - \frac{m}{NT}\right)T, N\right]
\end{aligned} \tag{7.15}
$$

where $\mathrm{dsinc}(x,N) = \sin(Nx)/\sin(x)$ is the digital sinc function. Equation (7.15) yields a non-zero value only at the delay-Doppler bin where the scatterer belongs to (and at its ambiguous Doppler location), i.e. $l = l_{\tau 0}$ and $\frac{m}{NT} = f_{D_0} + \frac{r}{T}$, $r \in \mathbb{Z}$. However, it can be observed that this mean value is identically equal to zero all over the delay-Doppler plane when $v_{B0} = 0$.

The second moment of the output in (7.14) for a stationary scatterer is evaluated as:

$$
\begin{aligned}
\mathbb{E}\{|z_0[l,m]|^2 | v_{B0} = 0\} &= |A_0|^2 \sum_{n=0}^{N-1} \sum_{p=0}^{N-1} f[l,n,p] \\
&\quad \cdot \exp\left(j2\pi \left(f_{D_0} - \frac{m}{NT}\right)(n-p)T\right)
\end{aligned} \tag{7.16}
$$

where

$$
\begin{aligned}
f[l,n,p] &= \mathbb{E}\{(g_n^{\mathrm{MF}}[l-l_{\tau 0}] - g_{n-k}^{\mathrm{MF}}[l-l_{\tau 0}]) \\
&\quad \cdot (g_p^{*\mathrm{MF}}[l-l_{\tau 0}] - g_{p-k}^{*\mathrm{MF}}[l-l_{\tau 0}])\} \\
&= \begin{cases}
2\,\mathrm{Var}\{g_0[l]\} & p = n \\
-\mathrm{Var}\{g_0[l]\} & p = n \pm K \\
0 & \text{elsewhere}
\end{cases}
\end{aligned}
\tag{7.17}
$$

And using (7.12) and (7.16) is then:

$$
\mathbb{E}\{|z_0[l,m]|^2 | v_{B0} = 0\} \cong 4|A_0|^2 \kappa^2 L \sigma^4 (\mu - 1) \mathrm{N} \sin^2\left(\pi \left(f_{D_0} - \frac{m}{N\mathrm{T}}\right) K\mathrm{T}\right)
\tag{7.18}
$$

where the border effects, which result from the exploitation of a limited number of slow-time samples, are neglected.

As can be seen in (7.18), it can be observed that the output power level for a stationary scatterer is equal to zero only at Doppler bin $m = f_{D_0} N\mathrm{T}$. In contrast, a stationary scatterer resulting from delay-Doppler bin $[l_{\tau 0}, f_{D_0} N\mathrm{T}]$ contributes to the delay-Doppler bin $[l, m]$, where the output power level is modulated by a $\sin^2$ shape in the Doppler dimension.

Although ideal conditions were assumed, a complete removal of the contribution from a stationary scatterer is not possible, due to the varying shape of the employed waveform.

Moving point-like scatterer
Here the contribution of a moving target at the delay-Doppler bin where it is expected to appear will be analyzed. The target has a delay of $l_{\tau_T} = \tau_T f_S$ and a bistatic velocity of v_{B_T}, yielding to a bistatic Doppler of f_{D_T} and a complex amplitude A_T. Using (7.14) yields:

$$
\begin{aligned}
z_T[l = l_{\tau_T}, m = f_{D_T} N\mathrm{T}] &= A_T \exp\left(-j2\pi \frac{v_P}{\lambda} \cos \alpha_T K\mathrm{T}\right) \\
&\quad \cdot \sum_{n=0}^{N} \left(g_n^{\mathrm{MF}}[0] - \exp\left(j2\pi \frac{v_{B_T}}{\lambda} K\mathrm{T}\right) g_{n-K}^{\mathrm{MF}}[0]\right)
\end{aligned}
\tag{7.19}
$$

As the mean value of the random variable $g_n[l]$ is independent of n, the expectation value and the quadratic expectation value of (7.19) are given as:

$$
\begin{aligned}
\mathbb{E}\{z_T[l = l_{\tau_T}, m = f_{D_T} N\mathrm{T}]\} &= A_T \exp\left(-j2\pi \frac{v_P}{\lambda} \cos \alpha_T K\mathrm{T}\right) N\mathbb{E}\{g_0[0]\} \\
&\quad \cdot \left(1 - \exp\left(2\pi \frac{v_{B0}}{\lambda} K\mathrm{T}\right)\right)
\end{aligned}
\tag{7.20}
$$

$$P_T^{(MF)} = |\mathbb{E}\{z_T[l = l_{\tau_T}, m = f_{D_T}NT]\}|^2$$

$$= \left| A_T N \mathbb{E}\{g_0[0]\} \exp\left(-j2\pi(\frac{v_P}{\lambda}\cos\alpha_T - \frac{v_{B_T}}{\lambda})KT\right) 2j\sin\left(\pi\frac{v_{B_T}}{\lambda}KT\right)\right|^2$$

$$= |A_T|^2 N^2 |\mathbb{E}\{g_0[0]\}|^2 \cdot 4\left|\sin\left(\pi\frac{v_{B_T}}{\lambda}KT\right)\right|^2 \quad\quad (7.21)$$

$$\overset{(7.12)}{=} |A_T|^2 N^2 |\kappa L\sigma^2|^2 \cdot 4\left|\sin\left(\pi\frac{v_{B_T}}{\lambda}KT\right)\right|^2$$

One observes that the highest gain of the output of (7.21) is for $v_{B_T} = \frac{\lambda(1+2k)}{2KT}$, $k \in \mathbb{Z}$, while the system will be blind for targets with some particular radial velocities (so-called 'blind velocities') at $v_{B_T} = \frac{k\lambda}{KT}$, $k \in \mathbb{Z}$.

7.3.2 Multiple stationary point-like scatterers

In the following subsection a multitude of stationary point-like scatterers will be considered, i.e. clutter is simulated according to (7.4). It is assumed, that the clutter contributions from different clutter patches are independent and identically distributed across all clutter patches. The range compressed outputs at the input of the cancellation stage are:

$$x_C^{(LA)}[l, n] = \sum_{r=1}^{N_R} \int_{\Phi_r} A_r(\alpha)g_n[l - l_{\tau_r}] \exp\left(j2\pi\frac{v_P}{\lambda}\cos\alpha nT\right) d\alpha$$

$$x_C^{(TA)}[l, n] = \sum_{r=1}^{N_R} \int_{\Phi_r} A_r(\alpha)g_n[l - l_{\tau_r}] \exp\left(j2\pi\frac{v_P}{\lambda}\cos\alpha nT\right) \quad\quad (7.22)$$

$$\exp\left(-j2\pi\frac{d}{\lambda}\cos\alpha\right) d\alpha$$

Figure 7.5 shows as an example of the result obtained with the approach for an input signal including clutter returns only, i.e., $r^{(\gamma)}[l] = r_C^{(\gamma)}[l]$ after matched filtering as described by (7.22) and Doppler filtering via DFT across batches. The map in Figure 7.5 was scaled to provide unitary processing gain for thermal noise. In this way direct comparison of clutter power levels at the input and at the output of the processing chain is enabled. The reported map is lower limited to 50 dB below the highest peak appearing with a value of 61.46 dB. It can be observed that at all the considered range cells across a Doppler extension of approximately 50 Hz strong clutter returns appear. The value of the Doppler extension is dependent on the platform speed, the wavelength, and the employed antenna pattern. Slow moving targets, i.e. targets appearing with Doppler frequency in endo-clutter region, can be masked by these clutter contributions. Furthermore, at higher Doppler regions a generally high clutter power level is observed, which is due to the uncontrollable characteristics (mainly the high sidelobes) of the AF of the employed signal. The authors in [28] provide an estimate of the power level of the observed range-Doppler map background for a single point-like scatterer. Furthermore, the presence of side peaks due to the periodic structures of the OFDM modulation (e.g., pilots, guard intervals) produces

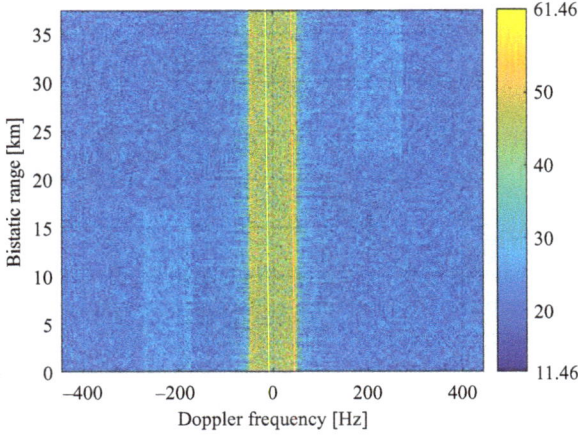

Figure 7.5   Range-Doppler map in dB for MF before DPCA processing for a simulated scenario including clutter echoes only. ©2019 IEEE. Reprinted, with permission, from [27].

regularly spaced regions with further increased clutter power levels (see regions at about ± 200 Hz) [29,30].

Under consideration of the DPCA condition $d = v_P K T$, the application of the cancellation stage leads to:

$$
\begin{aligned}
y_C[l, n] &= x_C^{(TA)}[l, n] - x_C^{(LA)}[l, n - K] \\
&= \sum_{r=1}^{N_{RC}} \int_{\Phi_r} A_r(\alpha) \left(g_n[l - l_{\tau_r}] - g_{n-K}[l - l_{\tau_r}] \right) \\
&\quad \exp\left(2\pi j \frac{v_P}{\lambda} \cos\alpha (n - K) T \right) d\alpha
\end{aligned}
\tag{7.23}
$$

and after a Doppler filter stage the output is:

$$
\begin{aligned}
z_C[l, m] &= \sum_n y_C[l, n] \exp\left(-j2\pi \frac{m}{NT} nT \right) \\
&= \sum_n \sum_{r=1}^{N_{RC}} \int_{\Phi_r} A_r(\alpha) \left(g_n[l - l_{\tau_r}] - g_{n-K}[l - l_{\tau_r}] \right)
\end{aligned}
$$

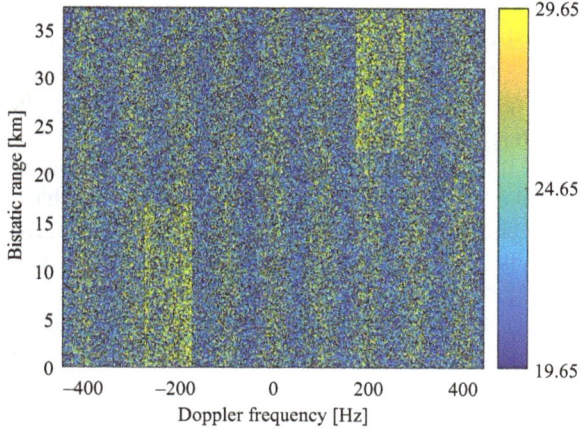

Figure 7.6 *Range-Doppler map in dB for MF after DPCA processing for a simulated scenario including clutter echoes only. ©2019 IEEE. Reprinted, with permission, from [27].*

$$\cdot \exp\left(2\pi j \frac{v_P}{\lambda} \cos\alpha(n-K)T\right) d\alpha \exp\left(-2\pi j \frac{m}{NT} nT\right)$$

$$= \sum_{r=1}^{N_{RC}} \int_{\Phi_C} A_r(\alpha) \exp\left(-j2\pi \frac{v_P}{\lambda} \cos\alpha KT\right)$$

$$\cdot \sum_n \left(g_n^{MF}[l - l_{\tau_r}] - g_{n-K}^{MF}[l - l_{\tau_r}]\right)$$

$$\exp\left(j2\pi \left(\frac{v_P}{\lambda} \cos\alpha - \frac{m}{NT}\right) nT\right) d\alpha$$

(7.24)

Figure 7.6 shows a range-Doppler map after DPCA clutter filtering. The highest peaks in Figure 7.5 have been correctly removed by the clutter filter. However, the map of Figure 7.6 is characterized by a generally high background level that is largely comparable to that observed in Figure 7.5, despite the fact that ideal conditions have been considered for the clutter returns (i.e., absence of ICM, perfect DPCA condition, no RX channel miscalibrations, and an error-free reference signal for range compression). Furthermore, no additional noise has been considered. Inside the clutter main ridge, a clutter cancellation limited to about 35 dB has been obtained. In the exo-clutter region the DPCA has basically no effect. In order to understand this limitation, it is useful to analyze the clutter output power using (7.18) and (7.24):

$$P_C^{(out)} = \mathbb{E}\left\{|z_C[l,m]|^2\right\}$$

$$= 4N \operatorname{Var}\{g_0[l]\} \sum_{r=1}^{N_R} \int_{\Phi_r} \sigma_{Cr}^2(\alpha) \sin^2\left(\pi \left(\frac{v_P}{\lambda} \cos\alpha - \frac{m}{NT}\right) KT\right) d\alpha$$

(7.25)

Analyzing (7.25), one can see, that there is a superposition of contributions from clutter patches on one particular range-Doppler bin. The contribution from the clutter patches comes from all considered range cells, which is encoded in the summation, and from all angles (i.e. direction of arrival), which is encoded in the integral. Each contribution undergoes an attenuation, which is encoded in the term $\sigma_{Cr}(\alpha)$, therefore being dependent on the angle of arrival α, and on the particular range cell r. The term $\sigma_{Cr}(\alpha)$ is modulated by the $\sin^2$ shape. Analyzing the $\sin^2$ shape reveals that the contribution of a particular delay-Doppler cell equals zero at that Doppler frequency f_D, which corresponds to the Doppler frequency of itself: $f_D = \frac{m}{NT} = \frac{vp}{\lambda}\cos\alpha$. That means, that a particular delay-Doppler cell $z_C[[l_{\tau_0}], m_0]$ does not contribute to delay-Doppler cells, that are modulated with the same Doppler frequency $f_D = m_0 N / T$.

The expression of (7.25) can be further developed by the assumption of omnidirectional antenna patterns with constant gain in azimuth. Furthermore, to simplify the analysis, it is assumed that homogeneous clutter returns are experienced within each range cell, which yields $\sigma_{Cr}^2(\alpha) = \sigma_{Cr}^2$. This assumption requires that the DSI from the TX is already suppressed. It must also be mentioned that this assumption cannot be strictly verified in a bistatic geometry as the Cassini's ovals intersect the bistatic range cells, so that the free-space propagation loss is not constant within each range cell:

$$P_C^{(\text{out})}[l, m] = 2N\kappa^2 L\sigma^2(\mu - 1)P_C^{(\text{in})}$$
$$\cdot \left\{ 1 - \cos\left(2\pi\frac{m}{N}K\right)J_0\left(2\pi\frac{d}{\lambda}\right)\right\} \tag{7.26}$$

$J_0(x)$ defines the Bessel function of the first kind of order zero, and $P_C^{(\text{in})}$ is the global clutter power at the input of the system. $P_C^{(\text{in})}$ can be written as:

$$P_C^{(\text{in})} = \mathbb{E}\left\{\left|r_C^{(\alpha)}[l]\right|^2\right\} = \pi\sigma^2\sum_{r=1}^{N_R}\sigma_{Cr}^2 \tag{7.27}$$

For a fair comparison of the power levels between the input and the output of the processing scheme of Figure 7.2, the multiplicative constant κ is chosen to provide unitary gain when only thermal noise as input of the processing scheme for the matched filter is considered. For κ this yields: $\kappa = \frac{1}{\sqrt{2NL_S\sigma^2\sigma^2}}$ and (7.26) becomes:

$$P_C^{(\text{out})}[l, m] = (\mu - 1)P_C^{(\text{in})}\left\{1 - \cos\left(2\pi\frac{m}{NT}KT\right)J_0\left(2\pi\frac{d}{\lambda}\right)\right\} \tag{7.28}$$

Analyzing (7.28) leads to two conclusions:

(i)   The clutter output power varies around the average value of $(\mu - 1)P_C^{(\text{in})}$. As μ is dependent on the constellation map (see (7.12)), the output power depends as well on the constellation map: for 16-QAM and 64-QAM μ is given as 1.32 and 1.38, respectively, which leads to a value of 4.9 dB and 4.2 dB below the clutter input power for the output power. For QPSK the output power becomes zero, as $\mu = 1$. However this is the result of neglecting the deterministic components, and in practice a lower bound on this average power level will be experienced, even in the QPSK case.

(ii)   The last factor $\left\{1 - \cos\left(2\pi\frac{m}{NT}KT\right)J_0\left(2\pi\frac{d}{\lambda}\right)\right\}$ varies in the interval from $(0, 2)$. For the employed simulation where $d = \lambda/2$, the factor is in the interval $[0.7, 1.3]$. The factor $\cos\left(2\pi\frac{m}{NT}KT\right)$ is dependent on the DPCA time shift $T_{\text{DPCA}} = KT = d/v_P$ and on Doppler bin m. This leads to a modulation of the clutter output power with period $\Delta f_D = \frac{1}{KT}$. The sinusoidal pattern is visible in Figure 7.6.

To compare the output of (7.28) to the outcome of Figure 7.6, the expected theoretical result of (7.28) is plotted in Figure 7.7 along with the estimated mean value for all Doppler bins for an extended range shown in Figure 7.6. The minima and the maxima, as calculated from (7.28), have a difference $\Delta_{P_{\text{out}}} = 10\log_{10}(1.3/0.7) \approx 2.7$ dB, which is well in line with the results from Figure 7.7.

Additionally one observes the modulation due to the DPCA delay time in Doppler dimension with period $f_P = \frac{1}{KT} \approx 99$ Hz.

DPCA cancellation ratio for MF
So far the DPCA performance was analysed under the background of the residuals (sidelobes, etc.) resulting from the range compression. However, DPCA is an effective clutter cancellation stage, although being limited here by the range compression residuals. Therefore it is worth to analyse how the clutter cancellation performs, despite the fact that it is not able to remove the clutter residuals in exo-clutter region, which as well result from the matched filtered processing. The cancellation ratio $CR[l, m]$ is given as:

$$CR[l, r] = \frac{P_C^{(\text{LA})}[l, m]}{P_C^{(\text{out})}[l, m]} \tag{7.29}$$

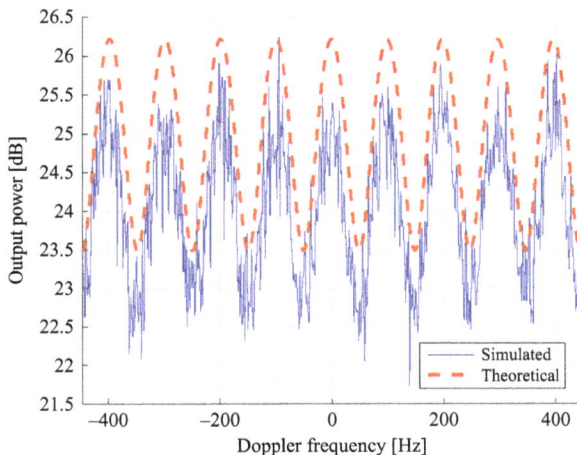

Figure 7.7   Clutter output power in dB versus Doppler frequency [Hz] after matched filtering and DPCA. The expected theoretical result from (7.28) is compared to the estimated mean value of Figure 7.6. The employed constellation map was 16-QAM. ©2019 IEEE. Reprinted, with permission, from [27].

The term $P_C^{(\text{LA})}[l, m]$ refers to the clutter power at the delay-Doppler bin $[l, m]$ without DPCA processing to be seen directly at Figure 7.5. The Doppler filtered output becomes:

$$z_C^{(\text{LA})}[l, m] = \sum_{r=1}^{N_R} \int_{\Phi_r} A_r(\alpha) \sum_n g_n^{(\text{MF})}[l - l_{\tau_r}] \exp\left(j2\pi \left(\frac{v_P}{\lambda} \cos\alpha - \frac{m}{NT}\right) nT\right) d\alpha$$

(7.30)

The output power is then:

$$P_C^{(LA)}[l, m] = \mathbb{E}\left\{\left|z_C^{(\text{LA})}[l, m]\right|^2\right\}$$

$$= (\mu - 1)P_C^{(\text{in})}$$

(7.31)

$$+ L\sigma^2 \sum_q^{N_R} \int_{\Phi_q} \sigma_{Cq}^2 \delta[l - l_{\tau_q}] \frac{1}{N} \left|\text{dsinc}\left[\pi\left(\frac{v_P}{\lambda}\cos\alpha - \frac{m}{NT}\right) T, N\right]\right|^2 d\alpha$$

Equation (7.31) is the summation of two terms. The first one $(\mu - 1)P_C^{(\text{in})}$ is dependent on the input clutter power and is constant for the complete range-Doppler map. It is a noise background and results from the matched filtering. It can be seen in Figure 7.5 outside the main clutter ridge. Usually it is smaller than the second term.

The second term represents the non-negligible main-lobe contributions of stationary scatterers defining the main clutter ridge. With the assumption of omnidirectional constant gain antenna elements, i.e. $\cos\alpha_{\max} = 1$, the clutter main ridge shows non-negligible values for $|f_D| = \left|\frac{m}{NT}\right| \le \frac{v_P}{\lambda}$. At each delay-Doppler bin $[l, m]$ the term (7.31) is dominated by the response of the scatterers belonging to the clutter patch defined by the corresponding range cell q, s.t. $l_{\tau_q} = l$, and angular sector $\Delta\bar{\alpha} = [\bar{\alpha} \pm w_\alpha]$ s.t. $\frac{v_P}{\lambda}\cos\bar{\alpha} = \frac{m}{NT}$ and $w_\alpha \cong \frac{\lambda}{2v_PNT}$. The output power can be written as: $P_C^{(LA)}[l, m] \cong LN\sigma_s^2\sigma_{C\bar{q}}^2 w_\alpha$, for $|f_D| \le v_P/\lambda$. The cancellation ratio CR$[l, m]$ can be approximated as:

$$\text{CR}[l, m] \cong \begin{cases} \dfrac{LN\left(\dfrac{\sigma_{C\bar{q}}^2}{\sum_{q=1}^{N_R}\sigma_{Cq}^2}\right)\dfrac{w_\alpha}{\pi}}{(\mu - 1)\left(1 - \cos\left(2\pi Km/N\right) J_0\left(2\pi d/\lambda\right)\right)}, & |f_D| \le \dfrac{v_P}{\lambda} \\[4ex] \dfrac{1}{1 - \cos\left(2\pi Km/N\right) J_0\left(2\pi d/\lambda\right)}, & |f_D| > \dfrac{v_P}{\lambda} \end{cases}$$

(7.32)

As example, in Figure 7.8 the cancellation ratio between Figures 7.5 and 7.6 is shown. The main clutter ridge appears to be filtered out, but the mean value is below 40 dB, thus far from the ideal cancellation ratio of being infinite. Furthermore, the suppression in exo-clutter region is not significant. The cancellation ratio values are just slightly higher than 0 dB.

In conclusion, this processing chain is not effective, as the limiting factor is the application of the MF for range compression of the uncontrollable transmitted waveform.

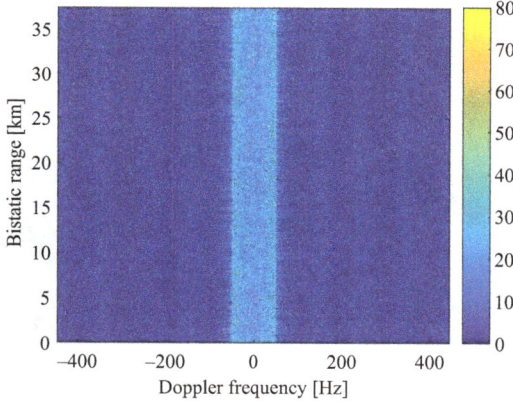

Figure 7.8   Cancellation ratio as a function of the delay-Doppler bin for the case study considered in Figures 7.5 and 7.6. ©2019 IEEE. Reprinted, with permission, from [27].

7.4   Range compression using reciprocal filter

The reason for the residual noise floor on the range-Doppler map, which prevents target detection, was referred to the time-varying content of the transmitted waveform: the payload data changes for each OFDM symbol, and therefore the crucial condition of transmission of time-invariant pulses for DPCA is not fulfilled here. Specifically, when using the MF the temporal variability is transferred to the output of the processing chain even for a point-like scatterer with constant amplitude.

As it is not possible to control the transmitted signal, the processing in order to filter the time-varying content of the waveform needs to be adapted. A method to remove the time-varying content is provided by application of the so-called RpF in the range compression stage. It performs basically a division of the received signal by the reconstructed copy of the transmitted signal. Although originally introduced in [31] and also addressed in subsequent papers [28,29,32,33] in order to control the level of unwanted sidelobes, it is effective in providing a time-invariant impulse response. This is followed by the DPCA clutter cancellation block, which operates on the output signals being waveform-independent after spectral equalization in the range-compression stage. The RpF in (7.5) is written as follows:

$$h_n[l] = h_n^{\text{RpF}}[l] = \kappa' \text{IDFT}\{S_n[m]^{-1}\} \tag{7.33}$$

The impulse response is then:

$$g_n^{\text{(RpF)}}[l] = s_n[l - nL] * h_n^{\text{(RpF)}}[l] = \kappa' \text{IDFT}\{\text{rect}[k]\} = \kappa' L \delta[l] \tag{7.34}$$

In Figure 7.9 the output of the RpF is reported. The high peaks and the constant noise floor have been removed, such that the impulse response is a digital sinc function, represented here using $\delta[l]$, which is typical for signals with a flat spectrum.

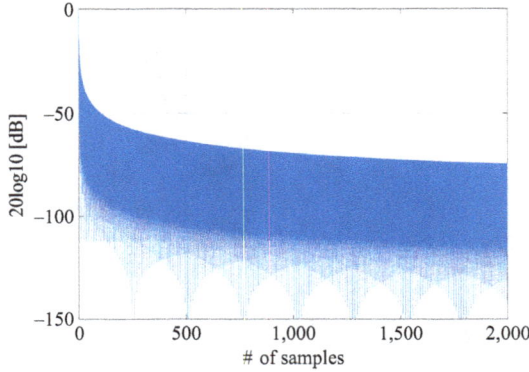

Figure 7.9   Impulse response of one reciprocal filtered DVB-T symbol

The RpF is not only beneficial in providing a time-invariant impulse response at the output of the range compression stage, it also provides a higher dynamic range due to lowering of the sidelobe level in the signal ambiguity function, which is beneficial for the detection of exo-clutter targets.

In order to avoid the issues due to overweighting possibly caused by the division by small or zero values [31], the application of the RpF is limited to non-zero subcarriers (i.e. the data- and pilot carriers in the considered DVB-T signal here). The guard carriers are excluded from the processing. It needs to be noted that if the transmitted signal is reconstructed following the reconstruction process [16,34], a division by zero (or by very small values) in practical applications will not happen, as all the data subcarriers are forced to represent a constellation symbol in the constellation map during the processing, therefore having a non-zero value.

In conclusion, it can be observed that due to the RpF strategy, the output of (7.34) loses its time dependence: the RpF provides an impulse response, which is deterministic and time-invariant so that: $g_n^{(\text{RpF})}[l] = g_p^{(\text{RpF})}[l]$, $\forall n, p$.

7.4.1   Single point-like scatterer

Stationary point-like scatterer

Analyzing the effects of the RpF on the return of a single point-like scatterer by recalling (7.8) and substituting in (7.13) $g_n^{(\text{MF})}$ with the impulse response for the RpF $g_n^{(\text{RpF})}$ from (7.34), the Doppler-filtered outputs at the end of the processing stage are:

$$
\begin{aligned}
\bar{z}_0[l, m] = & A_0 \exp\left(-j2\pi \frac{v_P}{\lambda} \cos \alpha_0 KT\right) g_0^{(\text{RpF})}[l - l_{\tau 0}] \\
& \cdot 2j \sin\left(\pi \frac{v_{B0}}{\lambda} KT\right) \text{dsinc}\left[\pi\left(f_{D_0} - \frac{m}{NT}\right)T, N\right]
\end{aligned}
\tag{7.35}
$$

which resembles the statistical expectation of the output obtained for the MF (see (7.15)). Obviously, the output of a stationary object $v_{B0} = 0$ becomes zero, therefore being perfectly suppressed. Furthermore, there is no contribution of a single point scatterer on other delay-Doppler cells, which is a substantial improvement compared to the MF case, compare to (7.14) and (7.24).

Moving point-like scatterer
The output peak power P_T for a moving target echo, measured at its expected range-Doppler location, i.e. $l = \lambda_T \tau_0$ and $\frac{m}{NT} = f_{D_0}$, can be evaluated as:

$$
\begin{aligned}
P_T^{(\mathrm{RpF})} &= |\mathbb{E}\{z_T[l_{\tau_0}, NTf_{D_0}]\}|^2 \\
&= 4|A_0|^2 N^2 \kappa'^2 L^2 |\sin\left(\pi \frac{v_{B0}}{\lambda} K T\right)|^2
\end{aligned}
\tag{7.36}
$$

where (7.34) has been used. The constant κ' is chosen such that at the outcome unitary gain is provided, in the case for thermal noise only as input signal, as was done for the MF evaluation. For κ' it is obtained as: $\kappa' = (2N\sigma_s^{-2}L\xi)^{-1/2}$, where ξ depends on the constellation map used by the TX: $\xi = \frac{1}{M_C}\sum_0^{M_C-1}\frac{1}{|c_m|^2}$. As the same scaling strategy for MF and RpF was used, the output power level for a single point-like scatterer at its expected position l_{τ_0} and f_{D_0} can be directly compared. Comparing (7.21) to (7.36) leads to:

$$
\Delta P_T = \frac{P_T^{(\mathrm{MF})}}{P_T^{(\mathrm{RpF})}} = \frac{\kappa^2 L^2 \sigma_s^4}{\kappa'^2 L^2} = \xi
\tag{7.37}
$$

Equation (7.37) represents the loss in signal power one can expect when replacing the MF with the RpF in the range compression stage. The expected loss is only dependent on the constellation map, which is given from the TX. For the three possible values of QPSK, 16-QAM and 64-QAM this can be evaluated as: $\xi = [0, 2.76, 4.29]$ dB for $M_C = [4, 16, 64]$. Apparently, no loss is observed for the QPSK case. The reason for this is that transmissions using the QPSK constellation map exhibits a rather flat spectral density, at least as long as only data carriers are regarded. The pilot carriers are transmitted at boosted power level, which might yield to an additional small loss. The transmissions using 16/64-QAM are more common, and then a non-negligible loss in Signal-to-Noise Ratio (SNR) has to be accepted, which is up to 4.3 dB for 64-QAM.

However, this loss needs to be compared to an improvement in terms of residual noise floor and ambiguity suppression, which can be expected from using the RpF for multiple stationary point-like scatterers.

7.4.2 Multiple stationary point-like scatterers

In the analysis of multiple stationary point-like scatterers, it is first observed that (7.35) equals zero for a stationary scatterer, i.e. $v_{B0} = 0$, under the assumption of ideal conditions, which are the absence of ICM, amplitude fluctuations, and a fulfillment of the DPCA condition. Consequently, for a superposition of clutter echoes from a multitude of stationary scatterers according to (7.1) and (7.4) the clutter output power evaluated as the superposition of the contributions from a multitude of stationary scatterers becomes at the end of the processing chain equally zero:

$$
P_C^{(\mathrm{out})}[l, m] = 0
\tag{7.38}
$$

which results in an ideal cancellation ratio CR[l, m] = ∞ all over the final range-Doppler map, as it is achieved for the case of an active radar which transmits a train of time invariant pulses.

Simulation results for the exchange of the MF with the RpF are shown in Figures 7.10 (before DPCA processing) and 7.11 (after DPCA processing).

Comparing Figure 7.5 to Figure 7.10 it can be seen that even before applying the clutter cancellation stage, a significant higher dynamic range is achieved for the RpF at exo-clutter regions. This means, targets experiencing a bistatic Doppler higher than the maximum of the clutter Doppler frequency: $f_{D,CI} > \frac{v_p}{\lambda}$ might already be detected without clutter removal. However, the clutter, visible in Figure 7.10, still blinds the systems for slow moving and/or weak targets. The peak value of the clutter ridge is at 58.61 dB, which is approximately 2.85 dB below the clutter peak power for MF application, visible in Figure 7.5. This loss in SNR results from the employed constellation map of 16-QAM used for the simulations. It is in accordance with (7.37), and holds for single stationary and multiple stationary scatterer returns as well as for a single target return, as both MF and RpF are linear filters, and target returns are formulated in the same manner as clutter returns.

The output of the cascade of the RpF and DPCA is reported in Figure 7.11. The same strategy as for scaling the results using the MF has been applied to represent the output map about its average value of −43 dB.

The clutter is completely removed, while the fluctuations are due to noise coming from border effects. The clutter cancellation is therefore only limited by noise present in the receiving system.

In Figure 7.12 the cancellation ratio evaluated between the range-Doppler maps of Figures 7.10 and 7.11 is reported.

Following can be observed:

1. The main clutter ridge has been effectively removed with either the MF or the RpF. This is due to the DPCA approach. Consequently, endo-clutter targets can

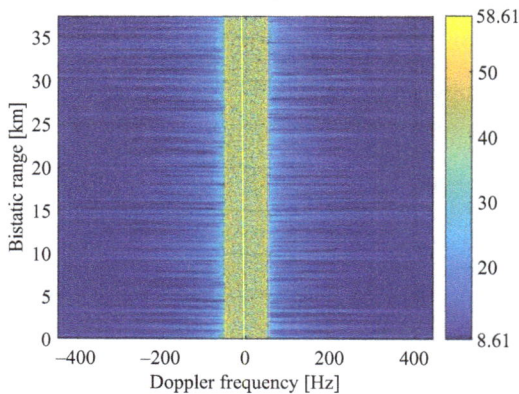

Figure 7.10   Range-Doppler map in dB for RpF before DPCA processing for a simulated scenario including clutter echoes only. ©2019 IEEE. Reprinted, with permission, from [27].

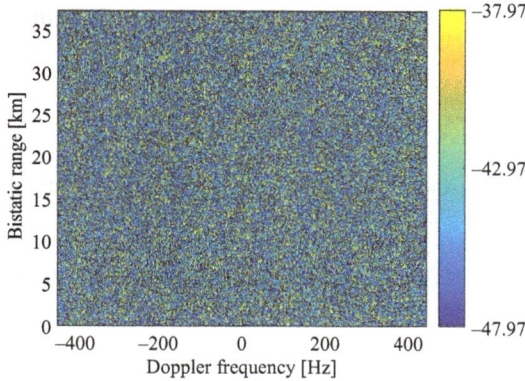

Figure 7.11   *Range-Doppler map in dB for RpF after DPCA processing for a simulated scenario including clutter echoes only. ©2019 IEEE. Reprinted, with permission, from [27].*

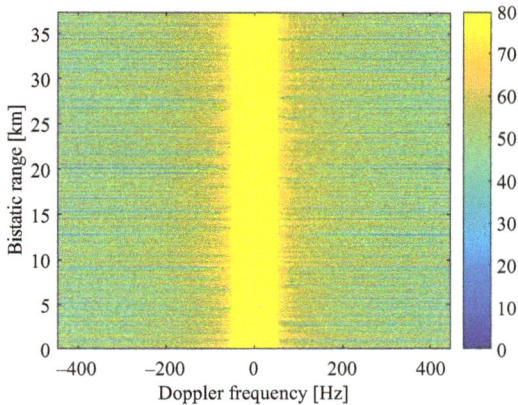

Figure 7.12   *Cancellation ratio in dB as range-Doppler map. Evaluated between the range-Doppler maps of Figure 7.10 and Figure 7.11. ©2019 IEEE. Reprinted, with permission, from [27].*

be potentially detected. In the RpF case, the cancellation ratio values obtained in the endo-clutter region are well above 80 dB (the figure has been upper limited to be directly compared to Figure 7.8).

2. In the cascade of the RpF and the DPCA clutter filter, the average level of the residual clutter contributions across the whole delay/Doppler map has been significantly reduced. In this way, the sensitivity of the radar against both fast and slowly moving targets is significantly improved. In fact, the average background

level moves from about 24.6 dB in Figure 7.6 to -43 dB in Figure 7.11. Correspondingly, cancellation ratio values in the range 4.0–60 dB are also obtained in the exo-clutter region.

7.4.3 Comparison of SCNR between MF and RpF

Cascading the MF and the DPCA approach yields the highest SNR output value, where the clutter residual power level at the output of the cancellation stage limits the detection performance. However, the cascade of the RpF and the DPCA processing ideally provides a perfect cancellation of the stationary clutter returns. The price to be paid is a limited loss in terms of SNR (see (7.37)) due to the non-flat spectral density for 16-/64-QAM. The output Signal-to-Clutter-and-Noise Ratio (SCNR) across the range-Doppler map can be studied with:

$$\text{SCNR}_{\text{MF\&DPCA}}[l, m] = \frac{P_T[l = l_{\tau_0}, m = NTf_{D_0}]}{P_C^{(\text{out})}[l, m] + 1} \tag{7.39}$$

Both the noise input power and the corresponding gain were set to unity, therefore a unitary power level for the noise at the output of the processing chain is given. With the expressions from (7.28), (7.21), (7.36), (7.37) and (7.38), we can write:

$$\text{SCNR}_{\text{MF\&DPCA}}[l, m] = \frac{|A_0|^2 N L \sigma_s^2 \left| \sin\left(\pi \frac{v_{B0}}{\lambda} KT \right) \right|^2}{(\mu - 1) P_C^{(\text{in})} \left\{ 1 - \cos\left(2\pi \frac{m}{N} K \right) J_0 \left(2\pi \frac{d}{\lambda} \right) \right\} + 1} \tag{7.40}$$

$$\text{SCNR}_{\text{RpF\&DPCA}}[l, m] = |A_0|^2 N \zeta^{-1} L \sigma_s^2 \left| \sin\left(\pi \frac{v_{B0}}{\lambda} KT \right) \right|^2 \tag{7.41}$$

A comparison leads to:

$$\Delta S[l, m] = \frac{\text{SCNR}_{\text{RpF\&DPCA}}[l, m]}{\text{SCNR}_{\text{MF\&DPCA}}[l, m]}$$

$$= \frac{(\mu - 1) P_C^{(\text{in})} \left\{ 1 - \cos\left(2\pi \frac{m}{N} K \right) J_0 \left(2\pi \frac{d}{\lambda} \right) \right\} + 1}{\zeta} \tag{7.42}$$

All variables can be regarded as constants, while only the input power varies, which is equal to the clutter power: $P_C^{(\text{in})} = \text{CNR}^{(\text{in})}$. Under these considerations, (7.42) shows that for the RpF to outperform the MF, the Clutterand-Noise Ratio (CNR) must be above a certain threshold. If stating that $\Delta S[l, m] > 1$, then:

$$P_C^{(\text{in})} = \text{CNR}^{(\text{in})} > \frac{\zeta - 1}{(\mu - 1) \left\{ 1 - \cos\left(2\pi \frac{m}{N} K \right) J_0 \left(2\pi \frac{d}{\lambda} \right) \right\}} \tag{7.43}$$

The inequality in (7.43) clearly reveals that the suggested approach of cascading RpF and DPCA provides an advantage in terms of the SCNR with respect to the cascade of the MF and the DPCA processing. This holds for any location of the range-Doppler map as far as the input CNR is above a proper threshold:

$$\text{CNR}^{(\text{in})} > \frac{\zeta - 1}{(\mu - 1) \left\{ 1 - \left| J_0 \left(2\pi \frac{d}{\lambda} \right) \right| \right\}} \tag{7.44}$$

The threshold for the reciprocal filtering stage to achieve a higher SCNR results is dependent of the employed constellation map and of the inter-element spacing d. Assuming that $d = \lambda/2$, then following threshold levels can be expected:

$$
\begin{aligned}
\text{CNR}^{(\text{in})} &> \infty & M_C &= 4 \\
\text{CNR}^{(\text{in})} &> 6.0\,\text{dB} & M_C &= 16 \\
\text{CNR}^{(\text{in})} &> 8.0\,\text{dB} & M_C &= 64
\end{aligned}
\tag{7.45}
$$

While for the QPSK case ($M_C = 4$) such improvement cannot be expected, the clutter cancellation stage of the RpF provides a significant advantage for 16-/64-QAM values for CNR values higher than 6/8 dB.

In practical cases, deviations from the results above might be due to a non-perfect DPCA condition, to a non-perfect interchannel calibration, and to the presence of ICM. Limitations due to the first issue are addressed in the next section, where the constraint of a perfect DPCA condition is relaxed. The effects of calibration errors [35] and ICM [7] on clutter cancellation capability have been largely investigated in the literature also for the case of conventional active radar and are not considered here. Furthermore, deviations in real data results can also be due to non-perfect reference signal reconstruction. As for example described in [12,36], the reference signal – reconstructed as copy from the transmitted signal – can deviate from the actual transmitted signal, which is due to noise, interferences, loss of orthogonality of the subcarriers due to platform motion, etc. Reconstruction errors lead to correlation artifacts after the range compression stage, limiting the clutter cancellation capability. However, these effects are not addressed here.

The analysis and comparison of matched filtering to reciprocal filtering, both followed by the clutter filtering stage DPCA is further extended. In the examples reported in the following, the antenna element spacing d was set to $d = \lambda/2$ and the platform carrying the RX is assumed to fly with velocity $v_P = 24.53\,\text{m/s}$. Correspondingly, the TA will occupy the same position of the LA after a time delay of $T_D = d/v_P = 0.01019\,\text{s}$. The target's echo is assumed to impinge the antennas from an angle $\alpha_G = \frac{\pi}{2}$, so that its bistatic Doppler frequency is only dependent on its velocity v_G. The CNR and the SNR are set to 30 dB and to -30 dB.

We note here that the DPCA condition is not met after an integer multiple of the duration of one DVB-T symbol, as: $K = \frac{d}{v_P \text{PRI}} = 9.0992$. In order to use the batching strategy based on single OFDM (or rather DVB-T) symbols, a co-registration of the sampled data is required [25,26], which can be achieved using the 'flexible DPCA' approach described in Section 7.2.

The result is shown in Figure 7.13. Both curves show the expected decay as the target bistatic velocity approaches $\frac{\lambda}{KT}p$, $p \in \mathbb{Z}$. However, at each Doppler value, the proposed approach allows a SCNR improvement of about 20 dB, thanks to the enhanced capability to remove clutter echoes in the final map. This is clearly apparent from Figure 7.14, where the ratio ΔS is reported of the SCNR results in Figure 7.13. The ΔS curve largely matches the theoretical expectation based on (7.42) despite small fluctuations are present due to the local estimation of the clutter+noise power

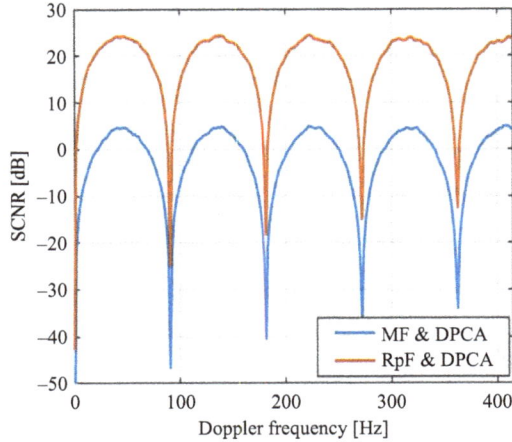

Figure 7.13   Performance comparison: SCNR vs. target Doppler frequency for RpF (red line) and MF (blue line) after clutter suppression application. ©2019 IEEE. Reprinted, with permission, from [27].

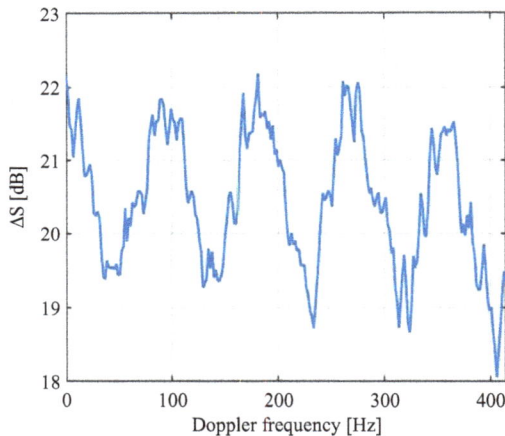

Figure 7.14   SCNR ratio corresponding to Figure7.13. ©2019 IEEE. Reprinted, with permission, from [27].

level at different Doppler bins of the map. It should be noticed that this improvement might be upper-bounded by the noise level since, for small CNR values, the SCNR at the output of the processing stages will be dominated by the output SNR that is only slightly higher when exploiting the MF with respect to the RpF (being the advantage dependent on the adopted modulation scheme). This effect was illustrated in (7.40)–(7.42) and is shown in Figure 7.15 where the output SCNR curves are reported as functions of the CNR at the input of the system for a fixed target Doppler value equal

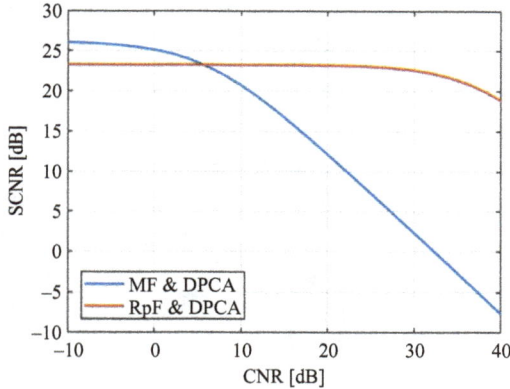

Figure 7.15   Performance comparison: SCNR vs. CNR for RpF (red line) and MF (blue line) after STAP application. ©2019 IEEE. Reprinted, with permission, from [27].

to -30 Hz. For CNR values below 5–6 dB the advantage of the proposed processing scheme disappears. Furthermore, according to (7.37), a loss of 2.76 dB has to be accepted in terms of output SCNR due to the exploitation of the RpF instead of the MF at the range compression stage. However, as the CNR increases, the RpF followed by clutter suppression stage outperforms the MF and clutter suppression approach. Up to 30 dB input CNR, the approach of RpF and DPCA is able to provide a constant input SCNR. We conclude that the result shown in Figure 7.15 is well in line with the calculation of (7.44).

7.5   Real data evaluation

7.5.1   *Measurement campaign*

The trials took place in the North of Poland close to the Baltic Sea and had been conducted in the context of a NATO Sensors & Electronics Technology (SET) work group. Figure 7.16 presents a map of the trial site. The blue line defines the trajectory of the airborne receiver, labeled in the following as airborne receiver (ARX). Its position at the time stamp of the evaluated data is marked with a white annulus.

Airborne receiver
The airborne receiving system is a dual super-heterodyne receiver, which was mounted in the cockpit of an Ultra Light Aircraft (ULA). It has four receiving channels, each with an effective bandwidth of 32 MHz [34].

   A pod mounted below one wing of the ULA carried four Yagi-Uda antenna elements, see Figure 7.17(a) and (b).

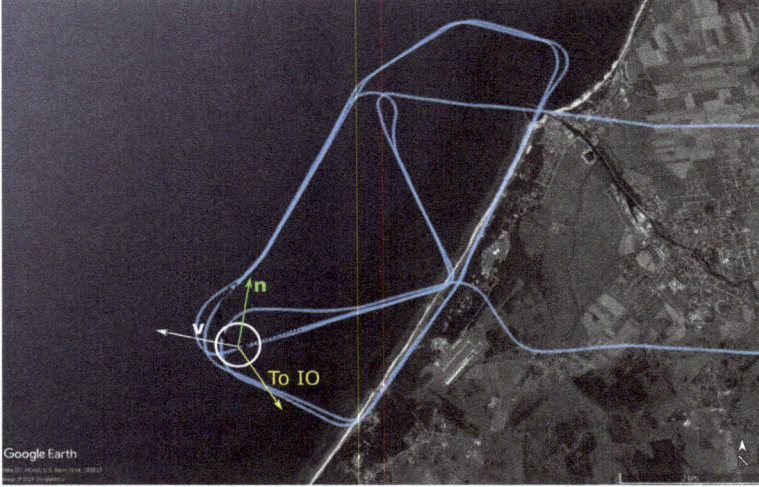

Figure 7.16   A map of the trial site in the northern part of Poland at the Baltic Sea. The yellow arrow defines the direction to the IO, while the white arrow marked with $\vec{v}$ defines the direction of movement. The green arrow marked with $\vec{n}$ defines the steering direction of the antenna array.

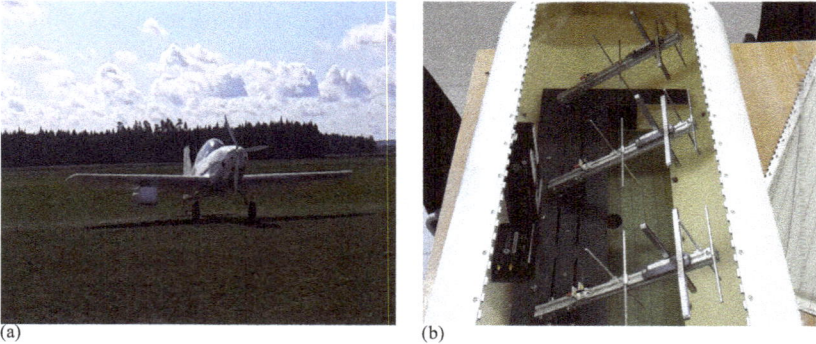

Figure 7.17   (a) The ULA Delphin with the pod mounted below the right wing. (b) The Yagi-Uda antenna elements mounted inside of the pod (only three of four in total are shown). On the left side the IMU and logging unit can be seen (without HF cables). ©2020 IEEE. Reprinted, with permission, from [12].

The signals received at the horizontal polarization of the two center elements were processed for the results presented here. Additionally, an IMU was mounted in the pod (see the black boxes in Figure 7.17) in order to measure accurately the position, velocity and attitude during data acquisition.

Stationary receiver

A ground-based stationary receiver (SRX) was deployed in a distance of approx. 55 km to the IO. The stationary receiving system consisted of a commercial Universal Software Radio Peripheral (USRP) receiver capable of sampling In-phase and Quadrature (IQ)-data with an effective bandwidth of 32 MHz at one input channel, to which a horizontally polarized antenna was connected. The SRX was employed to additionally capture a copy of the transmitted signal. By using the raw data captured from the SRX together with the data captured at the four receiving channels from the ARX, it is possible to improve the reference signal estimation by using the Maximal Ratio Combining (MRC) algorithm presented in [12].

Transmitter

As IO the DVB-T–TX *Koszalin-Gołogora* was chosen. During the trials, it transmitted five DVB-T signals, two of them transmitted on center frequencies 658 MHz and 682 MHz in horizontal polarization with a transmit power of 100 kW [Effective Radiated Power (ERP)].

7.5.2 Results

A CPI of $512 \cdot T_{OFDM} = 0.516$ s of data obtained at a single surveillance channel is first processed. The range-Doppler maps are reported in Figure 7.18(a) and (b) for MF and RpF, respectively, in the range compression stage. The scaling strategy introduced in the simulated analysis is also applied here for scaling of the range-Doppler maps. That means, the estimated noise floor level is represented by the 0 dB level in each map. It is kept constant at the output of each employed processing chain by properly scaling the filters applied so that different results can directly be compared.

The DSI peak comes with high sidelobes affecting the first range cells. Additionally, strong returns appear at Doppler bins different from the Doppler frequency of the DSI. They correspond to Doppler spread clutter echoes. The detection of both exo- and endo-clutter target echoes is impeded, as all these considered signals contribute to increase the background level of the final map when exploiting the MF in the processing chain. Given the platform velocity and the beamwidth of the employed antennas, targets can be considered as slowly moving when their bistatic Doppler frequency falls approximately within the interval $[-70, 70]$ Hz, which corresponds to an equivalent radial velocity smaller than 15 m/s.

In Figure 7.18 (a) and (b), range-Doppler maps after filtering with MF (in Figure 7.18 (a)) and after reciprocal filtering (in Figure 7.18) are reported. In both Figures (a) and (b) two regions are highlighted, which refer to targets of opportunity in the endo-clutter region. The target labeled 'T1' in the first region is isolated so that it can be reasonably identified. However, its detection must compete with a generally high background level. This target yields an SCNR value of about 45 dB. However, the local SCNR estimated at its location is lower due to the disturbance background that is much higher than the estimated noise floor. In contrast to target T1, the target labeled 'T2' in the second region cannot be discriminated among the very high peaks appearing in those areas. When the RpF is used in the range compression stage, the sidelobe level of the signal AF is significantly reduced so that targets appearing outside the

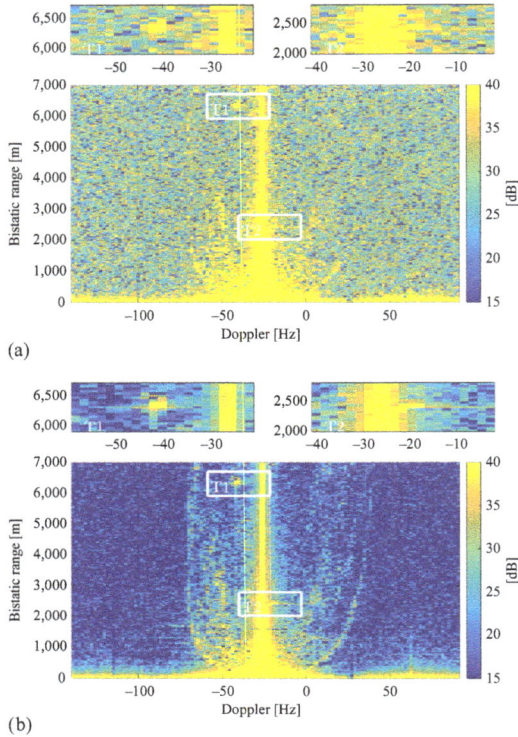

Figure 7.18   Range-Doppler map obtained for experimental data after (a) MF and (b) RpF

Doppler area of the main clutter contributions are clearly visible against the background. This is shown in Figure 7.18 (b). T1 appears at a level of about 43 dB above noise in this case. As analyzed before, there is a loss due the 64-QAM modulation scheme with respect to the MF case.

One can observe that the presence of slowly moving targets are still masked by main clutter structures in the final map. The next step in the processing scheme involves the Doppler spread clutter suppression by using the suggested spatio-temporal approach "flex-DPCA". To this purpose, a time shift is applied at the LA prior to subtraction from the TA [3,27]. The required time shifts, to be applied in the slow-time domain and the frequency domain, can be easily calculated using the velocity recorded by the IMU and the formulas given in (7.6) and (7.7).

In practical cases, the receive channels including the antenna elements might show differences regarding amplitude and phase. These differences can severely limit the cancellation stage [35]. To achieve appropriate clutter cancellation these effects have to be compensated. In order to achieve adaptivity against amplitude and phase imbalances between the receiving channels and despite the fact that more sophisticated approaches could be exploited to this purpose, for example refer to [37] and Chapter 8, a simple approach is applied here. The simple approach is based on calibrating the two receive channels by using the DSI [3,27]. Depending on trial geometry and employed antenna elements, this signal appears usually as the strongest contribution in the received signals. Therefore it is a strong reliable source for channel calibration. The amplitude and phase correction parameters between the two channels are calculated as

$$A_{cal} = \frac{|z^{(LA)}[\bar{l}, \bar{m}]|}{|z^{(TA)}[\bar{l}, \bar{m}]|} \tag{7.46}$$

$$\varphi_{cal} = \arg\left\{ z^{(TA)}[\bar{l}, \bar{m}] \left(z^{(LA)}[\bar{l}, \bar{m}] \right)^* \right\} \tag{7.47}$$

where $z^{(\gamma)}[\bar{l}, \bar{m}]$ is the complex value of the range-Doppler map at the location, which corresponds to the direct signal for the γ channel. Due to the synchronization process shown in Figure 7.2, the DSI appears in the first range cell with index $\bar{l} = 1$, while the Doppler index $\bar{m}$ is dependent on the relative motion of the RX to the TX.

The results obtained with the clutter filtering and calibration are shown in Figure 7.19 (a) after MF and (b) after RpF, respectively. It can be observed that a significant reduction is obtained for the main clutter contributions appearing at small Doppler values in both cases. These results clearly demonstrate the need for

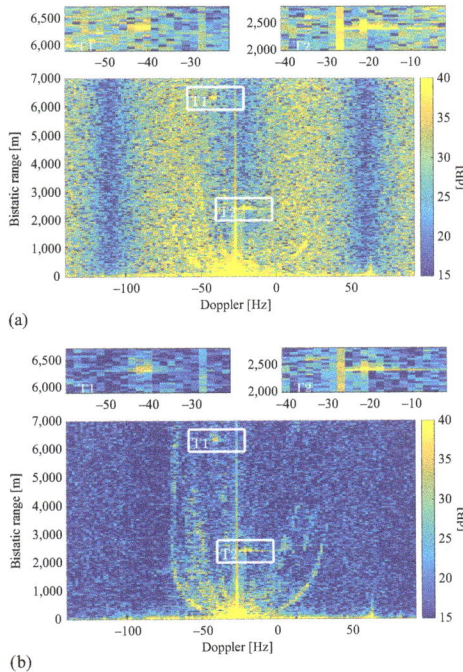

Figure 7.19 *Range-Doppler map obtained for experimental data after (a) STAP after MF and (b) STAP after RpF*

a clutter cancellation stage in the spatio-temporal domain. However, as for the case of simulated scenarios, the STAP after MF does not guarantee a complete removal of the clutter sidelobe structures: even after the cancellation stage (see Figure 7.19(a)) a generally high background level is provided in the final map. One can also note that the output power level shows a similar periodic modulation across Doppler frequencies as the output power level observed in Figures 7.6 and 7.11. As described in the previous analysis, this effect occurs due to the impossibility to effectively cancel the clutter sidelobe structures because of the random behavior of the waveform.

The slowly moving targets navigating on the sea are now visible, but they cannot be easily discriminated with respect to the background (see, for example, T1 in the upper portion of Figure 7.19(a)). In addition, it can be observed that T2's SNR is significantly reduced because this target falls within the cancellation notch of the DPCA filter, which is ambiguous in Doppler. But the cancellation notch of the clutter filter is shifted, such that the center of the notch aligns with the Doppler frequency of the DSI, as the DSI was used for phase calibration.

Both the stationary scatterers' main peaks and their sidelobe structures are filtered out after application of the RpF-based STAP approach, so that the targets appear now as isolated peaks. Despite the lower SNR values (see Figure 7.19(b)), they can be easily detected against the residual background. This holds especially for target T2.

7.6   Summary

In this chapter, we addressed the problem of GMTI with passive radar on a moving platform. Precisely, we focused on the impact of the random behavior of the waveform on subsequent clutter suppression. We analyzed the outcome of the range compression stage and the subsequent clutter filtering stage, if a matched filter is used in the range compression. Furthermore, we compared the outcome of the filtering stages, if the matched filter is exchanged with a reciprocal filter in the range compression stage. The detailed analysis and simulations have shown that – other as the matched filter – the reciprocal filter is able to restore the paradigm of a time-invariant impulse response at the output of the range compression stage. In this way, the reciprocal filter gives the passive radar the possibility to perform an ideal clutter suppression.

The developed algorithms were also applied against real data, which was acquired during trials with an airborne PCL system verifying the findings from the simulations and analytical derivation.

Other developed algorithms refer to performing a co-registration of the employed antenna elements, and applying a digital calibration of the receiving chain.

While this chapter is an important contribution to GMTI from a passive radar on a moving platform, it is far from being comprehensive. Many further improvements are expected to be developed in the next time regarding topics such as clutter suppression, digital calibration, GMTI using multi-static passive radar, among others. A further step in this direction is described in the next Chapter 8, which is concerned with improved STAP techniques for clutter suppression.

Acknowledgements

The authors would like to thank the Warsaw University of Technology for the support and the organization during the measurement campaign 'NATO Active-Passive-Radar-Trials—Ground-borne-Airborne-Stationary (APART-GAS)'.

References

[1] Dawidowicz B, Samczynski P, Malanowski M, *et al.* Detection of moving targets with multichannel airborne passive radar. *IEEE Aerospace and Electronics Systems Magazine.* 2012;27(11):42–49.

[2] Dawidowicz B, Kulpa KS, Malanowski M, *et al.* DPCA detection of moving targets in airborne passive radar. *IEEE Transactions on Aerospace and Electronic Systems.* 2012;48(2):1347–1357.

[3] Wojaczek P, Colone F, Cristallini D, *et al.* The application of the reciprocal filter and DPCA for GMTI in DVB-T – PCL. In: *International Conference on Radar Systems (Radar 2017)*, 2017. p. 1–5.

[4] Tan DKP, Lesturgie M, Sun H, *et al.* Space—time interference analysis and suppression for airborne passive radar using transmissions of opportunity. *IET Radar, Sonar & Navigation.* 2014;8(2):142–152.

[5] Yang PC, Lyu XD, Chai ZH, *et al.* Clutter cancellation along the clutter ridge for airborne passive radar. *IEEE Geoscience and Remote Sensing Letters.* 2017;14(6):951–955.

[6] Neyt X, Raout J, Kubica M, *et al.* Feasibility of STAP for passive GSM-based radar. In: *2006 IEEE Conference on Radar*, 2006. p. 546–551.

[7] Klemm R. *Principles of Space-Time Adaptive Processing.* 3rd ed. The Institution of Engineering and Technology IET, 2006.

[8] Melvin WL. A STAP overview. *IEEE Aerospace and Electronics Systems Magazine.* 2004;19(1):19–35.

[9] Colone F. Spectral slope-based approach for mitigating bistatic space-time adaptive processing clutter dispersion. *IET Radar, Sonar Navigation.* 2011;5(5):593–603.

[10] Palmer J, Ummenhofer M, Summers A, *et al.* Receiver platform motion compensation in passive radar. *IET Radar, Sonar Navigation.* 2017;11(6):922–931.

[11] Ummenhofer M, Schell J, Heckenbach J, *et al.* Doppler estimation for DVB-T based Passive Radar systems on moving maritime platforms. In: *2015 IEEE Radar Conference (RadarCon)*, 2015. p. 1687–1691.

[12] Wojaczek P, Cristallini D, Schell J, *et al.* Polarimetric antenna diversity for improved reference signal estimation for airborne passive radar. In: *2020 IEEE Radar Conference (RadarConf20)*, 2020.

[13] Lombardo P and Colone F. Advanced processing methods for passive radar systems. In: Melvin WL, Scheer JA, editors. *Principles of Modern Radar: Advanced Techniques.* SciTech Publishing, Inc., 2012.

[14]   Moscardini C, Petri D, Capria A, *et al*. Batches algorithm for passive radar: a theoretical analysis. *IEEE Transactions on Aerospace and Electronic Systems*. 2015;51(2):1475–1487.

[15]   Searle S, Howard S, and Palmer J. Remodulation of DVB-T signals for use in Passive Bistatic Radar. In: *2010 Conference Record of the Forty Fourth Asilomar Conference on Signals, Systems and Computers*, 2010. p. 1112–1116.

[16]   Poullin D. Passive detection using digital broadcasters (DAB, DVB) with COFDM modulation. *IEE Proceedings – Radar, Sonar and Navigation*. 2005;152(3):143–152.

[17]   Digital Video Broadcasting (DVB); Framing structure, Channel Coding and Modulation for Digital Terrestrial Television. 650 Route des Lucioles, F-06921 Sophia Antipolis Cedex – France, 2015.

[18]   Colone F, Palmarini C, Martelli T, *et al*. Sliding extensive cancellation algorithm for disturbance removal in passive radar. *IEEE Transactions on Aerospace and Electronic Systems*. 2016;52(3):1309–1326.

[19]   Colone F, O'Hagan DW, Lombardo P, *et al*. A multistage processing algorithm for disturbance removal and target detection in passive bistatic radar. *IEEE Transactions on Aerospace and Electronic Systems*. 2009;45(2):698–722.

[20]   Schwark C and Cristallini D. Advanced multipath clutter cancellation in OFDM-based passive radar systems. In: *2016 IEEE Radar Conference (RadarConf)*, 2016. p. 1–4.

[21]   Garry JL, Baker CJ, and Smith GE. Evaluation of direct signal suppression for passive radar. *IEEE Transactions on Geoscience and Remote Sensing*. 2017;55(7):3786–3799.

[22]   Searle S, Gustainis D, Hennessy B, *et al*. Cancelling strong Doppler shifted returns in OFDM based passive radar. In: *2018 IEEE Radar Conference (RadarConf18)*, 2018. p. 0359–0354.

[23]   Kulpa K. The CLEAN type algorithms for radar signal processing. In: *2008 Microwaves, Radar and Remote Sensing Symposium*, 2008. p. 152–157.

[24]   Dickey FR, Labitt M, and Staudaher FM. Development of airborne moving target radar for long range surveillance. *IEEE Transactions on Aerospace and Electronic Systems*. 1991;27(6):959–972.

[25]   Ender JHG. Space–time processing for multichannel synthetic aperture radar. *Electronics & Communication Engineering Journal*. 1999;11:29–38.

[26]   Gierull CH. *Digital Channel Balancing of Along-Track Interferometric SAR Data*. Defence Research and Development Canada, 2003.

[27]   Wojaczek P, Colone F, Cristallini D, *et al*. Reciprocal-filter-based STAP for passive radar on moving platforms. *IEEE Transactions on Aerospace and Electronic Systems*. 2019;55(2):967–988.

[28]   Gassier G, Chabriel G, Barrére J, *et al*. A unifying approach for disturbance cancellation and target detection in passive radar using OFDM. *IEEE Transactions on Signal Processing*. 2016;64(22):5959–5971.

[29]   Palmer JE, Harms HA, Searle SJ, *et al*. DVB-T passive radar signal processing. *IEEE Transactions on Signal Processing*. 2013;61(8):2116–2126.

[30] Colone F, Langellotti D, and Lombardo P. DVB-T signal ambiguity function control for passive radars. *IEEE Transactions on Aerospace and Electronic Systems*. 2014;50(1):329–347.

[31] Glende M. PCL-signal-processing for sidelobe reduction in case of periodical illuminator signals. In: *2006 International Radar Symposium*; 2006. p. 1–4.

[32] Searle S, Palmer J, Davis L, *et al*. Evaluation of the ambiguity function for passive radar with OFDM transmissions. In: *2014 IEEE Radar Conference*, 2014. p. 1040–1045.

[33] Chabriel G and Barrére J. Adaptive target detection techniques for OFDM-based passive radar exploiting spatial diversity. *IEEE Transactions on Signal Processing*. 2017;65(22):5873–5884.

[34] Heckenbach J, Kuschel H, Schell J, *et al*. Passive radar based control of wind turbine collision warning for air traffic PARASOL. In: *2015 16th International Radar Symposium (IRS)*, 2015. p. 36–41.

[35] Nickel U. On the influence of channel errors on array signal processing methods. *AEÜ – International Journal of Electronics and Communications*. 1993;47(4):209–219.

[36] Berthillot C, Santori A, Rabaste O, *et al*. BEM reference signal estimation for an airborne passive radar antenna array. *IEEE Transactions on Aerospace and Electronic Systems*. 2017;53(6):2833–2845.

[37] Blasone GP, Colone F, Lombardo P, Wojaczek P, and Cristallini D. Passive radar DPCA schemes with adaptive channel calibration. *IEEE Transactions on Aerospace and Electronic Systems*. 2020;56(5):4014–4034.

Chapter 8

Passive STAP approaches for GMTI

Giovanni Paolo Blasone[1], Fabiola Colone[1], Pierfrancesco Lombardo[1] and Philipp Markiton[2]

This chapter is concerned with the study of signal processing techniques and operational strategies for passive radar systems mounted onboard moving platforms and aimed at ground moving target indication (GMTI) applications. Specifically, the attention is focused on space–time adaptive processing (STAP) methodologies for systems equipped with multiple channels on receive. The spatial degrees of freedom are adaptively exploited for space–time clutter cancellation and slow-moving target detection and direction of arrival estimation. Specific solutions are devised and integrated into the typical signal processing architecture of passive radar to address the main limitations deriving from the passive bistatic framework.

Acronyms

ABPD	adjacent bin post-Doppler
AMF	adaptive matched filter
Cal-GLRT	calibrated generalized-likelihood ratio test
CFAR	constant false alarm rate
COTS	commercial off the shelf
CPI	coherent processing interval
CRB	Cramér–Rao lower bound
DFT	discrete Fourier transform
DOA	direction of arrival
DPCA	displaced phase centre antenna
DVB-T	digital video broadcasting – terrestrial
ECA	extensive cancellation algorithm
FM	frequency modulation

[1]Department of Information Engineering Electronics and Telecommunications, Sapienza University of Rome, Italy
[2]Fraunhofer Institute for High Frequency Physics and Radar Techniques, Germany

GLRT	generalized-likelihood ratio test
GMTI	ground moving target indication
ICM	internal clutter motion
MDV	minimum detectable velocity
ML	maximum likelihood
MLE	maximum-likelihood estimator
NC-GLRT	non-coherent generalized likelihood ratio test
OFDM	orthogonal frequency division multiplexing
PD	probability of detection
PFA	probability of false alarm
PRF	pulse repetition frequency
QAM	quadrature amplitude modulation
RAM	radiation absorbing material
RMS	root mean square
RX	Receiver
SAR	synthetic aperture radar
SCNR	signal to clutter plus noise power ratio
SNR	signal-to-noise power ratio
STAP	space–time adaptive processing
TX	transmitter
UHF	ultra-high frequency
ULA	uniform linear array
VHF	very high frequency

List of symbols

x	boldface and lower-case letter denotes vector
X	boldface and upper-case letter denotes matrix
x	letter in normal font denotes a scalar
I_N	$N \times N$ identity matrix
j	the imaginary unit
$\hat{(.)}$	estimate of $(.)$
$(.)^T$	transpose of $(.)$
$(.)^*$	conjugate of $(.)$
$(.)^H$	Hermitian (or conjugate transpose) of $(.)$
$|.|$	modulus of $(.)$ for scalars; determinant for matrices
$\angle(.)$	phase of $(.)$
$Re(.), Im(.)$	real and imaginary part of $(.)$
$(.)!$	factorial of $(.)$
$\exp(.)$	exponential function
$\log_{10}(.)$	logarithm of $(.)$ relative to base 10

$E\{.\}$ statistical expectation of (.)
$\otimes$ Kronecker product
c speed of light
λ wavelength
f_D Doppler frequency
d array element spacing
v_p platform velocity
v_b bistatic radial velocity
R_b bistatic range
φ Azimuth angle of receiver to scatterer line of sight
ϑ depression angle of receiver to scatterer line of sight
ψ angle between platform velocity and array end-fire
φ angle between array endfire and receiver to scatterer line of sight
B signal bandwidth
M number of pulses
L number of Doppler bins
N number of spatial channels
K number of training data

8.1 Introduction

In the last few years, passive radar technology has been gaining a renewed and increasing attention in the scientific community. The recent developments in this field have opened new perspectives and innovative areas of research [1–5]. The application of passive radar to airborne or ground moving platforms is one of the most interesting and challenging. Mobile passive radar may offer a number of strategic advantages compared to stationary ground-based solutions and extend the functionalities of passive sensors to applications such as synthetic aperture radar (SAR) imaging [6–9] and ground moving target indication (GMTI) [10–27].

The lack of the transmitter unit and the adoption of commercial-off-the-shelf (COTS) hardware components for the receiving system lead to potential reductions of weight, energy consumption and cost. For these reasons, mobile passive radar could be easily installed on small vehicles, boats, ultralight aircrafts, or unmanned aerial vehicles (UAVs), representing an appealing solution for covert and/or low-cost monitoring operations over wide areas (see Figure 8.1).

However, the potential benefits of passive radar onboard moving platforms are paid by the presence of Doppler distortions on the received signals, induced by the receiver motion, which can adversely affect the performance of the system.

In surveillance applications, the detection of moving targets with small radial velocity components is hindered by the Doppler spread of clutter echoes, due to the relative motion of the receiver with respect to the stationary scene. This

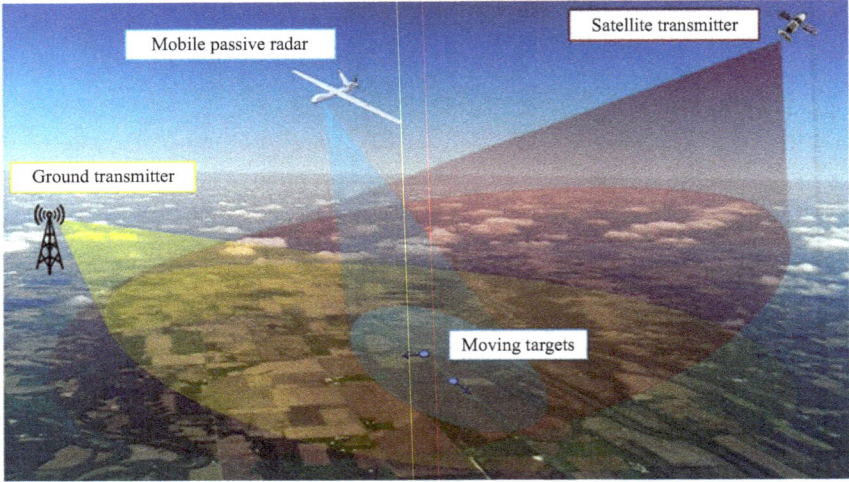

Figure 8.1   Sketch of an airborne passive radar for GMTI surveillance

effect, well known in standard airborne radar, tends to be even more stressed at the VHF/UHF bands of the most widely used illuminators of opportunity, due to the typical broad antenna beams available. As a result, the detection of slow-moving targets requires a proper suppression of clutter returns, which can be achieved by exploiting space–time processing applied to the signals collected by multiple receiving channels.

The techniques traditionally employed for direct signal and clutter suppression in passive radar, such as extensive cancellation algorithm (ECA) [47,48], and other time-domain filtering techniques [49], are not suited for the suppression of the Doppler-spread clutter observed by a moving receiver. Although adaptations of such techniques are possible [24,25], they reveal some limitations for wide range-Doppler clutter extensions and would not be directly able to discriminate the presence of moving targets beneath the clutter.

Conversely, space–time adaptive processing (STAP) and similar techniques borrowed from the airborne active radar literature allow an effective suppression of clutter by operating a two-dimensional filtering in the angle-Doppler domain. Such techniques have been widely studied for standard active radar in the last decades and many adaptations have been proposed [37–40].

However, providing GMTI capability to mobile passive radar poses several challenges. Some of them are due to the impact of receiver motion on the passive radar operations, others are associated to the non-straightforward application of standard methodologies from active moving radar in the passive bistatic scenario. Although the feasibility of this concept has been preliminarily demonstrated, such technology is still far from being mature and there is still great field for research.

Some of the main issues to be addressed are summarized in the following:

- The accurate replica of the reference signal must be properly collected and/or reconstructed by the system regardless of the platform motion.
- The continuous and time-varying waveforms of the most typically exploited illuminators of opportunity are not suitable for the conventional space–time clutter suppression techniques, rather conceived for standard pulsed radar.
- The typical ill-suited ambiguity functions may compromise the target detection capability, due to the presence of undesired peaks, and reduce the separation between range cells, generally required by the adaptive techniques.
- The direct signal interference from the exploited illuminator, frequently the dominant component in the received signal, must be properly handled and suppressed along with the Doppler-spread clutter echoes.
- The low directivity of practical receiving antennas, due to the typical exploited wavelengths, is paired with a bistatic illumination from broadcast transmitters, which do not focus the signal energy only in the direction of interest. This further increases the clutter Doppler bandwidth and easily raises channel calibration issues, due to the simultaneous reception of non-negligible clutter echoes from a very wide angular sector.
- The common use of low-cost COTS components and the long wavelengths compared to the practical antenna sizes may pose some limitations to the accuracy (or the feasibility) of preliminary system calibrations, resulting in non-negligible angle-dependent imbalances across the receiving channels.
- The limited number and low directivity of the receiving antennas make target angular localization a difficult task, thus requiring the available spatial degrees of freedom to be properly exploited both for space–time clutter filtering and for target DOA estimation.
- The bistatic illumination offered by ground-based transmitters is easily prone to shadowing phenomena and to generate non-homogeneous clutter scenarios, which may limit the effectiveness of adaptive techniques.

All the above issues contribute to making GMTI from passive radar on moving platforms a challenging application. The need for practical and effective solutions to tackle these issues and thus contribute to the maturity of mobile passive radar technology is at the base of this research. Specifically, the attention is focused on developing signal processing algorithms and operational strategies for space–time clutter filtering and slow-moving target detection and localization, while possibly preserving the paradigm of a simple system architecture.

The use of space–time processing for moving target detection in passive radar was first considered in [10,11]. The authors generalized STAP for noise-like signals and proposed it for passive radar exploiting GSM and DVB-T transmission, showing simulated results from a moving receiver.

The first proof of concept of an airborne passive radar was given in [12], exploiting FM transmission. The detection of targets from mobile passive radar was demonstrated and the behaviour of stationary ground clutter was analysed.

The low-cost characteristic of passive radar, the typical size of the antennas at VHF/UHF bands and the high data rate of digital broadcast transmissions suggest the use of only few spatial channels and simple processing architectures for ground clutter suppression. For this reason, the displaced phase centre antenna (DPCA) approach has been primarily considered in [13–19].

DPCA ideally suppresses the clutter echoes by a non-adaptive subtraction of properly delayed signals collected by two along-track displaced receiving antennas [37]. However, the cancellation performance can be affected by several factors, such as internal clutter motion (ICM), non-perfect alignment of antenna phase centres, waveform temporal variability, and possible channel imbalances.

In [13,14], the authors considered the use of DPCA against experimental data from an airborne FM-based passive radar and against simulated DVB-T data. The continuous-wave characteristic of the opportunity signal is exploited to apply the required temporal delay between the receiving channels, being only limited by the sampling frequency. This mitigates the typical constraints posed by DPCA on the pulse repetition frequency (PRF) of standard pulsed radar. However, the temporal variability of the waveform can severely limit the clutter cancellation performance, even assuming a perfect DPCA condition.

In [15], a scheme was proposed based on the reciprocal range compression strategy in conjunction with a flexible DPCA approach. The flexible DPCA relaxes the constraint on the equivalent PRF, also when a batch processing architecture is adopted, which is typical of passive radar exploiting digital waveforms. The use of the reciprocal filter [34–36], in lieu of a conventional matched filter, has a twofold role: it controls the undesirable structures and sidelobes arising in the signal ambiguity function, and it removes the performance limitations deriving from the temporal variability of the opportunity waveform. The effectiveness of this scheme was demonstrated for an experimental DVB-T-based mobile passive radar.

Other contributions addressed the impact of the receiver motion on reference signal estimation and passive radar operations. In [28], the effects of the signal mismatch on the correlation process in moving passive radar are analysed. In [29], the authors investigate the impact of the motion induced Doppler shift on the reference signal reconstruction, providing potential solutions. In [30], an improved estimation capability is achieved by exploiting antenna and polarimetric diversity.

Despite the effectiveness of the techniques devised and the progress that has been made, the performance and the operational capability of mobile passive radar might still be strongly limited. The work described in this chapter is aimed at addressing the still open issues and overcome them by devising specific solutions.

8.1.1   *Key points*

The goal of this chapter is to contribute to research on multichannel mobile passive radar, developing signal processing techniques and operational strategies aimed at GMTI applications. Specifically, the attention is focused on STAP methodologies and suitable solutions are devised to address the main limitations deriving from the passive bistatic framework and related to the problem of inter-channel imbalance.

As observed, several aspects brought in by the passive bistatic operation make channel imbalance a critical issue in mobile passive radar, limiting the applicability of conventional strategies and compromising the system performance. These aspects must be properly tackled to guarantee an effective clutter cancellation and slow-moving target detection and angular localization capability.

The proposed solutions make use of some peculiar characteristics of passive radar systems, such as the typical long integration times. Moreover, they are designed to preserve the paradigm of limited complexity and simple system architecture. The main innovative contributions are listed in the following.

- The design of a post-Doppler STAP approach for passive radar, which operates after the Doppler processing on a clutter subspace accounting for a limited angular sector, making use of the long integration time and the resulting fine Doppler resolution to reduce the size of the adaptive problem and compensate for the angle-dependent channel errors.
- The introduction and performance assessment of novel space–time generalised-likelihood ratio test (GLRT) schemes specifically devised for mobile passive radar in the presence of unknown imbalances affecting the receiving channels.
- The characterisation of the target angular localisation problem in mobile passive radar featuring few wide beam antennas, which requires a convenient exploitation of the available spatial degrees of freedom (DOF) both for space–time clutter filtering and for target direction of arrival (DOA) estimation.
- The definition of a calibration strategy for spatial steering vector in the desired direction, which exploits the one-to-one relationship between angle of arrival and Doppler frequency of stationary scatterers and enables an accurate target DOA estimation.
- The introduction of a dual cancelled channel STAP approach, which allows to further reduce the systems computational complexity, simplify the target DOA estimation process, and increase the robustness against adaptivity losses, typical of the highly non-homogeneous clutter scenario experienced in bistatic passive radar exploiting ground-based transmitter.

Finally, it is worth noting that the processing techniques developed in this work, which define a set of effective methodologies for GMTI in mobile passive radar, are duly tested against simulated and real data. Specifically, the experimental validation exploits the data acquired by a DVB-T-based multichannel receiver developed by Fraunhofer FHR and mounted on a ground moving platform.

8.1.2 Outline

The remainder of this chapter is organized as follows.

Section 8.2 proposes a post-Doppler STAP approach for clutter cancellation and moving target detection and localization in multichannel mobile passive radar, in the presence of an angle-dependent imbalance affecting the receiving channels. While the clutter suppression capability is guaranteed by the adaptive space–time filtering, different solutions are devised aimed at mitigating the impact of channel calibration errors on target detection and localization performance. A spatially non-coherent detection scheme is compared with a fully coherent detector, where clutter echoes are exploited for steering vector calibration. Finally, the target localization problem is addressed and the STAP scheme is employed for maximum likelihood (ML) estimation of target DOA against the impact of channel imbalance.

Section 8.3 proposes a dual cancelled channel STAP scheme aimed at reducing the system computational complexity, as well as the number of required training data, compared with a conventional full-array solution. The proposed scheme is shown to yield comparable performance with respect to the equivalent full array case, both in terms of target detection capability and DOA estimation accuracy, despite the lower cost. Moreover, it offers more robustness against adaptivity losses, operating effectively even in the presence of a limited set of training data.

Section 8.4 summarises the main results and provides an outlook for further works.

8.2   Passive radar STAP detection and DOA estimation under antenna calibration errors

In [15–19], a DPCA approach has been considered to provide GMTI capability in mobile passive radar. DPCA involves a simple system architecture and a limited computational load, which make it attractive for the passive radar application.

However, significant performance limitations may come from the presence of amplitude and/or phase imbalances affecting the receiving channels, which can severely compromise the clutter cancellation capability of a non-adaptive scheme. Such imbalance can, in general, be a function of the angle of arrival due to several factors (dissimilarities between the receiving antenna patterns, mutual coupling effects, interaction with near-field obstacles) and its impact is made more severe by the characteristics of the passive bistatic scenario.

The low directivity of the typically available receiving antennas for VHF/UHF bands, paired with a bistatic illumination from omnidirectional transmitters, results in non-negligible clutter contributions simultaneously received from a wide angular sector and therefore easily affected by different amplitude and phase responses across the receiving channels, as also verified against experimental data in [19].

On the other hand, the typical long integration time of passive radar provides a fine Doppler frequency resolution, which can be usefully exploited. In [18,19], effective solutions were developed for channel calibration in passive radar DPCA. The proposed strategies took the advantage of the clutter angle-Doppler dependence for the estimation of the angle-dependent imbalance. The accurate calibration of the received data proved to be largely required to preserve the clutter suppression capability. Moreover, the effects of channel errors on target signal had to be neglected, in favour of an effective clutter cancellation. In fact, target signal may experience a different imbalance compared with the clutter contributions appearing at the same Doppler frequency, since belonging to a different angle, thus possibly affecting the detection performance despite an effective suppression of clutter. Finally, with two channels on receive, only the detection of target echoes can be sought whereas the problem of its angular localization within the broad antenna beam remains unsolved.

The intrinsic limitations of DPCA and its reliance on an adaptive calibration stage for the compensation of localized errors suggest moving towards a space–time adaptive processing (STAP) approach [20]. At the expense of a higher computational load, STAP offers more flexibility and adaptation capability, thanks to a higher number of adaptive degrees of freedom. The use of STAP for clutter rejection in passive radar was first considered in [10,11]. Preliminary experimental results of STAP in a DVB-T-based mobile passive radar are presented in [23]. In [24], sparse Bayesian learning is employed for accurate estimation of clutter covariance matrix based on few secondary samples. In [27], a three-dimensional model is proposed to integrate clutter modelling and waveform impact in passive STAP. Further applications of STAP in passive radar were considered; in [31], for improved target detection exploiting spatial diversity with OFDM waveforms; in [32], for direct signal interference suppression; and in [33], for clutter rejection in airborne bistatic inverse-SAR (ISAR) imaging.

In this section, we propose a STAP scheme for mobile passive radar and analyse its effectiveness in terms of clutter cancellation and moving target detection and localization, in the case of an unknown angle-dependent imbalance affecting the receiving channels. The set of adopted methodologies takes advantage of some peculiar characteristics of the passive radar scenario.

First, we point out the key role of a post-Doppler STAP approach in preserving clutter cancellation capability also in the presence of channel calibration errors, identifying it as particularly suitable for mobile passive radar. It makes use of the long integration time and the resulting fine Doppler resolution to considerably reduce the size of the adaptive cancellation problem and compensate for the angle-dependent channel errors, by operating on a clutter subspace accounting for a limited angular sector.

Therefore, two detection schemes are proposed, aimed at mitigating the effects of the channel imbalance on the target signal, whose DOA and related experienced imbalance may differ from those of clutter in the surrounding range/doppler cells, possibly affecting the detection and localization performance despite the effective clutter suppression. The first scheme consists of a partially non-coherent space–time generalized-likelihood ratio test (GLRT), which performs a non-coherent integration of target signal across the receiving channels, to cope with losses due to spatial steering vector mismatch. Entirely excluding the presence of a calibration stage, it provides a simple but effective solution for target detection in systems featuring few receiving channels. The second scheme consists of a fully coherent space–time GLRT, where the echoes from the stationary scene are exploited to estimate the channel errors in the desired search direction and calibrate the spatial steering vector. At the expense of an additional calibration stage, it provides slightly better detection performance and preserves the target DOA estimation capability.

Finally, the target angular localization problem is addressed. The limited number and low directivity of receiving antennas, due to the typically exploited wavelengths, make the accurate estimation of target DOA a critical task. Such capability requires the availability of multiple spatial degrees of freedom, which must be properly exploited both for space–time clutter filtering and for target DOA estimation and can be severely compromised by the presence of channel imbalance. For this purpose, the proposed STAP scheme is adopted for a ML DOA estimation, assessing the key role of the steering vector calibration in mitigating the negative impact of the unknown channel errors.

The moving target detection and localization performances of the proposed solutions are analysed and compared against a simulated clutter scenario. Moreover, some results are shown against experimental data collected by a DVB-T based multichannel passive radar mounted on a ground moving platform. The obtained results clearly demonstrate the effectiveness of the proposed strategies.

8.2.1 STAP scheme for passive radar

Let us consider an N channel passive radar mounted on a moving platform and exploiting a stationary transmitter as illuminator of opportunity (see Figure 8.2). The platform moves at constant velocity v_p on a straight-line trajectory, assumed without loss of generality along the x-axis. Angles φ and ϑ indicate respectively the azimuth and depression angle of the receiver to scatterer line of sight.

Recalling the signal model adopted in [19] and extending it to a multichannel case, for a linear array of elements equally spaced by d in a side-looking configuration, the discrete time baseband signal received at the ith antenna from a moving target at angles (φ_0, ϑ_0) and relative bistatic range R_b can be expressed as:

$$r_0^{(i)}[l] = G_i(\varphi_0, \vartheta_0) A_0 \sum_n s_n[l - nL - l_{\tau_0}] e^{j2\pi f_D nT} e^{-j2\pi i \frac{d}{\lambda} \cos \varphi_0 \cos \vartheta_0} \tag{8.1}$$

where

- the time index l denotes the lth sample of the signal, sampled at frequency f_s;
- the transmitted signal is partitioned in batches of duration T and $s_n[l]$ denotes the nth batch, including $L = Tf_s$ samples; notice that the Doppler induced phase term within each batch has been neglected;
- $l_{\tau_0} = f_s R_b / c$ is the bistatic propagation delay, assumed constant in the CPI;
- A_0 is the target complex amplitude and $G_i(\varphi_0, \vartheta_0)$ the complex gain of the ith channel at target angular direction; the latter represents the overall receiver chains, including the antenna pattern, and accounts for the possible imbalance between channels;
- f_D is the bistatic Doppler frequency, which can be expressed as the sum of two contributions:

$$f_D = \frac{v_p}{\lambda} \cos \varphi_0 \cos \vartheta_0 - \frac{v_b}{\lambda} \tag{8.2}$$

the first is associated to the platform motion and the receiver to scatterer geometry, the second is due to the target's own bistatic velocity v_b, given by the sum of the projections of its velocity vector on the target to receiver and the target to transmitter lines of sight; λ denotes the signal carrier wavelength.

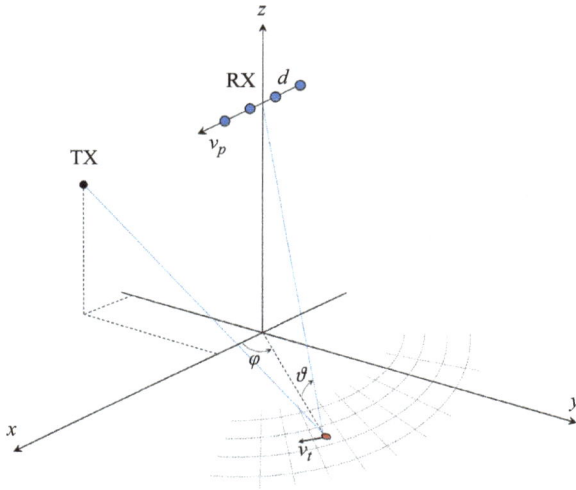

Figure 8.2   System geometry of a multichannel mobile passive radar exploiting a ground-based stationary transmitter as illuminator of opportunity

The signal representing the clutter contribution before range compression can be expressed as the superposition of echoes from a distribution of stationary scatterers $(v_b = 0)$ at different bistatic ranges $R_q (q = 1, \ldots, N_R)$ and different angles φ:

$$r_C^{(i)}[l] = \sum_{q=1}^{N_R} \int_\varphi G_i(\varphi, \vartheta) A_q(\varphi) \sum_n s_n[l - nL - l_{\tau_q}]$$

$$\times e^{j2\pi \frac{v_p}{\lambda} \cos\varphi \cos\vartheta nT} e^{-j2\pi i \frac{d}{\lambda} \cos\varphi \cos\vartheta} d\varphi \qquad (8.3)$$

where $A_q(\varphi)$ is the complex amplitude and τ_q the bistatic propagation delay of echo from clutter patch at angle φ and range R_q.

The adopted processing scheme for the application of STAP to the passive radar framework is sketched in Figure 8.3. It is obtained by extending the DPCA scheme presented in [19] to the general case of an N channel receiver. After a preliminary stage, including synchronization to the transmitter and reference signal reconstruction, the range compression is performed based on a batch processing architecture, which establishes the conventional fast-time/slow-time framework of a pulsed radar system operating at an equivalent PRF given by the inverse of the batch duration $(PRF = 1/T)$. The range compression stage is performed for each batch by means of a reciprocal filter [34–36], which has the dual role of controlling the signal ambiguity function and removing the temporal variability of the employed opportunity waveform [15]. For this purpose, a perfect reconstruction of the reference signal is supposed available by means of a decode/recode approach. Notice that the use of the reciprocal filter, which is fundamental in a non-adaptive approach like DPCA, also plays an

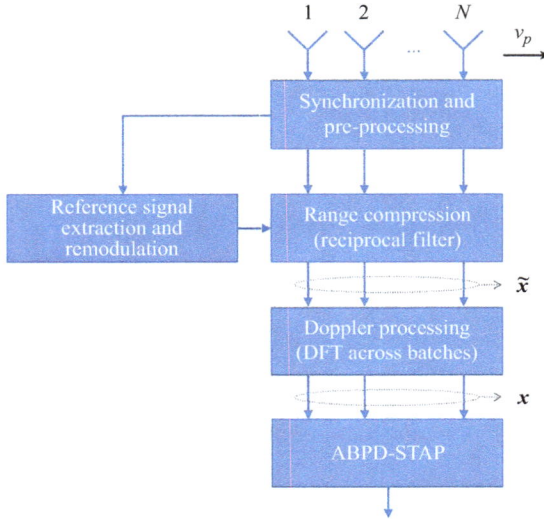

Figure 8.3   Processing scheme for post-Doppler STAP in passive radar

important role in a STAP approach. By providing a time-invariant impulse response and improving the separation between range cells, it allows an ideal estimation of the adaptive filter.

After range compression stage, by collecting the samples associated to the range cell under test, from the N receiving channels and the M batches in the coherent processing interval (CPI), the $(NM \times 1)$ space–time data vector $\tilde{x}$ is obtained. In the presence of a target with Doppler frequency f_D, DOA ϕ and complex amplitude $\bar{A}$, the data vector can be written as $\tilde{x} = \bar{A}\tilde{s} + \tilde{d}$, where $\tilde{d}$ represents the disturbance (clutter plus noise) component, assumed Gaussian with space–time covariance matrix $\tilde{Q} = E\left\{\tilde{d}\tilde{d}^H\right\}$. The space–time steering vector $\tilde{s}$ can be expressed as $\tilde{s}(f_D, \phi) = s_t(f_D) \otimes s_s(\phi)$, where $s_t(f_D) = \left[1, e^{-j2\pi f_D T}, \ldots, e^{-j2\pi f_D MT}\right]^H$ is the temporal steering vector and $s_s(\phi) = \left[1, e^{j2\pi d/\lambda \cos\phi}, \ldots, e^{j2\pi Nd/\lambda \cos\phi}\right]^H$ is the spatial steering vector, $\otimes$ denoting the Kronecker product and H the Hermitian transpose. Notice that ϕ denotes the angle between the antenna end-fire and the receiver to scatterer line of sight ($\cos\phi = \cos\phi\cos\vartheta$).

As known, STAP adaptively combines spatial and temporal samples of the signal in order to suppress clutter in the angle-Doppler domain and maximize the detection probability of potential moving targets [37,38]. It is based on the inversion of the space–time disturbance covariance matrix $\tilde{Q}$, which is usually not available in practical applications and has to be estimated based on proper training data. In order to reduce the computational effort and the amount of training data required for an effective estimation of $\tilde{Q}$, a number of reduced-order STAP approaches have been suggested, where adaptive processing is applied after a non-adaptive projection of data in a proper subspace [40].

Particularly suitable for the passive radar case, characterized by relatively long integration times, are the post-Doppler STAP approaches, where adaptation occurs on a subset of Doppler-processed data. The sufficiently long CPIs ensure proper decoupling of different clutter Doppler components. Specifically, we consider an adjacent-bin post-Doppler (ABPD) approach, which adaptively combines N spatial samples from a subset of L adjacent Doppler bins centred at the cell under test.

Let T be a transformation matrix, which consists, in this case, of L adjacent columns of a discrete Fourier transform (DFT) matrix. The resulting vectors in the space-Doppler domain are given by $x = T^H \tilde{x}$ and $s = T^H \tilde{s} = s_d \otimes s_s (\phi)$ of size ($NL \times 1$). The corresponding ($NL \times NL$) space-Doppler disturbance covariance matrix will be $Q = T^H \tilde{Q} T$.

The well-known space–time GLRT detector can be easily derived following the approach in [49] and applied after the ABPD transformation:

$$\frac{\left| s^H \hat{Q}^{-1} x \right|^2}{s^H \hat{Q}^{-1} s (1 + x^H \hat{Q}^{-1} x)} \gtreqless \eta_1 \tag{8.4}$$

where Q is substituted by its ML estimate $\hat{Q} = X_k X_k^H$, being $X_k = [x_1, \ldots, x_K]$ a set of training data of size ($NL \times K$) from K adjacent range cells, assumed as statistically independent, identically distributed and target-free. The detection threshold η_1 is selected according to the desired value of false alarm probability (PFA), whose known analytical expression is:

$$PFA = \left(\frac{1}{l} \right)^{K-NL+1} \tag{8.5}$$

where $l = 1/(1 - \eta_1)$.

The STAP solution can handle a higher number of degrees of freedom, offering more flexibility and adaptation capability compared to a DPCA approach. On the one hand, it involves a higher computational effort, since it requires the estimation and inversion of a covariance matrix, potentially for each range-Doppler bin. However, the ABPD approach limits the computational cost and the amount of required training data by significantly reducing the size of the covariance matrix. This also plays a key role in a real scenario, where the effectiveness of STAP would be subject to the potential non-homogeneity of clutter, which might not offer a sufficient number of homogeneous training data for matrix estimation.

In order to analyse the effectiveness of the considered space–time processing scheme in a controlled environment, we test it against a simulated clutter scenario for a multichannel moving passive radar.

We assume a ground moving receiver exploiting a stationary transmitter in a quasi-monostatic geometry. An $8k$ mode DVB-T signal sequence is generated as a reference signal. Details on the DVB-T signal parameters can be found in Table 8.1. Clutter returns are generated according to the model in (8.3), for a scene spanning $N_R = 1{,}000$ range cells. Amplitudes $A_q(\varphi)$ associated with different clutter patches are assumed independent and identically distributed complex Gaussian variables, thus resulting in a homogeneous clutter scenario. We assume the availability of $N = 3$ perfectly balanced receiving channels, arranged in the along-track direction, in a side-looking configuration. Omnidirectional antennas are considered, within an angular sector $\varphi = [0, \pi]$ (no back-lobe contributions). Carrier frequency is set to 690 MHz, platform velocity $v_p = 13$ m/s and antenna element spacing $d = \lambda/2$. We consider a CPI length of 512 OFDM symbols, corresponding to approximately half a second duration ($\sim$0.57 s), and we assume absence of ICM. Notice that, the DVB-T elementary period (7/64 µs) defines the fast-time sampling rate, while the batch duration, deliberately selected as equal to the OFDM symbol duration, defines the slow-time sampling rate (equivalent PRF $\cong$ 893 Hz).

The generated input signal includes clutter returns and thermal noise and is scaled so that the overall clutter contribution has an assigned power level of 20 dB above the noise level, at the input of each Rx channel. The echo from a moving target is also included, with bistatic range $R_b = 4$ km, azimuth angle of arrival $\phi_0 = 90°$ and bistatic radial velocity $v_b = 5$ m/s. Target signal-to-noise ratio (SNR), defined as the ratio of target signal power level with respect to noise, at the input of each channel before range compression, is set to −43 dB.

The range-Doppler map obtained from a single channel is reported in Figure 8.4(a). As apparent, clutter returns appear across a Doppler extension of approximately $\pm v_p/\lambda \cong \pm 30$Hz, while target signal-to-clutter-plus-noise ratio (SCNR) is −26 dB. In the simulation, the target SCNR is measured by taking the power level at the target range-Doppler location, when the processing is fed with target echoes only, and disturbance power level estimated over a proper area surrounding target location, in the maps containing only clutter and noise.

Applying the ABPD-STAP scheme in Figure 8.3, the resulting range-Doppler map at the output of the adaptive filter (numerator in (8.4)) is shown in Figure 8.4(b). In particular, $L = 3$ adjacent Doppler bins are used, for a total on $NL = 9$ degrees of freedom. The number L of Doppler bins is generally selected as small as possible, to reduce the computational complexity, but sufficient to guarantee good clutter cancellation capability. In our case, the value $L = 3$ proved to be a suitable trade-off between required cost and effectiveness of the adaptive filter. The amount of training data is set to $K = 6NL = 54$. This large sample support is selected in order to minimize the undesirable adaptivity losses.

Notice that all the range-Doppler maps are scaled to provide unitary processing gain for thermal noise, thus allowing a direct comparison of results. As expected, the clutter background is effectively cancelled and the resulting target SCNR is 15 dB, with an overall improvement of 41 dB.

8.2.2   Limitations due to channel calibration errors

To analyse the effects of channel imbalance, we included in the simulation process the presence of a deterministic but unknown angle-dependent imbalance between the receiving channels. This allows to emulate realistic channel error conditions. We denote by $\Gamma_{ij}(\phi) = G_i(\phi)/G_j(\phi)$ the complex imbalance between channel i and j.

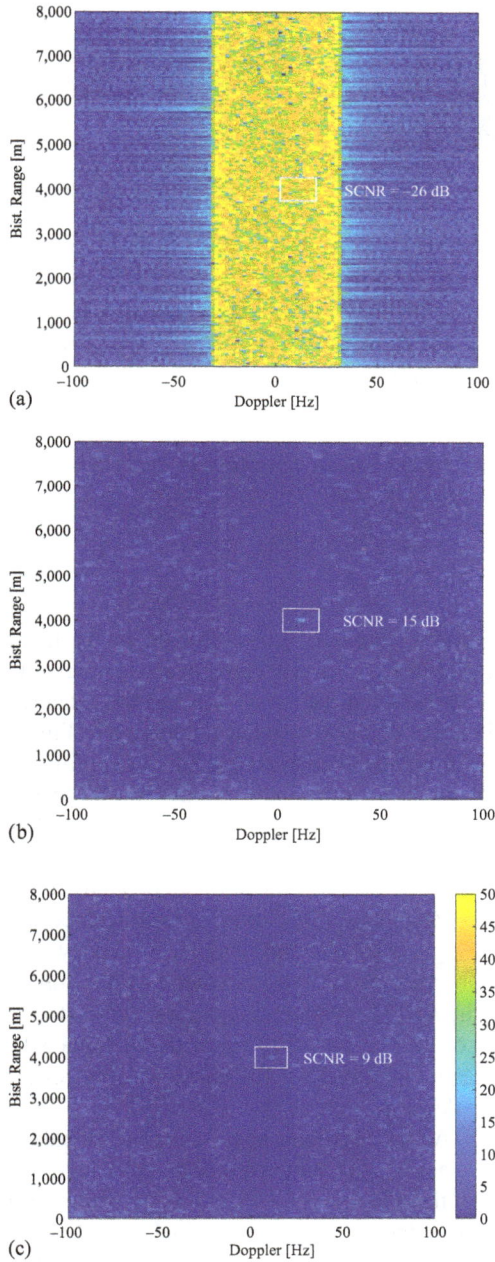

Figure 8.4   Range-Doppler maps from simulated clutter scenario: (a) single channel; (b) after ABPD-STAP with perfectly balanced channels; (c) after ABPD-STAP in the presence of channel imbalance

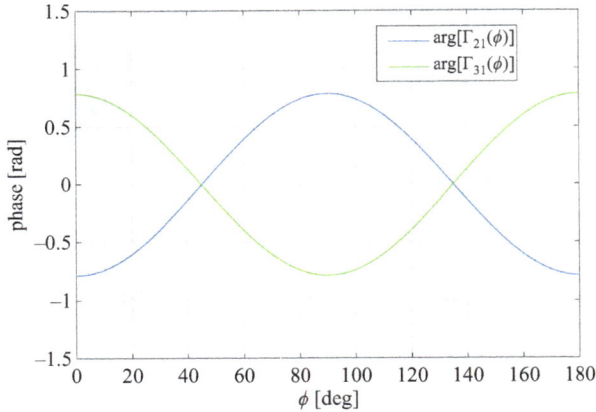

Figure 8.5   Simulated phase imbalance between the receiving channels as a function of the angle of arrival

Specifically, we assume a sinusoidal phase imbalance as illustrated in Figure 8.5, where channel 1 is arbitrarily taken as reference. Channel error is modelled as a function of the angle ϕ between the array line and the receiver to scatterer line of sight. Notice that the level of the simulated imbalance in Figure 8.5 is of the same order of magnitude of the imbalance experienced on experimental data in [17].

The range-Doppler map resulting at the output of the STAP scheme in the presence of channel imbalance is reported in Figure 8.4(c). The ABPD approach with $L = 3$ and $K = 54$ is again applied. As expected, the adaptation capability of the STAP filter allows to intrinsically compensate for the angle dependent channel errors and keeps providing an effective cancellation of clutter echoes. In fact, the known one-to-one relationship between angle of arrival and Doppler frequency of stationary scatterers, as well as the fine Doppler resolution guaranteed by the long integration times of passive radar, allow the ABPD approach to operate on a clutter subspace accounting for a limited angular sector. This allows to significantly reduce the number of degrees of freedom of the adaptive filter, by still being able to compensate for the local variation of the angle-dependent channel imbalance.

However, while STAP proves robust against channel imbalance for what concerns clutter suppression capability, the same cannot be said for the corresponding target steering vector used in the adaptive filter. In fact, the presence of unknown channel errors may cause a mismatch between the nominal spatial steering vector $s_s(\varphi)$ and the actual target vector affected by channel imbalance, which can be modelled as $diag\{[1, \Gamma_{21}(\phi), \ldots, \Gamma_{N1}(\phi)]\}s_s(\phi)$, by taking channel 1 as reference without loss of generality. This may result in target signal gain losses. In the case of Figure 8.4(c), the final target SCNR is in fact limited to 9 dB, with a loss of 6 dB with respect to the case without channel imbalance.

To evaluate the detection performance losses associated to the presence of channel imbalance, a Monte Carlo analysis has been performed, assuming the same simulated scenario and processing parameters adopted in the previous example. The space–time GLRT detector in (8.4) has been considered, with a desired PFA set to 10^{-6}. A Swerling 0 target model has been assumed.

In Figure 8.6(a), the results are shown in terms of probability of detection (PD) as a function of the target input SNR. Two targets are assumed, at DOA $\phi_0 = 90°$ and with a bistatic velocity of 7 and 3 m/s. They are representative of target condition sufficiently far and close to the clutter notch, respectively. The dashed curves represent

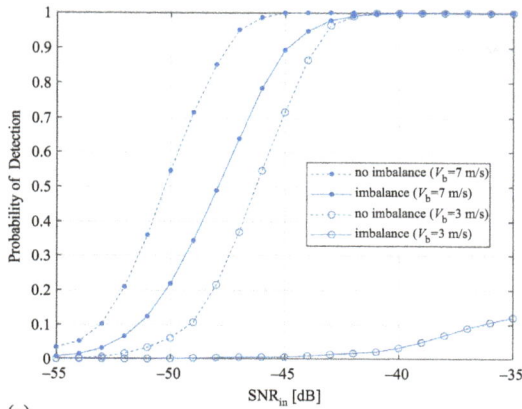

(a)

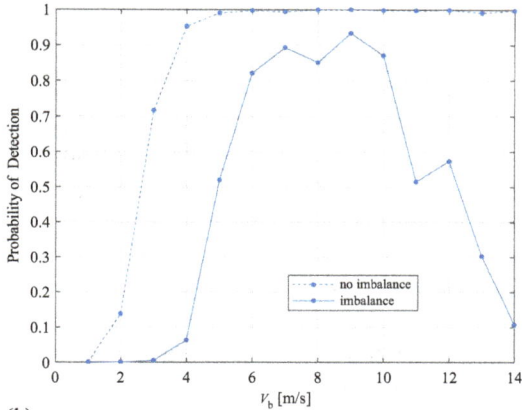

(b)

Figure 8.6 *Effects of channel imbalance on target detection performance. (a)*
Probability of detection as a function of target input SNR, for target
bistatic velocities 3 and 7 m/s. (b) Probability of detection as a function
of target bistatic velocity, for target input SNR = −45 dB. Desired false
alarm probability is set to 10^{-6}.

the detection performance achievable in the absence of channel imbalance, while the solid curves show the corresponding performance when receiving channels are affected by the imbalance modelled in Figure 8.5.

A significant loss can be observed, especially for a target velocity of 3 m/s, corresponding to a slow target close to the clutter notch. This indicates that the presence of channel mismatch may increase the clutter rank and produce an enlargement of the clutter notch (as from [38]), thus raising the target minimum detectable velocity (MDV). This is also confirmed by looking at Figure 8.6(b), where PD is evaluated as a function of the target bistatic velocity, for a fixed SNR of −45 dB. Notice that a reduction in PD performance is present also for higher velocity values. This is mostly caused by a dispersion of the clutter energy in the angle-Doppler domain due to the angle-dependent imbalance. This may result in losses also far from the clutter notch, as an effect of the cancellation filter.

8.2.3   Solutions for target detection

We have shown how the presence of an unknown imbalance affecting the receiving channels can impact on moving target detection performance of a mobile passive radar. In particular, when adopting a STAP approach on a multichannel system, the resulting mismatch of the target steering vector may result in undesirable gain loss or even partial suppression of the target signal at the output of the adaptive filter.

It is worth noting that applying digital channel calibration techniques such as those proposed in [17] to the received data is essential to ensure clutter cancellation in a non-adaptive approach like DPCA, while it is not strictly required in the STAP case. In fact, the adaptation capability of STAP can intrinsically compensate for localized channel errors, thus preserving clutter suppression capability. Moreover, in the case of angle-dependent channel errors, the target signal may experience a different imbalance compared to clutter echoes appearing at the same Doppler frequencies, since belonging to a different angular direction. As a result, a Doppler-based calibration strategy would not be useful against target steering mismatches.

In the following, two possible strategies for target detection are proposed, aimed at recovering the performance losses associated to channel calibration errors in a mobile passive radar exploiting a STAP scheme.

8.2.3.1   Spatially non-coherent GLRT (NC-GLRT)

If the knowledge of channel imbalance in each desired steering direction is not possible, a simple potential solution can be to renounce to a coherent integration in the spatial domain. If a small number of receiving channels is available, typically true in passive radar systems, this would produce a limited loss in terms of final signal to disturbance ratio.

We propose a partially non-coherent space–time GLRT detector (referred to as NC-GLRT), where the steering vector is specified in the Doppler domain but not specified in the spatial domain, resulting in a non-coherent integration of the spatial target echoes. Such a detector, first introduced in [20], can be derived along the line of [51], where a polarimetric adaptive detection scheme is addressed. Considering the spatial component of the steering vector s_s as a vector of unknown parameters, we replace it with its ML estimate during the derivation process. By defining the $(NL \times N)$ matrix $\boldsymbol{\Sigma} = s_d \otimes \boldsymbol{I}_N$, where $\boldsymbol{I}_N$ is the N-dimensional identity matrix, the resulting GLRT detector is given by:

$$\frac{x^H \hat{Q}^{-1} \boldsymbol{\Sigma} \left(\boldsymbol{\Sigma}^H \hat{Q}^{-1} \boldsymbol{\Sigma} \right)^{-1} \boldsymbol{\Sigma}^H \hat{Q}^{-1} x}{\left(1 + x^H \hat{Q}^{-1} x \right)} \gtrless \eta_2 \tag{8.6}$$

This detector keeps the CFAR property and the PFA follows (see [51]):

$$PFA = \frac{(1 - \eta_2)^{K-NL+1}}{(K - NL)!} \sum_{j=1}^{N} \frac{(K - NL + N - j)! \, \eta_2^{N-j}}{(N - j)!} \tag{8.7}$$

This approach still performs an adaptive space–time filtering of data, aimed at whitening of the clutter returns. While, at the expense of a limited loss in terms of maximum integration gain (and directivity), it is robust against losses due to spatial steering vector mismatches, thanks to a non-coherent integration of target echoes across the receiving channels.

8.2.3.2 GLRT with steering vector calibration (Cal-GLRT)

An alternative and more sophisticated approach consists in maintaining a fully coherent detection scheme and exploiting the information from stationary scene echoes to estimate the angle-dependent channel errors and define the correct spatial steering vector towards the desired target search direction.

The basic idea is to make use of the Doppler spread that characterises clutter returns seen from a moving receiver and the corresponding relationship between angle of arrival and Doppler frequency. In a similar fashion to the approaches in [17], Doppler frequency resolution can be exploited to isolate contributions from scatterers belonging to specific angular directions, thus allowing an estimation of the angle-dependent imbalance affecting the receiving channels. In this regard, the fine Doppler resolution provided by the typically long integration times of passive radar is an additional asset.

This principle was also exploited in some previous works for array calibration in airborne radar [41–45]. The estimation of channel errors and the correct spatial steering vector in a specific direction can be made in different ways: for instance, selecting the principal eigenvector from a clutter sample covariance matrix formed at the corresponding Doppler bin [42,45]; or through a least square estimation comparing returns at the same Doppler bin from co-registered range-Doppler maps (aligned by delay compensation) [44]. In this work, the latter approach is adopted.

Let us assume $z^{(i)}[l, m]$ to be the complex value at the generic range-Doppler bin of the range-compressed and Doppler-processed channel i, after proper temporal co-registration. The imbalance estimated at the mth Doppler bin between channels i and j (the latter assumed as reference) is given by:

$$\hat{\Gamma}_{ij}[m] = \frac{\sum_{l=l_1}^{l_2} z^{(i)}[l,m] z^{(j)*}[l,m]}{\sum_{l=l_1}^{l_2} \left| z^{(j)}[l,m] \right|^2} \tag{8.8}$$

where the average is evaluated over consecutive range cells spanning indices from l_1 to l_2.

The imbalance at the specific direction of interest $\hat{\Gamma}_{ij}(\phi_0)$ can be obtained by selecting or interpolating the imbalance values estimated as a discrete function of Doppler bins in (8.8), at Doppler frequency $f_D = v_p/\lambda \cos\phi_0$.

By defining $\boldsymbol{\Lambda}(\phi_0) = \boldsymbol{I}_L \otimes diag\left\{\left[1, \hat{\Gamma}_{21}(\phi_0), \ldots, \hat{\Gamma}_{N1}(\varphi_0)\right]\right\}$, the space–time coherent GLRT detector with calibrated steering vector (referred to as Cal-GLRT) is given by:

$$\frac{\left| \boldsymbol{s}^H \boldsymbol{\Lambda}^H \hat{\boldsymbol{Q}}^{-1} \boldsymbol{x} \right|^2}{\boldsymbol{s}^H \boldsymbol{\Lambda}^H \hat{\boldsymbol{Q}}^{-1} \boldsymbol{\Lambda} \boldsymbol{s}(1 + \boldsymbol{x}^H \hat{\boldsymbol{Q}}^{-1} \boldsymbol{x})} \gtrless \eta_1 \tag{8.9}$$

At the expense of the additional cost required for steering vector calibration, this solution is expected to maximise the spatial integration gain on the target signal, provided that a good estimation of channel imbalance can be achieved. In addition, as shown in the following, this scheme preserves phase information between receiving channels, being suitable for the target DOA estimation purpose.

8.2.4   Detection performance analysis

In order to test and compare the performance of the proposed detection strategies in a controlled environment, we consider the same simulated clutter scenario of the previous section, with $N = 3$ receiving channels affected by the simulated angle-dependent imbalance in Figure 8.5. The ABPD-STAP approach is applied, with $L = 3$ Doppler bins and the number of training data is set to $K = 54$.

Detection performance is analysed by means of Monte Carlo analyses, for moving targets in the endo-clutter region. The NC-GLRT scheme in (8.6) and the Cal-GLRT scheme in (8.9) are compared with the standard GLRT detector in (8.4), where a mismatched steering vector is considered due to channel imbalance. We refer to this latter case as mismatched GLRT. The case without imbalance is also considered as a reference. Specifically, in Figure 8.7(a), the results are shown in terms of estimated PD as a function of target input SNR, for the same target parameters of Figure 8.6. The desired PFA is set to 10^{-6}.

As expected, both the solutions proposed in Section 8.2.3 allow to mostly prevent the partial suppression of the target signal due to the channel imbalance and the resulting steering vector mismatch, largely recovering the detection performance losses with respect to the case of no imbalance. A significant difference can be noticed, compared to the mismatched GLRT case (solid blue curves), in terms of minimum SNR required for a given PD, especially for lower target velocity. The performances of the ideal case (blue dashed curves) are almost restored. Note that, although the considered solutions largely recover the performance losses caused by imbalance, the condition of perfectly balanced channels in all directions cannot be completely re-established.

The above considerations are also confirmed by looking at Figure 8.7(b), where the PD is shown as a function of target bistatic velocity, for a fixed SNR of -45 dB. Both NC-GLRT and Cal-GLRT allow a considerable reduction of target MDV, by reducing the width of the clutter notch. The performance at higher velocity values is also considerably improved.

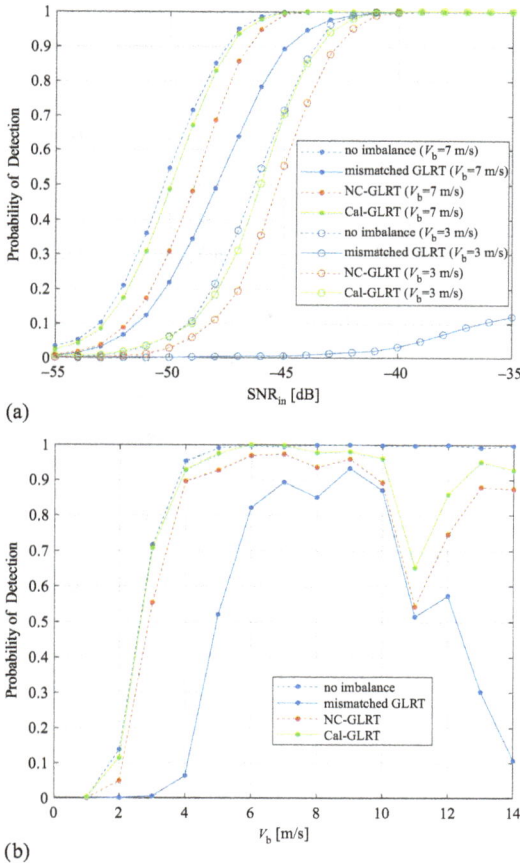

(a)

(b)

Figure 8.7   *Performance comparison of the considered detection schemes. (a) Probability of detection as a function of target input SNR, for target bistatic velocities 3 and 7 m/s. (b) Probability of detection as a function of target bistatic velocity, for target input SNR = -45 dB. False alarm probability is set to 10^{-6}.*

The Cal-GLRT detector (green dash-dot curves) yields the best performance. In fact, the steering vector calibration allows to maximise the coherent integration gain of target signal in the spatial domain. Small losses far from clutter notch are still present, due to the mentioned dispersion of clutter energy in the angle-Doppler domain. They are mostly associated to side-lobe residual clutter contributions and depend on the specific angular behaviour of the considered imbalance.

Notice that, in the assumed simulated scenario, the homogeneous distribution of clutter facilitates the estimation of channel imbalance and the corresponding steering vector correction at the desired angular direction. In a real environment, strategies accounting for potential variation of imbalance as a function of range or for robustness of estimation against outliers could be adopted, as described in [17].

The NC-GLRT approach (red dashed lines) only shows slight losses (in the order of 1–2 dB) compared to the Cal-GLRT case. These are mostly associated to a minor loss of non-coherent integration in the spatial domain (when using a small number of receiving channels). Nevertheless, it proves that an effective disturbance rejection capability is still preserved, despite the lower complexity due to the absence of an online calibration stage.

Therefore, this last approach proves to be a suitable and simple solution for the purpose of target detection in mobile passive radar with few receiving channels, being robust against significant imbalance possibly affecting the channels.

8.2.5 Target DOA estimation

Typically, passive radar exploiting VHF/UHF bands are characterized by broad antenna beams and a limited number of array elements. Therefore, an accurate target localization represents a critical task and interferometric approaches are commonly used to estimate target DOA, by exploiting the phase information across the available receiving channels.

Specifically, we refer to an ML DOA estimation approach for radar employing STAP (see [52]). Like the detection schemes in previous sections, DOA estimation can be performed after a non-adaptive transformation, aimed at containing the adaptivity losses and the required computational complexity.

By operating after the ABPD transformation of Figure 8.3, the target angle estimate is obtained by maximizing the log-likelihood function with respect to ϕ:

$$\hat{\phi}_t = \underset{\phi}{\operatorname{argmax}} \left\{ \frac{\left| s^H(\phi) \hat{Q}^{-1} x \right|^2}{s^H(\phi) \hat{Q}^{-1} s(\phi)} \right\} \tag{8.10}$$

where the unknown ϕ_t represents the actual target DOA.

In practice, the ML estimate can be viewed as the location of the peak in a dense grid of adaptive matched filters (AMF) [53]. Typically, the radar performs detection tests over a bank of filters coarsely spaced by the nominal beamwidth (BW) in angle. Once a target is detected, refined angle measurement is achieved through (8.10). The desired level of RMS error is typically chosen to be about one-tenth or one-twentieth of the BW.

The presence of an unknown imbalance affecting the receiving channels is a major problem for target detection but even more for their accurate angular localization. Inter-channel imbalance is expected to produce a significant degradation of DOA estimation accuracy. Although a non-coherent integration strategy in the spatial domain represents an effective solution for target detection, a coherent approach paired to a proper calibration of spatial steering vector is required when interested in target DOA estimation.

The same strategy adopted in Section 8.2.3.2 for calibration of target steering vector based on returns from stationary scene can be exploited to achieve a corrected ML DOA estimator.

By defining $\Lambda(\phi) = I_L \otimes diag\left\{\left[1, \hat{\Gamma}_{21}(\phi), \ldots, \hat{\Gamma}_{N1}(\phi)\right]\right\}$, $\hat{\Gamma}_{ij}(\phi)$ being the estimated imbalance between channels i and j for each angle ϕ in a dense grid around the direction of target detection, the ML DOA estimator with calibrated steering vector, referred to as calibrated maximum-likelihood estimator (MLE), is given by:

$$\hat{\phi}_t = \underset{\phi}{\mathrm{argmax}}\left\{\frac{\left|s^H(\phi) \Lambda^H(\phi) \hat{Q}^{-1}x\right|^2}{s^H(\phi) \Lambda^H(\phi) \hat{Q}^{-1}\Lambda(\phi) s(\phi)}\right\} \tag{8.11}$$

$\hat{\Gamma}_{ij}(\phi)$ for each test direction can be obtained by interpolating the imbalance estimated as a function of Doppler bins $\hat{\Gamma}_{ij}[m]$, at frequency $f_D = v_p/\lambda \cos\phi$.

The DOA estimation accuracy of the standard ML estimator in (8.10), referred to as mismatched MLE, and of the calibrated version in (8.11) is evaluated by means of a Monte Carlo analysis, against the same simulated clutter scenario of the previous sections. An ABPD-STAP approach is again applied, with $N = 3$ receiving channels, possibly affected by the angle-dependent imbalance shown in Figure 8.5, and $L = 3$ Doppler bins. The number of training data for the disturbance covariance matrix estimation is set to $K = 54$. Both the estimators operate with a bank of filters equally spaced in angle by $\delta\phi = 1°$ within the array nominal BW ($\sim 57°$), centred at $\phi_0 = 90°$. The simulated target DOA is $\phi_t = 85°$.

First, to appreciate the effect of steering vector calibration on DOA estimation performance when in the presence of channel imbalance, we show in Figure 8.8, the likelihood function of (8.10) and (8.11), obtained in the ideal case of known covariance matrix and in absence of noise. Notice that the phase imbalance generates a bias error in the mismatched case (blue curve), which is instead removed by the steering vector calibration (green curve). Moreover, especially for slower targets, also the width of the clutter notch plays a fundamental role.

In Figure 8.9, the accuracy of the DOA estimators is compared as a function of target input SNR, for bistatic velocity 7 and 3 m/s. Results are shown in terms of standard deviation and bias of the estimation, both normalised to the nominal BW.

First, we notice that the presence of channel imbalance significantly degrades the accuracy of the mismatched MLE (solid blue curves), by increasing both the standard deviation and the bias error, with respect to the case where no imbalance is present (dashed blue curves). In particular, the resulting mismatch tends to polarize the DOA estimate, as clearly visible in Figure 8.9(b), for high SNR values. Conversely, it is

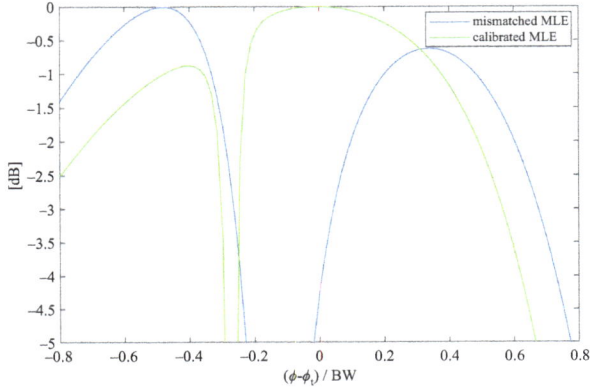

Figure 8.8   *Example of the likelihood functions of the ML DOA estimators, in the
known covariance matrix case, when in presence of channel imbalance.
Target set at DOA 85° with a bistatic velocity of 3 m/s.*

evident that the calibrated MLE (dash-dotted green curves) can mostly prevent the
performance losses due to the channel imbalance, almost recovering the estimation
accuracy of the ideal case (i.e., absence of imbalance).

To provide an additional reference for the DOA estimation accuracy of the considered scheme in the ideal case and to better evaluate the impact of channel errors and of
the proposed calibration approach, we also consider the Cramér–Rao bound (CRB).
Following [54,55], and assuming known target Doppler frequency and covariance
matrix Q, the CRB after the non-adaptive ABPD transformation can be expressed as:

$$\sigma_\phi^2 = [J^{-1}] = \left\{ 2|\bar{A}|^2 \left[(\dot{s}^H(\phi)\,Q^{-1}\dot{s}(\phi)) - \frac{|\dot{s}^H(\phi)\,Q^{-1}s(\phi)|^2}{s^H(\phi)\,Q^{-1}s(\phi)} \right] \right\}^{-1} \quad (8.12)$$

where σ_ϕ is the standard deviation of the DOA estimation error, J is the Fisher
information matrix, $\bar{A}$ is the target complex amplitude and $\dot{s}(\phi) = \partial s(\phi)/\partial\phi$.

In Figure 8.9(a), the corresponding CRB is reported for both the considered
target velocities (solid and dotted grey curves). For high SNR values, there is a good
match between theoretical results and simulation. This is due to the ML nature of the
estimator, which guarantees the condition of asymptotic efficiency. A little departure
of the simulated results from the theory occurs for low SNR, when the standard
deviation of the estimate becomes comparable with the BW. This is however a case
of limited interest. Also notice that, for higher SNR, the ML estimator is subject to
a saturation effect due to the use of a discrete set of angles (bank of filters) for DOA
estimation.

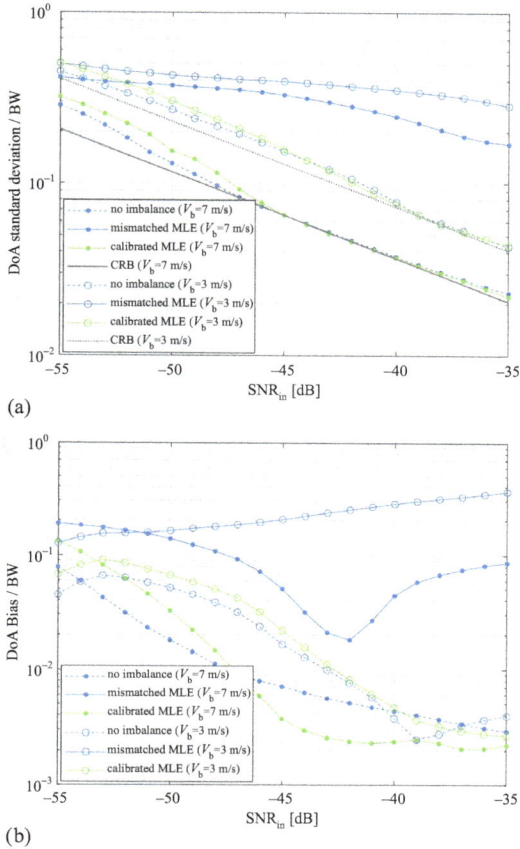

(a)

(b)

Figure 8.9   Comparison of ML DOA estimation accuracy as a function of the target input SNR: (a) standard deviation normalized to BW; (b) bias error normalised to BW

8.2.6   Experimental results

In this section, the effectiveness of the proposed strategies for target detection and DOA estimation is demonstrated against a set of experimental data acquired by a DVB-T based multichannel passive radar mounted on a ground moving platform.

The acquisition campaign was conducted by Fraunhofer FHR in a rural area of the Eifel region, in western Germany. The selected DVB-T illuminator of opportunity was the Eifel/Scharteberg transmitter, shown in Figure 8.10(a). The receiving system consisted of two PARASOL units [46], each providing two receiving channels, and was mounted on a trailer behind a van [see Figure 8.10(b)]. The four channels were connected to discone antennas and arranged to form a uniform linear array (ULA) in

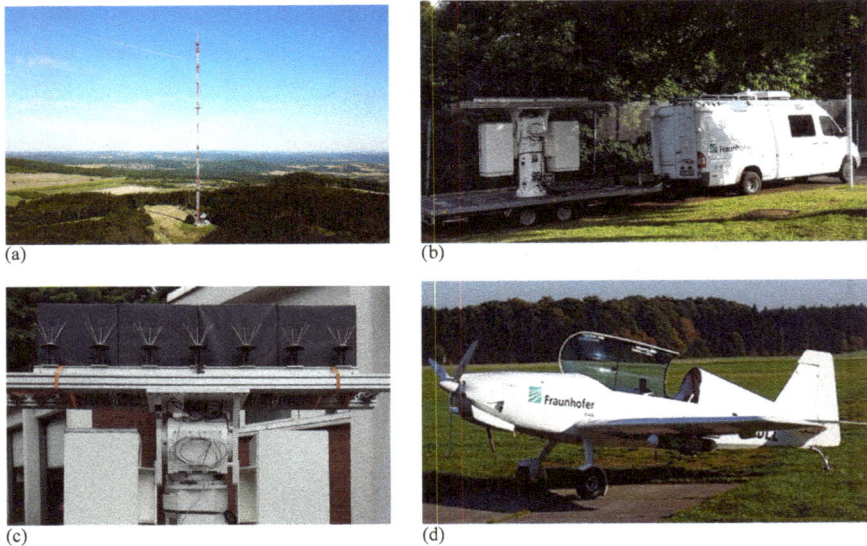

Figure 8.10   (a) The Eifel/Scharteberg transmitter of opportunity. (b) The experimental multi-channel receiver mounted on the back of a van in side-looking configuration. (c) A detailed view of the receiving array. (d) The aircraft (Delphin) employed as cooperative aerial target.

side-looking configuration. A radiation absorbing material (RAM) was placed on the one side, to attenuate the back lobe contributions. A close-up of the array is shown in Figure 8.10(c). Notice that only four central elements were connected to the receiving units, while the external ones were dummy elements resistively loaded. The receiving channels served as surveillance channels, while the reference signal was reconstructed from one of them. The parameters of the exploited DVB-T transmission are reported in Table 8.1, as well as the main system and processing parameters.

An ultralight aircraft from Fraunhofer FHR (Delphin) has been employed as a cooperative target during the acquisition campaign [see Figure 8.10(d)]. Figure 8.11 shows the acquisition scenario, characterized by a bistatic geometry where the transmitter is located approximately in direction opposite to the observed scene.

As thoroughly described in [19], where data from the same acquisition campaign are used, the four receiving channels are affected by a considerable angle dependent imbalance. Differently from the DPCA case analysed in [19], such an imbalance is not expected to compromise the clutter suppression capability by adopting the ABPD-STAP approach. Nevertheless, as mentioned, it can have an impact on target detection and localization performance.

To prove the effectiveness of the proposed solutions in preventing the performance degradation due to steering vector mismatch, four simulated moving targets

Table 8.1   Parameters of experimental test

Symbol	Description	Value
	DVB-T signal parameters	
	DVB-T standard	8k16QAM
f_c	Carrier frequency	690 MHz
N_c	Number of useful carriers	6,817
T_s	OFDM symbol duration	1,120 μs
B	Bandwidth	7.61 MHz
	System and processing parameters	
v_p	Platform velocity	~13.8 m/s
d	Antenna spacing	0.36 m
CPI	Coherent processing interval	512 T_s

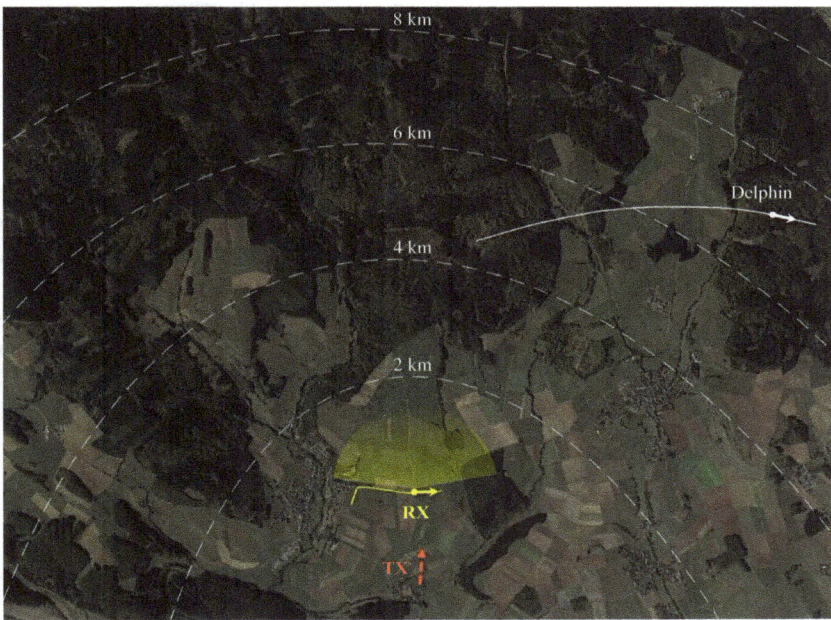

Figure 8.11   Optical image of the acquisition scenario. Yellow and white arrows indicate the position and the direction of motion of the receiver and aerial target, respectively. Red arrow indicates the transmitter direction of arrival. Dashed lines represent the bistatic iso-range curves.

are injected into the real data, according to the model in (8.1), in addition to the real target Delphin. Proper imbalance is applied to the generated target echoes across the receiving channels, according to the imbalance estimated at the corresponding target DOA. The parameters of the real and the simulated targets are reported in Table 8.2. Notice that the direction of Delphin target is known only with a certain approximation.

Figure 8.12 shows the range-Doppler maps resulting from the considered experimental data set. Specifically, Figure 8.12(a) represents the range-Doppler map obtained from a single channel, namely before STAP processing, scaled to the estimated noise power level. As evident, the clutter returns from the stationary scene extend over a Doppler bandwidth compatible with the platform velocity ($v_p/\lambda \cong \pm 32$Hz) and are characterized by a strong heterogeneity in terms of power levels across the map. Moreover, a strong direct signal contribution appears at first bistatic range bin and low Doppler frequency, being the Rx–Tx line of sight approximately orthogonal to the platform velocity vector.

The considered moving targets, due to their bistatic radial velocity and DOA, fall within the clutter Doppler bandwidth and appear as buried into clutter, being hardly detectable. Their positions are indicated by white boxes and the corresponding SCNR values are reported.

Figure 8.12(b) shows the resulting range-Doppler map at the output of the space-time adaptive filter [numerator of (8.9)]. The ABPD-STAP scheme is applied, with $N = 4$ channels and $L = 3$ Doppler bins. In this case, the amount of training data is limited to $K = 3NL = 36$, in order to account for the heterogeneity of the real clutter scenario. In fact, this can limit the number of available homogeneous secondary data, thus reducing the effectiveness of STAP.

The steering vector of the adaptive filter is selected towards the DOA of the targets T1 (i.e., $\varphi_0 = 37°$) and calibrated based on the imbalance estimated from clutter at the corresponding Doppler bin. With such steering, also the real target Delphin is expected to be included in the BW. For this reason, the SCNR values after STAP filtering are reported in Figure 8.12 only for targets T1 and Delphin. The other targets do not reach their maximum SCNR, since not included in the main beam, however they are still visible in the final map. The output SCNR values achievable when the proper steering vector is applied for each target are reported in Table 8.2.

The above result clearly demonstrates the effective clutter suppression capability of the proposed STAP approach, which proves to be robust against the presence of angle dependent calibration errors affecting the receiving channels.

Table 8.2 Target parameters

	T1	T2	T3	T4	Delphin
R_b	4,700 m	3,700 m	6,850 m	2,500 m	6,596 m
v_b	−6 m/s	6 m/s	11 m/s	4 m/s	−24.5 m/s
ϕ_t	37°	157°	109°	70°	∼37°
$SCNR_i$	−1 dB	−6 dB	−2 dB	−12 dB	2 dB
$SCNR_o$	17 dB	16 dB	12 dB	17 dB	15 dB

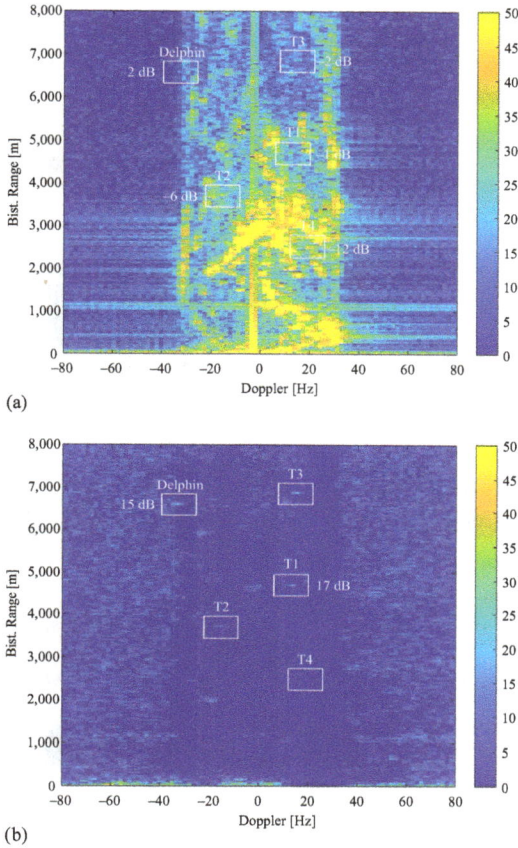

Figure 8.12 *Range-Doppler maps obtained from the experimental data: (a) single channel; (b) after ABPD-STAP with calibrated steering vector towards DOA of targets T1 and Delphin. Target positions are indicated by white boxes. SCNR values after STAP are reported only for targets included in the main beam.*

To analyse more in detail the role of the solutions proposed in Section 8.2.3 for target detection and to compare their performance, Figure 8.13(a) and (b) reports the results obtained with the NC-GLRT detector in (8.6) and with the Cal-GLRT detector in (8.9), respectively.

Specifically, for each solution, we report the test statistics over the bistatic range-Doppler map before the application of a proper threshold, selected according to a desired value of nominal PFA. For a fair comparison, the test statistic is mapped into the PFA setting that would allow to exceed the corresponding threshold. In other words, each pixel in the map has been scaled so that it represents the minimum value

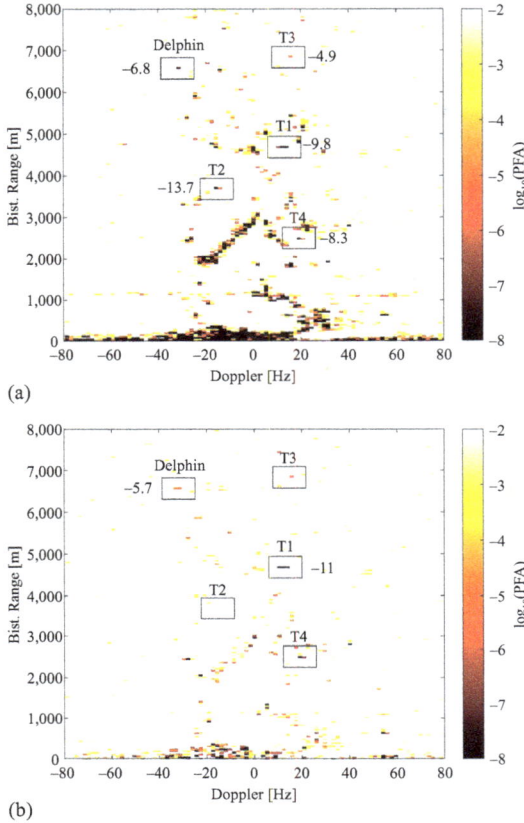

Figure 8.13 *Minimum nominal PFA to be set to detect each bin, using (a) the NC-GLRT scheme; (b) the Cal-GLRT scheme with steering towards DOA of targets T1 and Delphin. Values are expressed as $\log_{10}$ (PFA). PFA values for the Cal-GLRT case are reported only for targets included in the main beam. Target positions are indicated by black boxes.*

of nominal PFA to be set for that pixel to yield a detection. Notice that results are reported as $\log_{10}$(PFA).

Target positions are indicated on maps by black boxes. Notice that, for the Cal-GLRT scheme in Figure 8.13(b), the calibrated steering vector is again steered towards the DOA of targets T1 and Delphin. For this reason, the PFA values are compared in the figure only for the two targets belonging to the steering direction.

For a complete comparison, in Table 8.3, we report the minimum nominal PFA values to detect each target with the different detection schemes. Specifically, for the Cal-GLRT scheme we consider the case where the selected direction for steering and

Table 8.3   Minimum nominal PFA for target detection [log₁₀(PFA)]

	T1	T2	T3	T4	Delphin
Mismatched GLRT	−1.4	−1.9	−3.4	−3.2	−1.8
NC-GLRT	−9.8	−13.7	−4.9	−8.3	−6.8
Cal-GLRT	−11.1	−15.3	−6.2	−9.7	−5.7
Cal-GLRT @+5°	−8.9	−14.7	−5.6	−9.4	−5.3
Cal-GLRT @+10°	−6.3	−13.5	−4.9	−5.2	−4.9

calibration is coincident with each target DOA, and the cases where it deviates by 5° or 10° with respect to target angular positions, with targets still included in the array nominal BW ($\sim 23°$). This in order to simulate realistic detection conditions of a target search stage. Notice that such different target steering conditions are not considered for the NC-GLRT detector, which does not include a spatial steering vector component, due to the non-coherent integration in the spatial domain. Finally, for comparison, the results obtained in the mismatched GLRT case are also reported.

From the above results, the following considerations are in order:

- The NC-GLRT scheme allows us to detect targets in all directions simultaneously (depending on the antenna element pattern), while the fully coherent scheme normally requires proper beam steering.
- The non-coherent integration yields a higher number of false alarms for the same PFA compared to the Cal-GLRT scheme, even at low PFA values, due to the enhancement of persistent clutter structures. However, notice that the NC-GLRT tends to integrate the false alarms accounting for all different angular directions.
- It is also worth noting that the presence of additional non-cooperative moving targets in the scene during the acquisition cannot be excluded.
- The Cal-GLRT scheme yields better results with respect to the NC-GLRT, when steering is aligned with target DOA, allowing target detection until lower values of PFA. The only exception is the Delphin target, where a slightly better result is achieved with the non-coherent approach. This is mostly because for simulated targets the imbalances were generated based on clutter data, therefore a perfect calibration can be reached with the adopted technique. While the real target, whose DOA may not exactly match the nominal one and which also features an elevation angle, may deviate from the imbalance estimated from clutter.
- Accordingly, when target DOA does not exactly coincide with the steering direction (last two rows of Table 8.3), but still falls within the BW and therefore detection is desirable, the NC-GLRT approach may outperform the Cal-GLRT also for the simulated targets. In this case, in fact, the estimated calibration may differ from that required by the target, due to the angular variation of channel imbalance. Nevertheless, both the considered solutions yield significantly better results compared to the mismatched GLRT case.

- Selecting a PFA of 10^{-4} all the considered targets would be detected by both the proposed detection approaches; conversely, none of them would be detected in the mismatched case.

In the coherent approach, the array nominal BW is in the order of 23°, thus offering poor target localization capability. An appropriate estimation of target DOA is then required.

To estimate the target angular position and verify the role of steering vector calibration in the target localization process, we consider the space–time ML DOA estimation approach proposed in Section 8.2.5. Specifically, it is applied for each target after the same ABPD transformation used for detection, exploiting the $N = 4$ receiving channels, $L = 3$ Doppler bins and $K = 36$ training data. A bank of filters equally spaced by $\delta\varphi = 0.1°$ within the nominal BW is adopted.

The results of DOA estimation for each target are reported in Table 8.4, for the mismatched MLE in (8.10) and the calibrated MLE in (8.11). The errors with respect to the corresponding true DOA values in Table 8.2 are reported in brackets.

It is evident that the steering vector calibration allows to significantly improve the estimation accuracy, preventing the negative impact of channel imbalance on target localization performance. In fact, a correct DOA estimated is achieved for all targets, with an average error below one-thirtieth of the nominal BW. Notice that a slightly less accurate estimation is achieved for T1 and Delphin targets, whose DOA is far from the broadside direction.

The above results clearly demonstrate the effectiveness of the proposed strategies, both in terms of target detection and DOA estimation, in the presence of channel calibration errors. From the above considerations, for a mobile passive radar equipped with a limited number of receiving channels, it seems reasonable to suggest the following operational strategy. A less demanding spatially non-coherent approach like in (8.6), which does not require an online calibration stage, could be adopted for the purpose of target detection. While an additional calibration stage could be applied in a fully coherent space–time scheme for DOA estimation like in (8.11), once a target has been detected.

8.2.7 Summary

In this section, we addressed the problem of clutter rejection and slow-moving target detection and localization from multichannel mobile passive radar in the

Table 8.4 Target DOA estimation results

	T1	T2	T3	T4	Delphin
Mismatched MLE	32.1°	145.5°	106.3°	81.5°	46.9°
	$(-4.9°)$	$(-11.5°)$	$(-2.7°)$	$(+11.5°)$	$(+9.9°)$
Calibrated MLE	35.7°	157.1°	108.6°	70.2°	35.4°
	$(-1.3°)$	$(+0.1°)$	$(-0.4°)$	$(+0.2°)$	$(-1.6°)$

presence of antenna calibration errors. Specifically, we considered the application of a post-Doppler STAP scheme and discussed the impact of channel calibration issues, proposing some practical solutions.

ABPD-STAP was suggested as a suitable approach for mobile passive radar and its effectiveness was tested in terms of clutter cancellation and moving target detection, against both a simulated and real clutter scenario. The scheme is especially tailored to address the case of an angle-dependent imbalance affecting the receiving channels. In this case, the clutter cancellation capability is guaranteed by the adaptive degrees of freedom of the STAP filter. Nevertheless, we showed that the resulting target steering vector mismatch and the estimated clutter filter coefficients may result in gain losses and partial suppression of target signal at the output of the adaptive filter, as well as in its inaccurate angular localization.

To address these points, the proposed STAP scheme includes solutions aimed at recovering the target detection loss and performing an accurate DOA estimation.

First, we considered a spatially non-coherent space–time GLRT (NC-GLRT) scheme, where the steering vector is not specified in the spatial domain, resulting in a non-coherent integration of target echoes across the receiving channels. Our analysis showed that, at the expense of a limited loss in terms of maximum integration gain, this approach is robust against a significant imbalance affecting the received signals, thus preventing large detection losses. By entirely excluding the presence of a calibration stage, it represents a simple but effective solution for moving target detection, especially when few receiving channels are available.

Then, we considered a calibrated space–time GLRT (Cal-GLRT) scheme, where the echoes from the stationary scene are exploited for an estimation of the angle-dependent channel errors and a proper correction of the spatial steering vector mismatch, making use of the clutter spread in Doppler and of the one-to-one relationship between angle of arrival and Doppler frequency of stationary scatterers. This approach is specifically devised for the case of passive radar STAP, where it is customary to operate with long coherent integration intervals and wide antenna beams, which allows the proposed calibration approach to operate effectively. Our analysis showed that at the expense of an additional calibration stage, this solution can provide slightly better detection performance and allows to preserve target DOA estimation capability. Alternatively, it can be applied only to the potential targets detected by a first stage based on the NC-GLRT scheme, to enable DOA estimation capability.

Finally, we addressed the problem of target angular localization by DOA estimation, which represents a critical task in mobile passive radar featuring few wide beam antennas. We capitalized on the introduced steering vector calibration and assessed its key role in providing an accurate space–time ML estimation of target DOA, by mitigating the negative impact of the unknown channel errors.

The effectiveness of the proposed solutions in terms of moving target detection and localization has been tested against both simulated and experimental data from a DVB-T-based multichannel mobile passive radar, showing an effective operation.

8.3   Dual cancelled channel STAP for target detection and DOA estimation in passive radar

The use of multiple channels on receive and the adoption of STAP techniques for GMTI in mobile passive radar was shown to overcome the intrinsic limitations of a DPCA approach. A complete processing scheme was proposed in Section 8.2, as particularly suitable for mobile passive radar. By resorting to an adjacent-bin post-Doppler (ABPD) STAP approach, it easily fits into the typical passive radar processing chain and takes advantage of the long integration times to reduce the size of the adaptive cancellation problem and intrinsically compensate for potential angle-dependent channel errors. Target detection is then based on a space–time detection scheme where a calibrated spatial steering vector can be exploited to prevent accidental target rejection arising from severe channel imbalance. Finally, a space–time MLE is employed to also provide the accurate target angular localization.

The passive radar STAP scheme was shown to be effective for GMTI. Large flexibility is gained thanks to the adaptive use of multiple spatial degrees of freedom (DOF). In fact, clutter cancellation, target detection and DOA estimation are performed according to a full-array strategy, i.e., by jointly exploiting all the N available channels on receive. This is paid in terms of an increased complexity of the final system, since it requires: (i) the estimation and inversion of an $NL \times NL$ space–time disturbance covariance matrix, L being the number of temporal DOF; (ii) the availability of a number of training data greater than $2NL$ in order to limit the adaptivity loss [38], which might be difficult to be guaranteed in the considered bistatic passive radar scenario; and (iii) the implementation of computationally expensive algorithms for the maximization of the DOA MLE-likelihood function, being the computational burden dependent on the desired estimation accuracy.

In this section, we address the above limitations and propose an alternative approach based on a dual cancelled channel STAP scheme. This approach takes inspiration from a family of sub-optimal STAP schemes reported in the active radar literature [56–59], where an adaptive transformation is first applied on received data to reduce the number of spatial DOF, which are then directly exploited for target detection and DOA estimation. For instance, the solution in [57] exploits the generalised monopulse estimator (GME), which relies on the sum and difference channels obtained after adaptive clutter cancellation. However, to also limit the computational cost of the preliminary adaptive transformation, we resort to the AB-STAP technique presented in [58].

This technique is first integrated into the ABPD-STAP scheme proposed in Section 8.2. The output of the first processing stages, separately applied at each receiving channel, is grouped into two spatially displaced antenna sub-apertures and the same STAP technique is applied to each sub-aperture, thus obtaining two clutter cancelled channels. These channels are then properly recombined for the purpose of target detection, while target DOA is estimated from the two outputs via a simple closed-form expression by exploiting their displaced phase centres. The AB-STAP approach allows us to further reduce the computational complexity of the system, by reducing the number of the adaptive DOF in the space–time clutter filtering (namely the size of the covariance matrices to be estimated and inverted). Despite the lower computational cost, the proposed scheme does not suffer from significant performance losses with respect to the conventional full-array solution. Moreover, due to the smaller number of adaptive DOF, it is more robust against adaptivity losses, operating effectively even in the case of a limited sample support. This plays a key role in a real passive radar scenario, where the effectiveness of STAP is subject to the potential non-homogeneity of clutter, which may limit the amount of relevant training data. Lastly, the estimation of target DOA does not involve a functional maximisation, thanks to the closed form expression of the estimator, thus requiring a lower computational cost, which in turn is independent of the desired estimation accuracy.

The moving target detection and localisation performance of the proposed strategy are tested and compared with those of the equivalent full array solution (assumed as a benchmark), against the experimental data collected by a DVB-T-based multi-channel passive radar mounted on a ground moving platform.

8.3.1 Dual cancelled channel STAP scheme

Let us consider again the system configuration and signal model adopted in Section 8.2.1. The formerly proposed scheme for passive radar STAP is here referred to as full array STAP, since jointly exploiting all the available spatial DOF.

Such a scheme involves a non-negligible level of complexity since it requires the estimation and inversion of an $(NL \times NL)$ disturbance covariance matrix, for potentially each range-Doppler bin. Moreover, target DOA estimation requires finding the maximum of the non-linear function in (8.10), either by using fast-converging techniques or by evaluating it for a set of values and selecting the one providing the maximum. This might be expensive in terms of computational load, the latter being dependent to the required estimation accuracy. Such cost may not be in line with the usual low complexity of passive radar, especially in the case of mobile systems offering onboard processing.

In addition, the availability of a sufficient number of uniform training data to limit the adaptivity loss may be difficult to be guaranteed in a passive bistatic radar scenario. In fact, the typical ground-based transmitters of opportunity are likely to produce a non-uniform illumination of the ground due to the masking effects of terrain. Moreover, due to its covert operation and potentially limited coverage, passive radar might be operated also at short range, where non-uniform clutter distributions occur, especially in airborne acquisition geometries.

To mitigate the above limitations, we propose an alternative approach for an operational mobile passive radar.

The dual cancelled channel STAP is based on the idea of adaptively forming two clutter cancelled channels which could be directly exploited for both target detection and DOA estimation. Specifically, we consider the AB-STAP technique, originally presented in [58]. This approach is particularly suitable for the passive radar case, thanks to its simple architecture, reduced computational load and robustness against highly non-homogeneous clutter scenarios.

The adopted processing scheme is sketched in Figure 8.14. This is obtained by modifying the scheme proposed in Section 8.2, so as to integrate the considered dual cancelled channel approach. The antenna array is split into two spatially displaced (possibly overlapped) sub-apertures. The same ABPD-STAP technique is applied to each sub-aperture, thus obtaining two clutter cancelled channels, namely channels A and B. These channels are then coherently recombined for the purpose of target detection. For range-Doppler cells where detection is declared, target DOA estimation can be performed from the A and B channel outputs by exploiting their different phase centres.

The processing steps are described in the following:

(i) *Creation of two spatially displaced channels*: the $(NL \times 1)$ data vector x is split into two $(N_0 L \times 1)$ subvectors x_A and x_B, corresponding to channels A and B, respectively. x_A contains the data collected from the first N_0 antennas $(N/2 \leq N_0 < N)$, while x_B contains those collected from the last N_0 antennas.

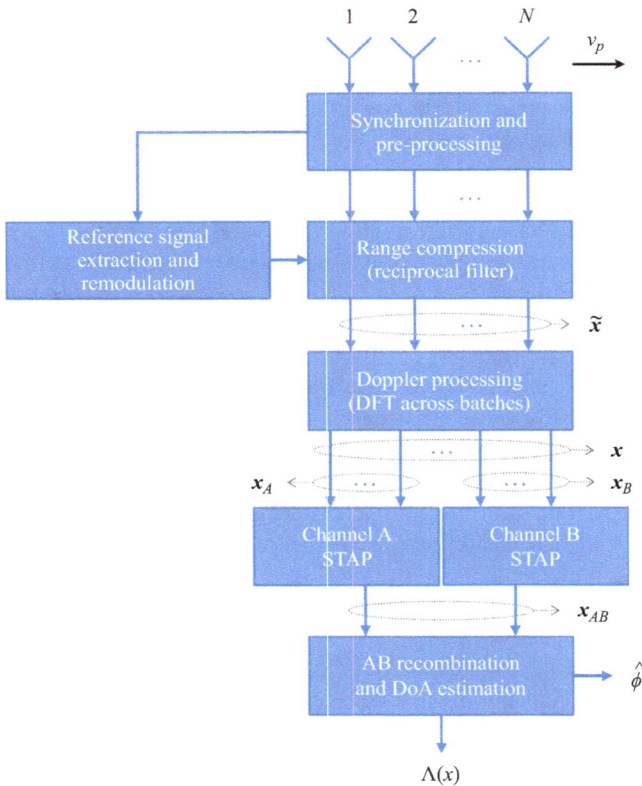

Figure 8.14   Processing scheme for post-Doppler AB-STAP in passive radar

(ii) *Cancellation of clutter on each channel*: the subvectors x_A and x_B are filtered with a sample matrix inverse scheme, giving outputs y_A and y_B

$$y_{A(B)} = w_{A(B)}^H x_{A(B)} = s_{A(B)}^H (\phi_L) \hat{\mathbf{Q}}_{A(B)}^{-1} x_{A(B)} \tag{8.13}$$

where $s_{A(B)} (\phi_L)$ and $\hat{\mathbf{Q}}_{A(B)}$ are the steering vector and the estimated covariance matrix of channel A (B), respectively. Notice that, while ϕ represents the generic unknown target DOA, we denote by ϕ_L the array look direction.

In a ULA, channels A and B exhibit disturbance with the same structure, since

$$s_B (\phi_L) = s_A (\phi_L) \exp \{j2\pi (N - N_0) d/\lambda \cos \phi_L\} \tag{8.14}$$

As a result, a better estimate can be obtained averaging the estimates made on the two channels, that is $\hat{\mathbf{Q}}_A = \hat{\mathbf{Q}}_B = \frac{1}{2} \left(X_{kA} X_{kA}^H + X_{kB} X_{kB}^H \right)$, being X_{kA} and X_{kB} the $(N_0 L \times K)$ training data of channels A and B, respectively. Although not totally independent (if $N_0 > N/2$), averaging the two estimates yields a more stable estimate of the true covariance matrix.

(iii) *Recombination of channels for target detection*: using only one of the two channels to detect targets may result in significant performance degradation in terms of cancellation capability, being the equivalent antenna length and the employed spatial DOF reduced. To recover this degradation and avoid detection losses, an optimal coherent recombination of the two channels is performed.

Arranging the two outputs in the (2×1) vector $x_{AB} = [y_A y_B]^T$ and applying the AMF detection scheme, we obtain

$$\frac{\left| s_{AB}^H \hat{\mathbf{Q}}_{AB}^{-1} x_{AB} \right|^2}{s_{AB}^H \hat{\mathbf{Q}}_{AB}^{-1} s_{AB}} \gtreqless \eta_{AB} \tag{8.15}$$

where η_{AB} is the detection threshold and the steering vector s_{AB} is given by

$$s_{AB} (\phi_L) = \begin{bmatrix} w_A^H s_A (\phi_L) \\ w_B^H s_B (\phi_L) \end{bmatrix} = w_A^H s_A (\phi_L) \begin{bmatrix} 1 \\ 1 \end{bmatrix}. \tag{8.16}$$

$\hat{\mathbf{Q}}_{AB}$ is the (2×2) estimated covariance matrix at the output of channels A and B. Notice that $\mathbf{Q}_{AB}$ is a Toeplitz Hermitian matrix having elements α (average disturbance power at the output of each channel) and ρ (cross-correlation between the outputs). This can also be used to obtain a more stable estimate from the data.

As shown in [58,59], the optimal recombination of the two channels allows to mostly recover the detection losses with respect to the full array case. Moreover, due to the smaller number of adaptive DOF used by each step of AB-STAP, smaller adaptivity losses are expected compared to full array STAP. This plays a fundamental role in a real scenario, where the potential non-homogeneity of clutter may limit the amount of relevant training data usefully exploitable.

(iv) *Estimation of target DOA from the two cancelled channels*: once a target is detected, its DOA can be estimated by applying the MLE to the two cancelled channels A and B. This results in

$$\hat{\phi}_{t,AB} = \arg \max_{\phi} \left\{ \frac{\left| s_{AB}^H (\phi) \, \hat{Q}_{AB}^{-1} x_{AB} \right|^2}{s_{AB}^H (\phi) \, \hat{Q}_{AB}^{-1} s_{AB} (\phi)} \right\} \tag{8.17}$$

The maximisation is performed only with respect to the last stage of the processing chain, which only involves 2-dimensional quantities. This allows to find a closed form expression for the ML estimate of target DOA.

By defining the quantities $z = [z_1 z_2]^T = Q_{AB}^{-1} x_{AB}$, $v = z_1 z_2^H / |z|^2$ and $u = \rho/\alpha$ the ML estimate of target DOA is obtained as (see [58,59] for more details)

$$\hat{\psi} = \arcsin \left(\frac{2 \mathrm{Im} \{ v u^H \}}{|2v + u|} \right) - \angle \, (2v + u) \tag{8.18}$$

$$\hat{\phi}_{t,AB} = \arccos \left\{ \frac{\lambda}{2 \pi d \, (N - N_0)} \hat{\psi} + \cos \phi_L \right\} \tag{8.19}$$

where $\mathrm{Im}\{\zeta\}$ and $\angle \, (\zeta)$ are the imaginary part and the phase of complex number ζ, respectively.

As evident, the DOA estimation does not require a functional maximisation and yields a simple formula to be implemented, limiting the computational cost.

It is worth noting that the value of N_0 defines the number of spatial DOF reserved for clutter cancellation on channels A and B, but it also affects their phase centres displacement. Better DOA estimation could be achieved with large phase centres displacement (i.e., small N_0), but this would be paid in lower clutter cancellation capability on both channels A and B. In other words, a trade-off exists between the theoretical sensitivity for DOA estimation and the SCNR measured at the output of the channels. Appropriate choices should be made according to the specific scenario. Notice that the SCNR at the output of the A and B channels, responsible for DOA estimation accuracy, is always lower than the SCNR available for target detection after channel recombination.

8.3.2 *Experimental results*

The effectiveness of the AB-STAP scheme for slow-moving targets detection and localization in mobile passive radar is demonstrated against a set of experimental data, acquired by a DVB-T-based multichannel receiver mounted on a ground moving platform. The experimental setup and the considered data set correspond to those formerly used in Section 8.2, where the full array STAP scheme has been successfully applied. In this section, the full array solution is compared with the AB-STAP approach, to verify the benefits of the latter in a real scenario.

We recall that the system features four receiving channels, arranged to form a side-looking uniform linear array (ULA), as shown in Figure 8.10. The parameters

Table 8.5 *Target parameters*

	T1	T2	T3	T4	Delphin
R_b	4,700 m	3,600 m	6,850 m	2,400 m	6,596 m
v_b	−6 m/s	6 m/s	11 m/s	7 m/s	−24.5 m/s
ϕ_t	37°	157°	109°	70°	∼ 37°
f_D	11.7 Hz	−14.9Hz	15.9 Hz	26.7 Hz	−31.1Hz

of the system and of the exploited DVB-T signal are the same reported in Table 8.1. Refer to Figure 8.11 for a sketch of the bistatic acquisition geometry.

In addition to the cooperative aerial target Delphin, four simulated moving targets have been injected into the acquired data. The parameters of the real and the simulated targets are reported in Table 8.5.

It is also worth noting that the same strategy presented in Section 8.2 for calibration of the spatial steering vector component has been adopted here, to mitigate the impact of the angle-dependent inter-channel imbalance on the target detection and localization performance. Specifically, this is applied to vector s in the full array STAP scheme and to vectors s_A and s_B in the AB-STAP scheme.

The resulting range-Doppler maps are shown in Figure 8.15. Specifically, Figure 8.15(a) represents the map obtained from a single channel, namely before STAP processing, scaled to the estimated noise power level. The considered moving targets, due to their radial velocity components and DOAs, fall within the clutter Doppler bandwidth, being mostly buried into clutter and therefore hardly detectable. The target positions are indicated by white boxes in the map and their corresponding SCNR values are reported.

The observed clutter scenario is characterized by a large heterogeneity, associated to the presence of densely vegetated and rural areas. This aspect is exacerbated by the bistatic geometry of the passive radar exploiting a ground-based transmitter, prone to non-uniform illumination and shadowing phenomena, due to the orography of terrain. In this context, the AB-STAP approach represents a suitable solution, since operating effectively even with a limited sample support.

Figure 8.15(b) shows the range-Doppler map obtained when applying the full array STAP scheme, jointly using all the $N = 4$ channels and $L = 3$ Doppler bins. The amount of exploited training data is set to $K = 16$. Notice that this number is less than twice the number of adaptive DOF for the full array solution ($2NL = 24$). As expected, the large adaptivity loss results in a high residual disturbance, which lower the achievable SCNR and may hinder the detection of targets.

For comparison, Figure 8.15(c) shows the corresponding range-Doppler map at the output of the adaptive filter, when applying the AB-STAP scheme employing the same amount of training data. In this case, the array has been split into two (non-overlapped) sub-arrays of $N_0 = 2$ elements, forming the A and B channels.

The spatial steering vector is oriented towards the DOA of targets T1 and Delphin ($\phi_L = 37°$), which appear clearly visible in the final maps. For this reason, the SCNR

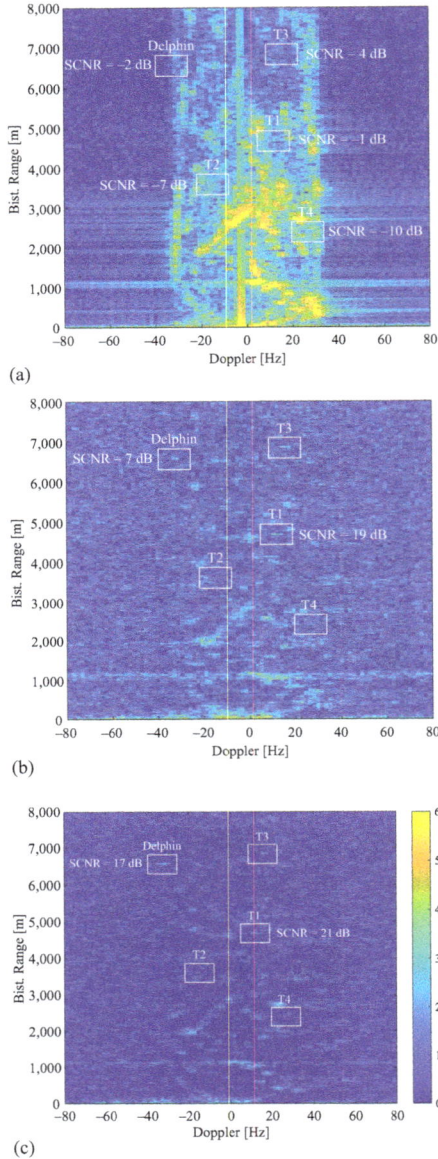

Figure 8.15 *Range-Doppler maps obtained from the experimental data: (a) single channel; (b) after full array STAP; (c) after AB-STAP. The number of training data is set to K=16. Steering is towards DOA of targets T1 and Delphin. SCNR after STAP is reported only for targets included in the main beam. Target positions are indicated by white boxes.*

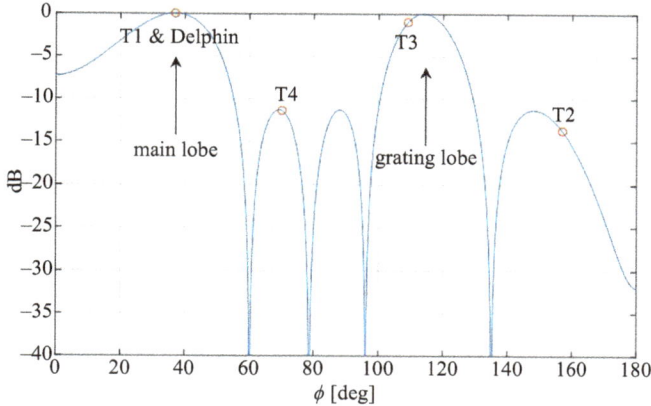

Figure 8.16   Theoretical array pattern with steering at φ=37° and target positions, indicated by red circles

values after STAP are reported in both figures only for these targets. The other targets, less visible, arrive from different DOAs and are not included in the main beam. The theoretical array pattern when steered at 37° is shown in Figure 8.16, with the corresponding target angular positions. Notice that, due to element spacing larger than $\lambda/2$, a grating lobe appears for steering directions far from broadside. In the present case, target T3 is also visible in the final map since located in the direction of the grating lobe.

The above result clearly demonstrates the effective clutter suppression and moving target detection capability of the AB-STAP approach, also in the presence of a limited sample support for the adaptive filter estimation.

For a more detailed analysis and comparison of the moving target detection performance, Figure 8.17 reports the results obtained with the full array STAP and with the AB-STAP detector in (8.15). The two detectors are compared under different conditions of availability of the secondary data: for $K = 24$ in Figure 8.17(a) and (b), and for $K = 16$ in Figure 8.17(c) and (d).

For each solution, we report the test statistics over the bistatic range-Doppler map before applying a proper threshold, according to a desired value of nominal PFA. For a fair comparison, the test statistic is mapped into the PFA setting that would allow to exceed the corresponding threshold. In other words, each pixel in the map has been scaled so that it represents the minimum value of nominal PFA to be set for that pixel to yield a detection. Notice that the results are reported as $\log_{10}(\text{PFA})$. Target positions are indicated by black boxes. Notice that the spatial steering vector is again steered towards the DOA of targets T1 and Delphin.

For a complete comparison, in Table 8.6 we report the minimum nominal PFA values which allow us to detect each target with the different detection schemes, when the selected steering direction coincides with each target DOA.

From the above results, the following considerations are in order:

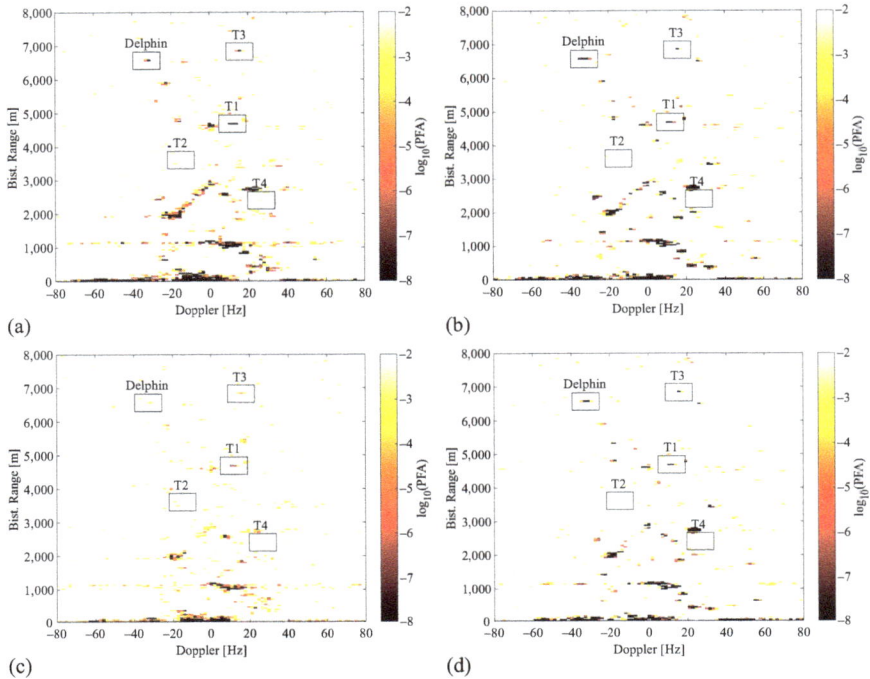

Figure 8.17   *Minimum nominal PFA to detect each bin: (a) full array STAP with K = 24; (b) AB-STAP with K = 24; (c) full array STAP with K = 16; (d) AB-STAP with K = 16. Steering is towards DOA of targets T1 and Delphin. Values are expressed as $\log_{10}$ (PFA). Target positions are indicated by black boxes.*

Table 8.6   Minimum nominal PFA for target detection [$\log_{10}(PFA)$]

	Full-array STAP					AB-STAP				
	T1	T2	T3	T4	Delphin	T1	T2	T3	T4	Delphin
K = 24	−11.6	−6.7	−10.5	−4.8	−7.5	−9.3	−7.0	−7.8	−5.1	−9.6
K = 16	−6.6	−2.2	−4.6	−3.3	−3.1	−8.6	−8.0	−6.5	−6.6	−9.3

- Comparing Figure 8.17(a) and (b), the full array STAP scheme seems yielding a higher number of false alarms compared to the AB-STAP scheme, for given PFA. False alarms are mostly associated to persistent clutter structures, especially in those areas where clutter power shows abrupt variations and filter adaptivity is more likely to fail.

- It is worth mentioning that the presence of additional non-cooperative moving targets in the observed scene during the acquisition cannot be excluded.
- When the training data are reduced to $K = 16$, Figure 8.17(c) and (d), the more localized adaptation capability allows the AB-STAP scheme to better handle the clutter discrete and reduce the number of false alarms, without compromising target detection. Conversely, in the full array STAP scheme this is paid in terms of higher adaptivity loss, which raises the detection threshold to prevent a general increase of the false alarms, thus compromising the detection capability.
- Looking at Table 8.6, when enough training data are available, both the detection schemes yield remarkable results, allowing target detection until low values of PFA. In the case of limited sample support, instead, the detection performance of the full array STAP drastically decreases, while the AB-STAP approach is able to mostly preserve or even improve its outcomes (see e.g. T2 and T4). As a result, the AB-STAP largely outperform the full array STAP for all targets in this case.
- By selecting a PFA of 10^{-4}, with $K = 24$ training data all the considered targets would be detected by both the detection schemes. For $K = 16$, the AB-STAP would still detect all the targets, while the full array scheme would miss the detection of three out of five targets.

After target detection, an appropriate estimation of target DOA is worthwhile for target localization purpose. In fact, the array nominal BW is in the order of $23°$ (when steered at broadside), thus offering poor target localization capability. This is achieved by resorting to a space–time ML DOA estimation strategy.

Specifically, in the full array STAP case, the target angular position is estimated by finding the maximum of (8.10) over a bank of filters equally spaced by $\delta\phi = 0.1°$ within the array nominal BW, corresponding to a step smaller than BW/100. In the AB-STAP approach, instead, the closed form expression in (8.18) and (8.19) is exploited. In both cases, the estimation is achieved after the ABPD transformation, using $N = 4$, $N_0 = 2$ and $L = 3$. For our analysis, we neglected the angular ambiguity resulting from the antenna spacing larger than $\lambda/2$ and, in the estimation process, we have used the known target position to identify the non-ambiguous angular sector. Note that this strategy does not affect our results as we are mostly interested in small estimation errors around the true target DOA.

The results of DOA estimation for each target are reported in Table 8.7, for an amount of training data equal to $K = 24$ and $K = 16$. The error with respect to the true DOA values in Table 8.5 is reported in brackets.

Analysing the results, we can notice that, for a large sample support, both the AB-STAP and the full array STAP provide an accurate DOA estimation for all targets, with an average error below one-twentieth of the nominal BW. When the sample support is limited to $K = 16$ instead, the estimation accuracy of the full array STAP is considerably reduced. On the other hand, the AB-STAP approach is able to mostly preserve its good localization performance, while requiring a lower computational complexity.

Finally, Figure 8.18 shows the DOA estimation results obtained for the real target Delphin over consecutive scans. Specifically, 24 CPIs were considered, each of length

Table 8.7 *Target DOA estimation results*

	Full-array STAP					AB-STAP				
	T1	T2	T3	T4	Delphin	T1	T2	T3	T4	Delphin
$K = 24$	36.3°	158.1°	108.1°	69.2°	34.3°	37.4°	157.8°	109.7°	69.1°	35.3°
	(−0.7°)	(+1.1°)	(−0.9°)	(−0.8°)	(−2.7°)	(+0.4°)	(+0.8°)	(+0.7°)	(−0.9°)	(−1.7°)
$K = 16$	32.8°	161.1°	110.5°	68.6°	48.6°	38.1°	156.2°	110.1°	68.5°	37.2°
	(−4.2°)	(+4.1°)	(+1.5°)	(−1.4°)	(+11.6°)	(+1.1°)	(−0.8°)	(+1.1°)	(−1.5°)	(+0.2°)

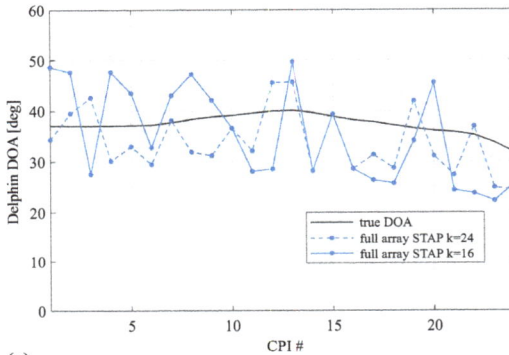

(a)

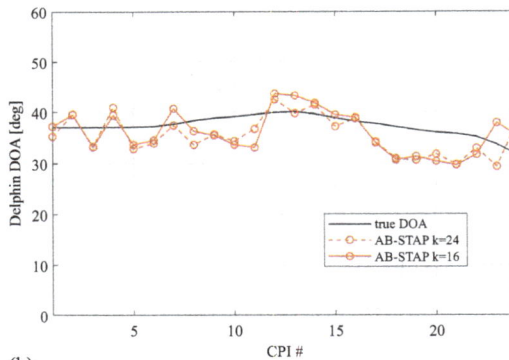

(b)

Figure 8.18 *DOA estimation results for real target Delphin over consecutive CPI: (a) using full array STAP scheme; (b) using AB-STAP scheme*

512 OFDM symbols and overlapped by 256 symbols, for an overall observation time of approximately 7 s. The results clearly show the advantage of the AB-STAP scheme compared to the full array scheme, in terms of accuracy and stability of the estimation, especially in the case of a limited sample support.

The results reported in this section clearly demonstrate the effectiveness of the AB-STAP strategy against a real clutter scenario and its advantage over the full array scheme. From the above considerations, the AB-STAP represents a suitable and convenient solution for target detection and DOA estimation in mobile passive radar equipped with multiple receiving channels.

8.3.3   Summary

In this section, we proposed a dual cancelled channel STAP scheme for clutter rejection and slowly moving target detection and localization in multi-channel mobile passive radar. The proposed scheme is aimed at reducing the computational complexity, as well as the number of required training data, compared to a conventional full-array solution.

Specifically, the AB-STAP technique was considered, combined with an ABPD strategy, and proved to be a suitable solution for multi-channel mobile passive radar. The reduction of computational complexity is obtained both by reducing the number of adaptive DOF in the space–time processing steps and by providing a simple closed-form expression for target DOA estimation.

Despite the lower computational cost, the proposed scheme was shown to yield comparable performance with respect to the equivalent full array solution, both in terms of target detection capability and DOA estimation accuracy. Moreover, it is more robust against adaptivity losses, operating effectively even in the presence of a limited sample support. This plays a key role in a real passive bistatic scenario, where the effectiveness of STAP is subject to the potential non-homogeneity of clutter, which may limit the amount of relevant training data usefully exploitable.

The effectiveness of the proposed scheme, both in terms of moving target detection and localization, has been tested against simulated and experimental data from a DVB-T based multi-channel mobile passive radar.

8.4   Conclusion and future work

The goal of this chapter is to contribute to research on mobile passive radar by addressing the main issues connected to the use of STAP methodologies for GMTI applications. The research has led to the development of specific signal processing techniques and operational strategies to tackle the limiting factors deriving from the passive bistatic framework to the capability of clutter cancellation and moving target detection and localization. The main novelties brought about by this chapter are summarised below, followed by some recommendations for future research.

8.4.1   Results and novelties

- First, the strict connection between mobile passive radar framework and channel imbalance problem has been highlighted. The low directivity of typical receiving antennas, paired with a bistatic broadcast illumination, results in the simultaneous reception of echoes from a very wide angular sector. Such echoes are likely to suffer from an angle-dependent imbalance across the receiving channels, made more severe by the use of low-cost components. This may affect the target detection performance and hinder their accurate angular localization within the broad antenna beams.

- We proposed an ABPD-STAP approach, which plays a key role in preserving the clutter cancellation capability also in the presence of an angle-dependent imbalance affecting the receiving channels. This approach proved particularly suitable for mobile passive radar. It takes advantage of the typical long integration times and the resulting fine Doppler resolution to: (i) significantly reduce the size of the adaptive problem, by exploiting the decoupling of the clutter Doppler components; (ii) allow the adaptive filter to intrinsically compensate for the angle-dependent channel errors, by operating on a clutter subspace accounting for a limited angular sector.

- Two novel detection schemes were developed, aimed at mitigating the effects of the channel imbalance on target signal:
  (i) A partially non-coherent space–time GLRT (NC-GLRT) scheme, where the steering vector is not specified in the spatial domain, resulting in the non-coherent integration of target echoes across the receiving channels. At the expense of a limited loss in terms of maximum integration gain, this approach proved to prevent large detection losses. By entirely excluding a calibration stage, it represents a simple but effective solution for moving target detection, especially when few receiving channels are available
  (ii) A calibrated space–time GLRT (Cal-GLRT) scheme, where the clutter echoes are exploited for estimation and correction of the spatial steering vector mismatch, making use of the fine Doppler resolution and of the one-to-one relationship between angle of arrival and Doppler frequency of stationary scatterers. At the expense of an additional calibration stage, it provides slightly better detection performance and preserves target DOA estimation capability. Alternatively, it can be applied only to the potential targets detected by a first stage based on the NC-GLRT scheme, for DOA estimation purpose.

- The problem of target angular localization in mobile passive radar has been addressed. The limited number and the low directivity of antennas make it a critical task and require an accurate estimation of target DOA. The multiple spatial degrees of freedom available must be properly exploited both for space–time clutter filtering and for target DOA estimation. To this purpose, the proposed STAP scheme was adopted, capitalizing on the introduced steering vector calibration to provide accurate ML DoA estimation, by mitigating the negative impact of the unknown channel errors.

- Finally, attention has been focused on further reducing the computational complexity of passive radar STAP in multichannel configurations, as well as on increasing its robustness against adaptivity losses. In fact, the heterogeneity of a real clutter scenario, exacerbated by the bistatic acquisition geometry of passive radar exploiting ground-based illuminators, may limit the availability of homogeneous training data, decreasing the effectiveness of STAP.

- We developed a dual cancelled channel STAP scheme, aimed at reducing the system computational complexity, as well as the number of required training data, compared to a conventional full-array solution. Specifically, the AB-STAP technique was considered, combined with the adjacent-bin post-Doppler strategy. The reduction of computational cost is obtained both by reducing the number of adaptive DOF in the space–time processing steps and by providing a simple closed-form expression for target DOA estimation. The proposed scheme showed to yield comparable performance with respect to the equivalent full array case. Moreover, it proved more robust against adaptivity losses, operating effectively even in the presence of a limited sample support.

• After the extensive performance assessment against simulated data, the effectiveness of the developed strategies in terms of moving target detection and localization has been demonstrated by experimental validations, against the data acquired by a DVB-T based multichannel mobile passive radar.

8.4.2 Future outlook

During this work, the following areas of further research were identified.

• The reliability and performance of the system could be improved by resorting to the exploitation of information diversity. For instance, the multi-frequency operation, i.e., the joint use of signals received at different carrier frequencies, typically available from broadcast transmitters, is expected to provide interesting and unexplored perspectives for mobile passive radar. The impact of exploiting such information diversity and its integration with STAP could be investigated and specific optimization criteria for the design of the receiving system could be derived.

• In addition to the multi-frequency and spatial diversity, also the information diversity conveyed in the temporal dimension by using the reciprocal range compression filter with different signal fragmentation strategies could be explored. The reciprocal filtering would allow the normalization of the signal ambiguity function, removing the temporal variability associated to waveform information content, without being bounded to the OFDM signal structure. This would further increase the flexibility of the system with respect to the characteristics of the waveform of opportunity.

• The possibility to exploit polarimetric diversity could also be considered. As known, a proper combination of signals received at different polarisations can enhance clutter cancellation and target detection capability of passive radar. In the case of mobile passive radar, multi-dimensional detection schemes exploiting space–time-polarimetric observations could be explored, adaptively exploiting the polarimetric differences between the target and the competing disturbance to improve the target discrimination capability.

• The effectiveness of the strategies proposed in this chapter, and of those foreseen for the future, could be extensively tested against larger experimental datasets with different system configurations and other categories of carrying platforms (e.g., airborne platforms). This will give the opportunity to confirm the above results and might put in evidence new aspects to be investigated. Finally, other types of illuminators of opportunity could be considered. Specifically, DVB-S transmissions offer very wide coverage and convenient geometry, compared to terrestrial illuminators.

References

[1]   P. Lombardo and F. Colone, "Advanced processing methods for passive bistatic radar systems", in W. L. Melvin and J. A. Scheer (Eds.), *Principles of Modern Radar: Advanced Radar Techniques*, SciTech, 2012, pp. 739–821.

[2]   J. Palmer, D. Cristallini, and H. Kuschel, "Opportunities and current drivers for passive radar research", in *Proceedings of the IEEE Radar Conference*, Johannesburg, South Africa, 2015, pp. 145–150.

[3]   R. Klemm (Eds.), *Novel Radar Techniques and Applications, Part III: Real Aperture Array Radar Imaging Radar and Passive and Multistatic Radar*, IET Publisher, 2017.

[4]   H. Griffiths and C. J. Baker, *An Introduction to Passive Radar*, Norwood, MA: Artech House, 2017.

[5]   M. Malanowski, *Signal Processing for Passive Bistatic Radar*, Norwood, MA: Artech House, 2019.

[6]   L. M. H. Ulander, P. Frölind, A. Gustavsson, R. Ragnarsson, and G. Stenström, "VHF/UHF bistatic and passive SAR ground imaging," in *2015 IEEE Radar Conference (RadarCon)*, Arlington, VA, 2015, pp. 0669–0673.

[7]   D. Gromek, K. Kulpa, and P. Samczyński, "Experimental results of passive SAR imaging using DVB-T illuminators of opportunity," in *IEEE Geoscience and Remote Sensing Letters*, vol. 13, no. 8, pp. 1124–1128, 2016.

[8]   K. Kulpa, M. Malanowski, and P. Samczyński, "Passive radar: from target detection to imaging," in *2019 IEEE Radar Conference (RadarConf)*, Boston, MA, 2019, pp. 1–286.

[9]   Y. Fang, G. Atkinson, A. Sayin, J, *et al.*, "Improved passive SAR imaging with DVB-T transmissions," in *IEEE Transactions on Geoscience and Remote Sensing*, vol. 58, no. 7, pp. 5066–5076, 2020.

[10]   X. Neyt, J. Raout, M. Kubica, *et al.*, "Feasibility of STAP for passive GSM-based radar," in *2006 IEEE Conference on Radar*, Verona, NY, 2006, pp. 1–6.

[11]   J. Raout, X. Neyt, and P. Rischette, "Bistatic STAP using DVB-T illuminators of opportunity," in *2007 IET International Conference on Radar Systems*, Edinburgh, UK, 2007, pp. 1–5.

[12]   J. Brown, K. Woodbridge, H. Griffiths, A. Stove, and S. Watts, "Passive bistatic radar experiments from an airborne platform," *IEEE Aerospace and Electronic Systems Magazine*, pp. 50–55, November 2012.

[13]   B. Dawidowicz, P. Samczyński, M. Malanowski, J. Misiurewicz, and K. S. Kulpa, "Detection of moving targets with multichannel airborne passive radar," *IEEE Aerospace and Electronic Systems Magazine*, vol. 27, pp. 42–49, 2012.

[14]   B. Dawidowicz, K. Kulpa, M. Malanowski, J. Misiurewicz, P. Samczyński, and M. Smolarczyk, "DPCA detection of moving targets in airborne passive

radar," *IEEE Transactions on Aerospace and Electronic Systems*, vol. 48, pp. 1347–1357, 2012.

[15] P. Wojaczek, F. Colone, D. Cristallini, and P. Lombardo, "Reciprocal-Filter-based STAP for passive radar on moving platforms," *IEEE Transactions on Aerospace and Electronic Systems*, vol. 55, no. 2, pp. 967–988, 2019.

[16] G. P. Blasone, F. Colone, P. Lombardo, P. Wojaczek, and D. Cristallini, "A two-stage approach for direct signal and clutter cancellation in passive radar on moving platforms," in *2019 IEEE Radar Conference (RadarConf)*, Boston, MA, 2019, pp. 1–6.

[17] P. Wojaczek, D. Cristallini, and F. Colone, "Minimum variance power spectrum based calibration for improved clutter suppression in PCL on moving platforms," in *2019 IEEE Radar Conference (RadarConf)*, Boston, MA, 2019, pp. 1–6.

[18] P. Wojaczek, D. Cristallini, D. W. O'Hagan, F. Colone, G. P. Blasone, and P. Lombardo, "A three-stage inter-channel calibration approach for Passive Radar on moving platforms exploiting the minimum variance power spectrum," *Sensors*, vol. 21, no. 1, pp. 69, 2021.

[19] G. P. Blasone, F. Colone, P. Lombardo, P. Wojaczek, and D. Cristallini, "Passive radar DPCA schemes with adaptive channel calibration," *IEEE Transactions on Aerospace and Electronic Systems*, vol. 56, no. 5, pp. 4014–4034, 2020.

[20] G. P. Blasone, F. Colone, and P. Lombardo, "Facing channel calibration issues affecting passive radar DPCA and STAP for GMTI," in *2020 IEEE International Radar Conference (RADAR)*, Washington, DC, 2020, pp. 31–36.

[21] G. P. Blasone, F. Colone, P. Lombardo, P. Wojaczek, and D. Cristallini, "Passive radar STAP detection and DOA estimation under antenna calibration errors," *IEEE Transactions on Aerospace and Electronic Systems*, vol. 57, no. 5, pp. 2725–2742, 2021.

[22] G. P. Blasone, F. Colone, P. Lombardo, P. Wojaczek, and D. Cristallini, "Dual cancelled channel STAP for target detection and DOA estimation in passive radar," *Sensors*, vol. 21, no. 13, pp. 4569, 2021.

[23] P. Wojaczek, A. Summers, and D. Cristallini, "Preliminary experimental results of STAP for passive radar on a moving platform," in *22nd International Microwave and Radar Conference (MIKON)*, Poznan, 2018, pp. 589–592.

[24] K. Kulpa, M. Baczyk, J. Misiurewicz, M. Malanowski, and D. Gromek, "Limits of ground clutter CLEAN based cancelation in mobile PCL radar," in *19th International Radar Symposium (IRS)*, Bonn, 2018, pp. 1–7.

[25] C. Berthillot, A. Santori, O. Rabaste, D. Poullin, and M. Lesturgie, "DVB-T airborne passive radar: clutter block rejection," in *2019 International Radar Conference (RADAR)*, Toulon, France, 2019, pp. 1–5.

[26] Q. Wu, Y. D. Zhang, M. G. Amin, and B. Himed, "Space–time adaptive processing and motion parameter estimation in multistatic passive radar using sparse Bayesian learning," *IEEE Transactions on Geoscience and Remote Sensing*, vol. 54, no. 2, pp. 944–957, 2016.

[27]   J. R. Lievsay and N. A. Goodman, "Modeling three-dimensional passive STAP with heterogeneous clutter and pulse diversity waveform effects," *IEEE Transactions on Aerospace and Electronic Systems*, vol. 54, no. 2, pp. 861–872, 2018.

[28]   J. Palmer, M. Ummenhofer, A. Summers, G. Bournaka, S. Palumbo, and D. Cristallini, "Receiver platform motion compensation in passive radar," *IET Radar, Sonar and Navigation*, pp. 922–931, 2017.

[29]   C. Berthillot, A. Santori, O. Rabaste, D. Poullin, and M. Lesturgie, "BEM reference signal estimation for an airborne passive radar antenna array," *IEEE Transactions on Aerospace and Electronic Systems*, vol. 53, no. 6, pp. 2833–2845, 2017.

[30]   F. Wojaczek, D. Cristallini, J. Schell, and D. O'Hagan, "Polarimetric antenna diversity for improved reference signal estimation for airborne passive radar," in *2020 IEEE Radar Conference (RadarConf)*, Florence, Italy, 2020.

[31]   G. Chabriel and J. Barrère, "Adaptive target detection techniques for OFDM-based passive radar exploiting spatial diversity," *IEEE Transactions on Signal Processing*, vol. 65, no. 22, pp. 5873–5884, 2017.

[32]   V. Navrátil, A. O'Brien, J. L. Garry, and G. E. Smith, "Demonstration of space-time adaptive processing for DSI suppression in a passive radar," in *2017 International Radar Symposium (IRS)*, Prague, 2017, pp. 1–10.

[33]   S. Gelli, A. Bacci, M. Martorella, and F. Berizzi, "Clutter suppression and high-resolution imaging of noncooperative ground targets for bistatic airborne radar," *IEEE Transactions on Aerospace and Electronic Systems*, vol. 54, no. 2, pp. 932–949, 2018.

[34]   M. Glende, "PCL-signal-processing for sidelobe reduction in case of periodical illuminator signals," in *2006 International Radar Symposium*, 2006, pp. 1–4.

[35]   S. Searle, J. Palmer, L. Davis, D. W. O'Hagan, and M. Ummenhofer, "Evaluation of the ambiguity function for passive radar with OFDM transmissions," in *IEEE Radar Conference*, Cincinnati, 2014.

[36]   G. Gassier, G. Chabriel, J. Barrère, F. Briolle, and C. Jauffret, "A unifying approach for disturbance cancellation and target detection in passive radar using OFDM," *IEEE Transactions on Signal Processing*, vol. 64, no. 22, pp. 5959–5971, 2016.

[37]   R. Klemm, *Principles of Space-Time Adaptive Processing*, 3rd edn., IET, 2002.

[38]   J. R. Guerci, *Space-Time Adaptive Processing for Radar*, Artech House Radar Library, 2003.

[39]   W. L. Melvin, "A STAP overview," *IEEE Aerospace and Electronics Systems Magazine*, vol. 19, no. 1, pp. 19–35, 2004.

[40]   J. Ward, "Space-time adaptive processing for airborne radar", Technical Report 1015, Lincoln Laboratory, Massachusetts Institute of Technology, Lexington, MA, 1994.

[41]   F. Le Chevalier, L. Savy, and F. Durniez, "Clutter calibration for space-time airborne MTI radars," in *Proceedings of International Radar Conference*, Beijing, China, 1996, pp. 82–85.

[42]   M. A. Koerber and D. R. Fuhrmann, "Radar antenna calibration using range-Doppler data," in *IEEE Seventh SP Workshop on Statistical Signal and Array Processing*, Quebec, City, Canada, 1994, pp. 441–444.

[43]   F. C. Robey, D. R. Fuhrmann, and M. A. Koerber, "Array calibration and modelling of steering vectors," in *Conference Record of Thirty-Fifth Asilomar Conference on Signals, Systems and Computers (Cat. No.01CH37256)*, vol. 2, Pacific Grove, CA, 2001, pp. 1121–1126

[44]   J. Ender, "The airborne experimental multi-channel SAR system AERII," in *Proceedings of the European Conference on Synthetic Aperture Radar (EUSAR)*, 1996, pp. 49–52.

[45]   C. H. Gierull, "Digital channel balancing of along-track interferometric SAR data," *Defence Research and Development Canada (DRDC)*, Tech. Rep. TM-2003-024, 2003.

[46]   J. Heckenbach, H. Kuschel, J. Schell, and M. Ummenhofer, "Passive radar based control of wind turbine collision warning for air traffic PARASOL," in *2015 16th International Radar Symposium (IRS)*, Dresden, 2015, pp. 36–41. doi: 10.1109/IRS.2015.7226394

[47]   F. Colone, D. W. O'Hagan, P. Lombardo, and C. J. Baker, "A multistage processing algorithm for disturbance removal and target detection in passive bistatic radar," *IEEE Transactions on Aerospace and Electronic Systems*, pp. 698–722, 2009.

[48]   C. Schwark and D. Cristallini, "Advanced multipath clutter cancellation in OFDM-based passive radar systems," in *IEEE Radar Conference (RadarConf)*, Seattle, 2016.

[49]   J. L. Garry, C. J. Baker, and G. E. Smith, "Evaluation of direct signal suppression for passive radar," *IEEE Transactions on Geoscience and Remote Sensing*, vol. 55, no. 7, pp. 3786–3799, 2017.

[50]   E. J. Kelly, "An adaptive detection algorithm," *IEEE Transactions on Aerospace and Electronic Systems*, vol. AES-22, no. 2, pp. 115–127, 1986.

[51]   D. Pastina, P. Lombardo, and T. Bucciarelli, "Adaptive polarimetric target detection with coherent radar. Part I: detection against Gaussian background," *IEEE Transactions on Aerospace and Electronic Systems*, vol. 37, no. 4, pp. 1194–1206, 2001.

[52]   J. Ward, "Maximum likelihood angle and velocity estimation with space-time adaptive processing radar," in *Conference Record of The Thirtieth Asilomar Conference on Signals, Systems and Computers*, vol. 2, Pacific Grove, CA, 1996, pp. 1265–1267.

[53]   F. C. Robey, D. R. Fuhrmann, E. J. Kelly, and R. Nitzberg, "A CFAR adaptive matched filter detector," *IEEE Transactions on Aerospace and Electronic Systems*, vol. AES-28, no.1, pp. 208–216, 1992.

[54]   J. Ward, "Cramér-Rao bounds for target angle and Doppler estimation with space–time adaptive processing radar" in *Proceedings of 29th ASILOMAR Conference on Signals Systems and Computers*, 1995, pp. 1198–1203.

[55]   R. Klemm, "Cramér-Rao analysis of reduced order STAP processors," in *IEEE International Radar Conference*, Alexandria, VA, 2000, pp. 584–589.

[56]   U. Nickel, "Monopulse estimation with adaptive arrays", *IEE Proceedings of Radar Signal Process*, vol. 140, no. 5, pp. 303–308, 1993.

[57]   U. Nickel, "An overview of generalized monopulse estimation", *IEEE AES Magazine*, vol. 21, no. 6, pp. 27–55, 2006, Part 2: Tutorials.

[58]   P. Lombardo and F. Colone, "A dual adaptive channel STAP scheme for target detection and DOA estimation," in *Proceedings of the International Conference on Radar (IEEE Cat.no.03EX695)*, Adelaide, SA, Australia, 2003, pp. 115–120.

[59]   F. Colone, D. Cristallini, D. Cerutti-Maori, and P. Lombardo, "Direction of arrival estimation performance comparison of dual cancelled channels space-time adaptive processing techniques," *IET Radar, Sonar & Navigation*, vol. 8, no. 1, pp. 17–26, 2014.

Chapter 9

SDR-based passive radar technology

Amerigo Capria[1], Anna Lisa Saverino[1], Elisa Giusti[1] and Marco Martorella[1,2]

The intrinsic flexibility offered by the SDR paradigm represents a very suitable implementation strategy for passive bistatic radars which operate over a wide range of frequencies with a variety of instantaneous bandwidth. SDR together with digital beamforming techniques allow the realization of passive radars able to achieve the best performance in different operative scenarios and RF environment. In this framework, a number of new demonstrators have been designed and realized in the last few years.

Acronyms

ADC:	analogue to digital converter
ADS-B:	automatic dependent surveillance broadcast
AIS:	automatic identification system
CapTechs:	capability technology areas
CFAR:	constant false alarm rate
CNIT:	National Inter-University Consortium for Telecommunications
CORA:	COvert Radar
COTS:	commercial off-the-shelf
CPI:	coherent processing interval
CPU:	central processing unit
CSSN:	Naval Support and Experimentation Center of the Navy
DAB:	digital audio broadcasting
DAC:	digital to analogue converter
DELIA:	dab experimental radar with linear array
DoA:	direction of arrival
DSP:	digital signal processor
DVB-S:	digital video broadcasting satellite

[1]Radar and Surveillance Systems (RaSS) National Laboratory, CNIT (National Inter-University Consortium for Telecommunications), Italy
[2]Department of Information Engineering, University of Pisa, Italy

DVB-T:	digital video broadcasting terrestrial
EDA:	European Defence Agency
EW:	electronic warfare
FHR:	Fraunhofer Institute for High Frequency Physics and Radar Techniques
FM:	frequency modulation
FMCW:	frequency modulated-continuous wave
FPGA:	field programmable gate array
GMTI:	ground moving target indicator
GPS:	global positioning system
GPU:	graphics processing unit
GSM:	global system for mobile communications
GUI:	graphical user interface
HDD:	hard disk drive
HPBW:	half power beamwidth
HW:	hardware
IF:	intermediate frequency
IO:	illuminator of opportunity
IPP:	image projection plane
ISAR:	inverse synthetic aperture radar
JCC:	Joint Italian Navy
LNA:	low-noise amplifier
LO:	local oscillator
LPDA:	logarithmic periodic dipole antenna
MIMO:	multiple-input and multiple-output
MoD:	Ministry of Defence
MSPS:	megasamples per second
NATO SET:	North Atlantic Treaty Organization Sensors and Electronics Technology
NCTR:	non-cooperative target recognition
NI:	national instruments
OFDM:	orthogonal frequency-division multiplexing
PaRaDe:	passive radar demonstrator
P-ISAR:	passive ISAR
PBR:	passive bistatic radar
PC:	personal computer
PCI Express:	peripheral component interconnect express
PCL:	passive coherent location
PETRA:	passive experimental Tv radar
PPS:	pulse per second
PXIe:	PCI eXtensions for instrumentation
RAID:	redundant array of independent disks
RaSS:	radar and surveillance systems
RCS:	radar cross section
RD:	range-Doppler
RF:	radio frequency

Rx:	receiver
SAR:	synthetic aperture radar
SDR:	software-defined radio
SMARP:	software-defined multiband array passive radar
SSD:	solid-state drive
SWOT:	strengths, weaknesses, opportunities, and threats
TV:	television
Tx:	transmitter
UAV:	unmanned aerial vehicle
UHF:	ultra-high frequency
UMTS:	universal mobile telecommunications service
USRP:	universal software radio peripheral
VHF:	very high frequency
WUT:	Warsaw University of Technology

9.1 Introduction

Software-defined radio (SDR)-based radars achieve advanced flexibility, scalability, robustness and more powerful functions over conventional radars. A SDR-based radar system is capable of changing its main specifications by reconfiguring its operational parameters on-the-fly like frequency, bandwidth, digital signal processing blocks and others.

Since passive bistatic radar (PBR) can utilize a number of transmitters of opportunity that operate in ground and space over a wide range of frequencies with a variety of instantaneous bandwidth; therefore, the intrinsic flexibility offered by the SDR technology represents a very suitable implementation strategy for this class of radars. The jointly exploitation of SDR and digital beamforming allow the realization of a flexible passive radar able to achieve the best performance in different operative scenarios and RF environment. The chapter is organized as follows. Section 9.1 introduces the SDR technology applied to radar applications. Section 9.2 is instead devoted to a brief introduction of the PBR application and to the advantages of implementing a PBR exploiting the SDR technology. The section also reports some reference SDR-based passive radars developed in the last years by the German Fraunhofer Institute for High Frequency Physics and Radar Techniques (FHR) and by the Polish Warsaw University of Technology (WUT). Finally, in Section 9.3, the Italian SDR-based passive radar developed by RaSS National Laboratory of CNIT is presented, the main functionalities are described and some experimental results are reported.

9.2 SDR technology for radar application

The growing demand for radar systems able to perform multiple functions like long- and short-range air-ground surveillance, target detection, target imaging, target recognition, clutter/interference cancellation and communication, leads to the development of highly performing radars in terms of agility, adaptivity, flexibility, interoperability, weight and cost. These aspects became fundamental with the advent of ground-breaking platforms (e.g. unmanned vehicles) equipped with heterogeneous sensors and with limited available space and weight.

Future radars must be designed to account these requirements with a focus on the requirements/cost trade-off.

The SDR concept enables radar technology to cope with these features. SDR applied to the radar concept means that some or all of the hardware layer processing, like waveform generation, filtering, up-and down conversion, etc. are performed by software and in the digital domain. Then, the required characteristics, performance and functionalities are achieved via software without any changes to the hardware layer. As a result, SDR-based radars achieve advanced flexibility, scalability, robustness and more powerful functions over conventional radars. In fact, conventional radars typically require long development cycle and high costs because of the dedicated application in a certain radar environment.

Among all these features, the flexibility and adaptivity represent the key points of the SDR technology. The software re-programmability guarantees them, thus reducing hardware components and consequently costs and weight. For instance, add-on suitable programs allow to manage different operations, each of which would require a different circuit in conventional radars, thus reducing the hardware sub-systems complexity. This will help to extend the system functionality to a wide range of applications. Working in the software domain also adds flexibility, predictability of performance, repeatability and reconfigurability. The reconfigurability allows the system to be reprogrammed, making its capabilities adjustable to different functions/missions/applications by re-using the same hardware components.

The objective of the SDR research is therefore to move the problem of radio engineering from the domain of the hardware to that of the software, where each radio signal is differentiated only by the software required to implement it.

A SDR-based radar system derives the same principles of a software-defined radio. Specifically, many hardware components (e.g. mixers, filters, modulators, demodulators, detectors, etc.) are implemented via software on a computer, using digital signal processor (DSP) or field programmable gate array (FPGA). The Universal Software Radio Peripheral (USRP) which is developed at a low cost is one of the most known SDR platforms [1]. USRPs and GNU radio have been getting attention by research, industry, academia, government and hobbyist environments, to implement very sophisticated SDRs based on. GNU radio is an open-source software framework that provides both signal processing functions and graphical design approach. In GNU radio, C++ blocks perform specific signal processing tasks, while Python applications connect the blocks together to realize a functional software radar. For this purpose, it can be used either with low-cost external RF hardware or without hardware in a simulation-like environment [2].

A basic SDR-based radar system consists of a wideband RF front-end, up/down converter, digital to analogue converter/analogue to digital converter (DAC/ADC) converter, a FPGA/SDR platform and a base station. Differences may arise at the transmitting and receiving functions [3].

The SDR-based radar can be used in the following two-fold configuration:

1. The computer is only devoted to manage the radar and display the results, while the whole processing is performed by the FPGA/SDR platform.
2. FPGA/SDR platform deals with high-speed processing like decimation and filtering while most of the radar signal processing is carried out by the computer.

The selection of the most suitable configuration depends on a number of variables, as the level of optimization of the radar signal processing code, the available FPGA computing power and the level of integration of the system under design.

The concept beyond the SDR-based radar system leads to many advantages such as [1–3]:

- The opportunity to design a multipurpose radar. Since, the radar platform is completely software defined, it can switch among several operative modes or RF applications on-the fly. This capability is performed by only acting on the transmitted waveforms and signal processing tasks. Multipurpose radars, which combines different RF applications in one transmitter–receiver (Tx–Rx) module and one antenna, are essential to fulfil the new radar field trends and the advent of new platforms (e.g. UAVs – unmanned aerial vehicle) where the on-board space and weight (e.g. for antennas) are very constrained.
- The opportunity to share the available hardware for different purposes.
- A great reduction of development times and production costs. In fact, the possibility to reuse already existing hardware (HW) components accelerates the radar system development reducing costs.

These features make SDR-based radars useful for many different applications as follows:

- Human motion detection area. In this area, SDR-based radar can be used to classify human motions exploiting micro-Doppler technique [4–5].
- Automotive area. SDRs perform a two-fold functionality, radar and communications [6]. These possibilities can be actuated by switching modes or through specific waveforms like orthogonal frequency-division multiplexing (OFDM) [7]. For instance, a measurement system making use of a hybrid radar scheme with continuous wave frequency modulation and a pseudo-random code pulse techniques is presented in [8]. Here, the ability to obtain high precision information concerning the velocity of a vehicle, the distance, the direction and other information useful to improve the security in the automotive field is depicted.
- Ground penetrating radar area. A SDR-based radar exploits its features of implementing hardware modules via software to reproduce a compact multiband radar [9].
- Military area. SDR-based radar enables multifunction radars as ground moving target indicator (GMTI) function, synthetic aperture radar (SAR) function, Inverse synthetic aperture radar (ISAR) function, weather surveillance, communication mode and electronic welfare (EW) mode, suitable to be implemented on UAVs or similar vehicles [9].

- Weather surveillance area. In this area, a radar network is fundamental, then costs reduction of implementing it is achieved through the SDR paradigm. A weather surveillance application based on GNU radio-based software-defined frequency modulated-continuous wave (FMCW) radar is described in [10].
- In passive bistatic radar area, SDR paradigm guarantees flexibility and agility to configure the receiver in order to select the most suitable available illuminator of opportunity (IO) according to the environment. The illuminators of opportunity fall into two main classes: analogue signal like frequency modulation (FM) radio, analogue television (TV) or digital signal like digital video broadcasting terrestrial (DVB-T), universal mobile telecommunications service (UMTS), and wireless fidelity (Wi-Fi). In [11], the capability of the USRP technology is demonstrated in the realization of a passive radar by designing a low-cost DVB-T software defined system for costal ship detection.

These attractive peculiarities make SDR-based radar systems the future trend in radar field. A SWOT (strengths, weaknesses, opportunities, and threats) analysis of this technology is provided in Table 9.1.

9.3   Passive bistatic radar

Bistatic radar may be defined as a radar in which the transmitter and receiver are at separate locations. The very first radars were bistatic, until pulsed waveforms and T/R switches were developed. Bistatic radars can operate with their own dedicated transmitters, which are specially designed for bistatic operation, or with transmitters of opportunity, which are designed for other purposes but found suitable for bistatic operation. When the transmitter of opportunity is from a non-radar transmission, such as broadcast, communications or radio-navigation signal, the bistatic radar are called: PBR and passive coherent location (PCL).

In recent years, there has been a growing interest in PBR using existing transmitters as illuminators of opportunity to perform target detection, localization and tracking [12,13].

The PBR system concept is of great interest in both civilian and military scenarios mainly due to a number of advantages offered by this system with respect to active radar. The main achievable benefits are:

- No need of frequency allocation.
- Enhanced target radar cross section (RCS) thanks to the lower frequency of operation (i.e.: VHF – very high frequency and UHF – ultra high frequency bands) and the bistatic configuration (possibility to detect stealthy targets).
- Low power absorption (receiver only): solar panel could be used in order to achieve an electrical independent system.
- Robustness against jammers.
- Possibility to operate in a covert manner due to the low probability of intercept (no emissions).
- Small size and low weight.
- Potentially low cost architecture (absence of transmitter).

Table 9.1 SDR-based radar technology SWOT analysis

Strengths	Weaknesses
✓ Software re-programmability	• Poor dynamic range in some SDR designs
✓ Decreased number of hardware components	• Analog/digital conversion is still a technical bottleneck
✓ Reduced complexity of the hardware subsystem	• The software to be developed must be kept separated from the hardware as far as possible to ensure portability and reusability of programs
✓ Reduced costs and weight	
Opportunities	**Threats**
✓ Flexibility and versatility guaranteed by the software re-programmability	• It could be difficult to find an effective digital signal processing implementation of receiver functions that achieves dynamic range sufficient for the frequency range of interest
✓ Predictability of performance and repeatability	• Physical/technological limitations
✓ Interoperability and multi-functionality	
✓ Adaptability to different functions/missions/applications	
✓ Reconfigurability: the system can be reprogrammed to extend its capabilities avoiding the intervention on the hardware components	
✓ Simplified maintenance	
✓ Possibility for updates or bug fixes after delivery	
✓ Possibility of adaptively choosing the best-suited operating frequency and mode	
✓ Opportunity to recognize and avoid interference with other communications channels	

In Table 9.2 a SWOT analysis applied to the PBR technology is presented.

There are a great variety of signals that can be used for PBR purposes. Their performance in PBR systems will vary significantly, depending on a variety of factors: (i) power density at target, (ii) coverage (both spatial and temporal), and (iii) ambiguity function shape depending both on the waveform and on the transmitter–target–receiver geometry. In particular, broadcast transmitters represent some of the most attractive choices for long-range surveillance application due to their excellent coverage. The most common signals used for PBR applications are FM radio and UHF television broadcasts as well as digital transmission such as digital audio broadcasting (DAB) and digital video broadcasting-terrestrial.

Since PBR can utilize a number of transmitters of opportunity that operate in ground and space over a wide range of frequencies with a variety of instantaneous bandwidth, therefore the intrinsic flexibility offered by the SDR technology represents

Table 9.2   PBR technology SWOT analysis

Strengths	Weaknesses
✓ Low electromagnetic pollution	• Non-cooperative IO (performances are driven by the TX waveform)
✓ Receiver only	• Waveforms are generally not designed for radar purposes
✓ Potential low cost architecture	• Spatial resolution and RX positioning are dependent on the geometry
✓ Low power consumptions	
Opportunities	**Threats**
✓ Counter-stealth	• Some functions/applications may not be feasible depending on the specific operative scenario and required performance
✓ Low probability of intercept	
✓ Covert, directed jamming is unlikely	

a very suitable implementation strategy for this class of radars [14]. Indeed, a SDR-based radar system is capable of changing its main specifications by reconfiguring its operational parameters on-the-fly like frequency, bandwidth, digital signal processing blocks and others.

Achieving an effective frequency agility is a key capability for PBR applications. Indeed even when the IO to be exploited is defined, for instance the DVB-T one, the broadcasting frequencies span over hundreds of MHz, therefore the most suitable carrier frequency can be chosen according to the signal level, the electromagnetic (e.m.) environments and others.

In order to achieve a high level of adaptability also on the antenna system, PBR can exploit phased arrays with digital beamforming in order to be able to form multiple adaptive antenna beams and to steer nulls. The jointly exploitation of SDR and digital beamforming allows the realization of a flexible passive radar able to achieve the best performance in different operative scenarios and RF environment.

It should be noted that the advantages of PBR and SDR can be obtained at the expenses of a huge signal processing load. However, the computational power of nowadays central processing unit (CPUs) and FPGAs are making the real-time realization of such a radar system possible.

In the framework of SDR-based passive radars, a number of new demonstrators have been designed and realized in the last few years. Some of the most advanced and innovative ones developed by European research institutes are briefly descripted in the following.

Fraunhofer is a German research organization with several divisions that deal with different branches of science and technology. One of the branches is the Fraunhofer Institute for High Frequency Physics and Radar Techniques (FHR), located in Wachtberg, near Bonn.

Fraunhofer FHR has been leading the research on the topic of PBRs since the beginning of the renewed interest in this technology started in the 2000s. FHR had developed and operated a number of experimental systems like COvert RAdar (CORA), Passive Experimental Tv RAdar (PETRA), and Dab Experimental radar with Linear Array (DELIA) [15].

FHR's more recent SDR-based passive radar for DAB/DVB-T exploits a scalable, modular receiver named ATLAS [16] (Figure 9.1). The main specification of the system are: (i) 12 RF receiver channel modules; (ii) 12 ADC (analogue-to-digital-converter) and 6 field programmable gate array (FPGA) modules; (iii) 12 high performance personal computer (PC) modules; (iv) one central local oscillator (LO) and ADC-clock generation unit.

Depending on the measurement task, the system modules can be configured such that the defined requirements are met.

In case of DVB-T operations, the RF front end filters from 470 to 870 MHz. The signal is then amplified and mixed to the first IF (intermediate frequency) at about 1,105 MHz. Then a second stage down-converts the signal to an IF of 80 MHz (bandwidth of 32 MHz). The usage of two mixer stages avoids the back-folding of subjacent channels into the desired measurement channels. The signal level is finally optimized for the input requirements of the ADC and adjacent spectral products are eliminated thanks to an amplifier and a further band filter.

The down-converted signals are then digitized with a 16-bit ADC at a sampling frequency of 64 MSPS (megasamples per second). The FPGA pre-processes the incoming data from up to two channels and transfers them to the high-performance PCs via Peripheral Component Interconnect Express (PCI Express).

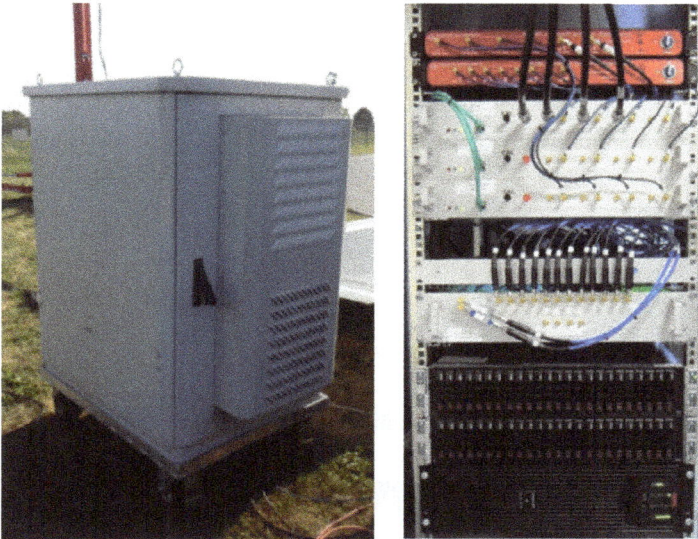

Figure 9.1   ATLAS front end [16]

The multi-channel RF front-end ATLAS has been successfully validated in combination with an 11-element uniform linear array and uniform circular array. Several measurements campaigns have shown the effectiveness of the system able to detect and tracks even low RCS targets like an ultralight aircraft and manoeuvring speed boats [17].

The Warsaw University of Technology (WUT) is one of the leading institutes of technology in Poland and one of the largest in Central Europe.

WUT has begun to work on passive radar technology since 2002 gaining a broad experience over the years. The first demonstrator called Passive Radar Demonstrator (PaRaDe) has been developed in 2007 [18,19]. This demonstrator exploits FM illuminators for airborne target detection and tracking. The hardware includes four sampling modules, a clock distribution module, a clock signal generator, a calibration signal generator, a power supplier, a global positioning system (GPS) receiver and an Automatic-Dependent Surveillance Broadcast (ADS-B) receiver (used as a source of the reference data).

Up to now, different versions of this first demonstrator have been developed. The PaRaDe system has been tested during several measurement campaigns [20]. Depending on the version, the demonstrator operates in many commercial bands e.g. FM, DVB-T, Global System for Mobile Communications (GSM), WiFi [21] as well as with signals transmitted by active radars as its sources of illumination. All demonstrators are designed according to the SDR paradigm which confers them flexibility and scalability [21]. These features allow for the use of the PaRaDe family in a wide range of configurations (e.g. ground based, airborne, MIMO-multiple-input and multiple-output) and applications (e.g.: detection, tracking, imaging) [22,23].

For instance, a passive SAR for Earth imaging purpose has been built using National Instruments (NI) commercial off-the-shelf (COTS) products; specifically NI PXIe-5663 (PCI eXtensions for Instrumentation-5663) has been used to receive and digitize signals and NI HDD-8264 (NI Hard Disk Drive-8264) RAID (Redundant Array of Independent Disks) to store a large amount of data with high-throughput streaming [24]. The used RAID system allows for 600 MB/s continuous streaming. Other NI solutions allow to improve the throughput up to 3.6 GB/s (e.g. HDD-8266 – solid-state drive (SSD) version). The field test and the hardware components are depicted in Figure 9.2.

9.4 Software-defined multiband array passive radar (SMARP)

The Radar and Surveillance Systems National Laboratory of the Italian National Interuniversity Consortium for Telecommunications developed a consolidated expertise in passive radar systems both from the point of view of analysis and design, and from the point of view of realization of experimental equipment and measurements in relevant environments. The passive radar research activity started at RaSS Lab in the 2006 by means of a university master thesis [25], then in 2009 the first proof of concept demonstrator has been realized [26] and in 2010 the first research project proposal on this topic obtains funds from European Defence Agency (EDA) [27].

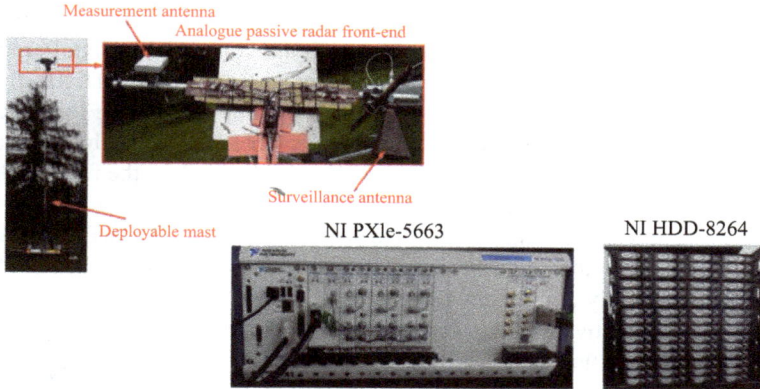

Figure 9.2   WUT passive SAR demonstrator [24]

The staff of the RaSS Lab has been actively engaged in the activities of numerous North Atlantic Treaty Organization Sensors and Electronics Technology (NATO SET) panel research groups relating to passive radar systems, as well as teachers for the NATO Lecture series on "Passive radar technology" and in the activities of organized CapTechs (Capability Technology Areas) by EDA. The RaSS Lab is strongly involved in the research and experimental activities in the field of passive radar conducted by the Joint Italian Navy (JCC) "Ugo Tiberio" laboratory, a joint laboratory between the CSSN (Naval Support and Experimentation Center of the Navy) and the CNIT, whose constitution was promoted by the RaSS Lab.

In the field of passive radars, the RaSS Lab team has published more than 40 works including several book chapters' contribution.

The most advanced PBR demonstrator developed by the RaSS Lab is named SMARP and it has been conceived in the framework of the Italian National Plan for Military Research founded by the Italian MoD (Ministry of Defence) [28]. The SMARP demonstrator peculiar features are:

- Multiband receiving array antenna (UHF and S band) with dual polarization reception.
- Software-defined multiband flexible receiver based on commercially available solutions.
- Digital array processing techniques and advanced radar signal processing algorithms implemented on COTS processing architectures (multicore CPUs and graphics processing unit (GPUs)).

Such features allow for a great flexibility such as the use of both DVB-T and UMTS IOs for the detection and tracking of targets in the range-Doppler map. Moreover, the use of array processing allows for the estimation of the target Direction of Arrival (DoA) thus allowing for accurate target localization.

As research in the passive radar field progresses, more radar techniques are added to PBRs to make them able to handle several tasks and to be applied in different scenarios. One of such tasks is the radar imaging of non-cooperative targets through ISAR imaging, which in turn may open the doors to Non-Cooperative Target Recognition (NCTR) capabilities. The RaSS Lab has been able to produce the first passive ISAR image starting from SMARP real data, a result obtained for the first time in the international scientific panorama [29,39].

Radar imaging functionality is activated on demand during SMARP demonstrator operation, by using a wider bandwidth signal. Signal with wide enough bandwidth are obtained by coherently combining adjacent DVB-T or UMTS channels. The ISAR capability may be activated on a specific target of interest or may be performed simultaneously on all the detected targets.

9.4.1 SMARP architecture: description and functionalities

The SMARP system architecture is reported in Figure 9.3 [30]. The main functional blocks are:

- The antenna system, namely, an array for the surveillance channel and a dedicated single element for the reference channel.
- The RF-Front-end #1 and the calibration network.
- The RF-Front-end #2 which includes the synchronization signal generator.
- Radar digital signal processing, control and display unit.

The SMARP demonstrator is installed in Livorno on the roof of the CSSN-ITE 'Istituto Vallauri'. Pictures of the system are shown in Figure 9.4 where the main functional blocks have been highlighted with different colours and will be briefly described in the following.

9.4.1.1 Antenna system

The single element receiving antenna is capable of operating in dual band and dual polarization mode. The main features of the surveillance antenna system are:

- Multiband receiving array antenna: UHF-band (470–790 MHz) and S-band (2,100–2,200 MHz), with dual polarization reception (H/V) [31].
- Four linear arrays, two for each band, composed of 8 logarithmic periodic dipole antenna (LPDA) patch antennas. The inter-element spacing is equal to 30 cm in the UHF band and 8 cm in the S-band.
- The antenna is equipped with two plane reflectors, one for each band (UHF-band: $3.1 \times 1.3 \times 0.86$ m and S-band: $1.1 \times 0.7 \times 0.27$ m).

The reference antenna is composed of four separate single receiving elements of the same type of the ones used for the surveillance antenna, one for each frequency band and polarization.

9.4.1.2 RF Front-end #1 and calibration network

The RF Front-end #1 is in charge of filtering out the interferences from the working frequency band and to amplify the desired signal by means of low noise amplifier

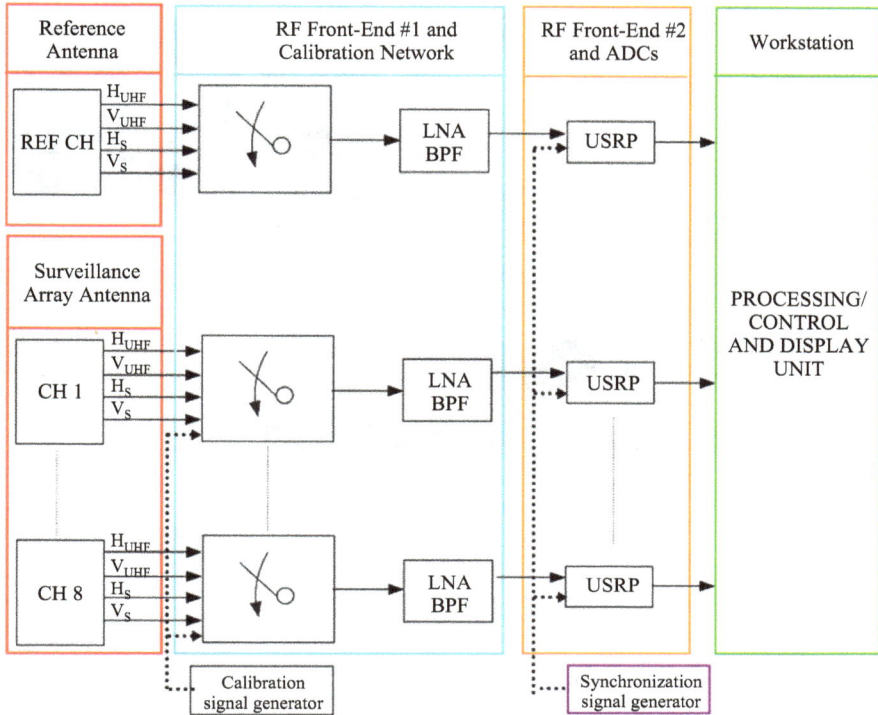

Figure 9.3 SMARP system architecture

(LNA) with a limiting effect on the system noise figure. A software-controlled switch allows to set the total gain given by the amplification chain which is dependent on the signal exploited and on the e.m. environment.

The internal calibration routine is also implemented at this stage. It estimates the constant phase offsets and amplitude imbalances, and generates the correction coefficients to be applied to the signal samples before digital beamforming. The selection between the calibration path and the antenna received signals path is performed by a switch which is remotely controlled via software. The signals coming from the antennas are the default input transferred to the RF Front-ends #1. On the contrary, when the switches are set in 'calibration mode', the RF Front-end #1 input changes to the calibration signal. During this phase, a board is used for generating and injecting, in a closed-loop mode, the calibration reference signal into the RF chains.

9.4.1.3 RF Front-end #2

The main receiver requirements to be satisfied by the SMARP demonstrator have been identified as:

- Flexile reception in the bandwidth 470–790 MHz and 2,100–2,200 MHz.

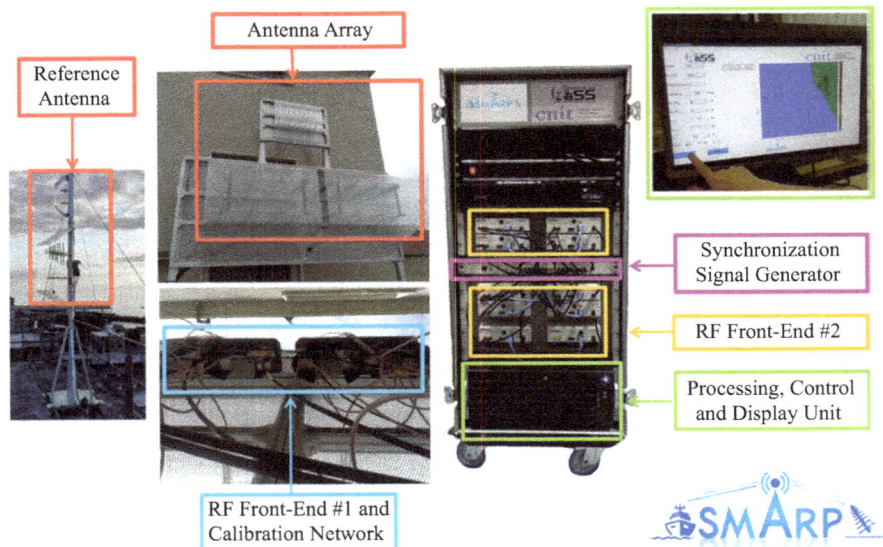

Figure 9.4   SMARP subsystems

- Receiving instantaneous bandwidth up to 7.61 MHz.
- Sampling resolution equal to 14 bits.
- Capability to operate in a coherent and synchronized multichannel configuration.

The selected solution for the SMARP receiver is based on a COTS Software Radio System from National Instruments, the NI USRP-2922. The Universal Software Radio Peripheral is an open design, low-cost, flexible board for the implementation of SDRs. The selected solution is configured as a general purpose versatile transceiver and it is able to fully meet the requirements. The board can operate on a wide frequency band, specifically 400 MHz–4.4 GHz and it is equipped with an FPGA in charge of dealing with high data rate processing (e.g. digital down conversion, digital filtering and decimation).

9.4.1.4   Synchronization signal generator

To perform digital array processing, a synchronous and coherent multichannel acquisition system is mandatory. When operating with several USRPs boards, the mentioned requirements can be achieved by feeding each board with two reference signals, specifically:

- A 10 MHz reference clock to provide a single frequency reference for all devices.
- A pulse-per-second (PPS) to enable a temporal synchronization of the acquired samples.

The generation of the reference signals has been achieved by using a clock distribution unit, produced by Ettus Research and named a OctoClock-G [32], which provides a suitable number of 10 MHz and 1 PPS outputs (i.e.: equal to the number of NI USRP-2922 boards to be synchronized).

9.4.1.5 Processing, control and display unit

The radar digital signal processing architecture implemented in SMARP is composed of the following blocks:

- Pre-processing to reduce the multipath effects and attenuate the spurious peaks caused by guard intervals, deterministic and pseudorandom pilot tones of the waveform [33].
- Range-Doppler (RD) map formation which aims at filtering out the direct path interference, its multipath and at generating the RD map for each receiving channel [34,35].
- Surveillance channel beamforming: this processing unit aims at forming one or more beams in the surveillance area. Moreover, this allows obtaining a raw estimation of the target DoA [36].
- Detection, which provides the target position in the RD domain. This step is accomplished by using constant false alarm rate (CFAR) detection techniques [37].
- Tracker, which aims at reconstructing the path of targets in the RD domain [38].

The SMARP demonstrator is provided with a GUI which is devoted to the user control of the main system parameters as well as the display of the radar outputs. The main user-defined system parameters are the operating mode, the frequency, the integration time as well as the amplification chain value applied on reference and surveillance channels. The radar display includes the range-Doppler maps and/or the visualization of target detections and tracking on a georeferenced map. The GUI has been conceived to guarantee a high level of flexibility and the possibility to choose the system parameters to be controlled, the outputs and the appearance of the radar plots.

9.4.2 SMARP experimental results

Different measurement campaigns have been carried out to perform detection of various vessels both arriving and departing from the Livorno harbour, by exploiting both DVB-T and UMTS signals. The surveillance antenna was directed towards an area of sea in front of the receiver site at an angular direction of about 260° (boresight) as shown in Figure 9.5. A building which limits the visible area toward North is located at the direction of about 275°. The exploited transmitters of opportunity are respectively a DVB-T transmitter located on 'Monte Serra' in Pisa (around 32 km far from the receiver) and a UMTS base station located at the stadium around 400 m far from the receiver. The used channels carrier frequencies were respectively 634 MHz and 2,147.5 MHz. For this experiment the system has been configured in UHF mode to acquire a wideband signal formed by three adjacent DVB-T channels for a total bandwidth of about 25 MHz. The wideband signal has been used for ISAR functionality, while detection and tracking functionality have been performed by using only one transmitted channel.

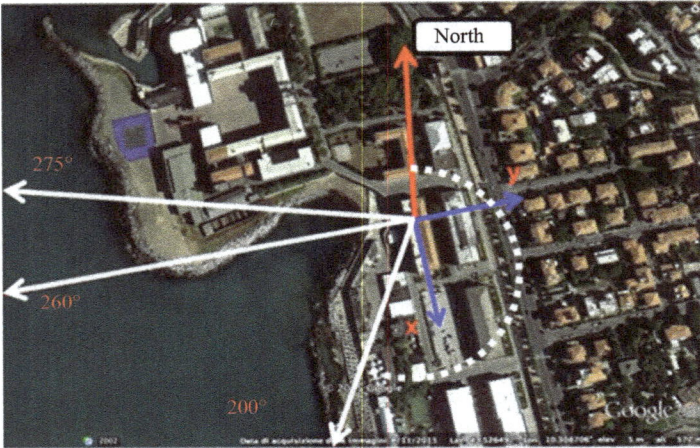

Figure 9.5   SMARP installation geometry

Table 9.3   Non-cooperative targets characteristics – UHF and S band experiments
(Figure 9.7 and 9.8)

MMSI	Name	Vessel type	Size	Exploited bandwidth
247294700	Eurocargo Malta	Cargo	200 m × 26 m	UHF
247334000	Espresso Catania	Passenger	150 m × 23 m	UHF
247552000	Corsica Marina	Passenger	120 m × 20 m	UHF
210090000	Vitality	Cargo	210 m × 30 m	UHF
636012570	Ceniris	Tanker	243 m × 42 m	UHF
247064930	Erpiu	Fishing	19 m × 5 m	UHF
247267800	Corrado Neri	Tug	35 m × 14 m	UHF
247103700	Iver Agile	Tanker	110 m × 17.8 m	UHF
247272700	Elba	Tanker	72.6 m × 13.2 m	S
247000400	Via Adriatico	Passenger	150 m × 23 m	S
247213700	Fratelli Neri	Tug	24 m × 12 m	S

Some results obtained during different measurement campaigns are presented hereinafter. The actual trajectories of non-cooperative vessels have been recorded with an automatic identification system (AIS) receiver to be used for validating the detection and tracking functionalities of the system. The results are obtained by separately exploiting both operating frequency bands (i.e., UHF and S-band). The integration time was 0.5 s and the total duration of each acquisition was about 30 min. A big variety of vessels have been observed during this trial, the main characteristics of the non-cooperative targets are shown in Table 9.3. The history of the range-Doppler map relative to the entire 30 min observation is reported in Figure 9.6 for the UHF band experiment. In Figures 9.7 and 9.8, the trajectories of the vessels acquired by

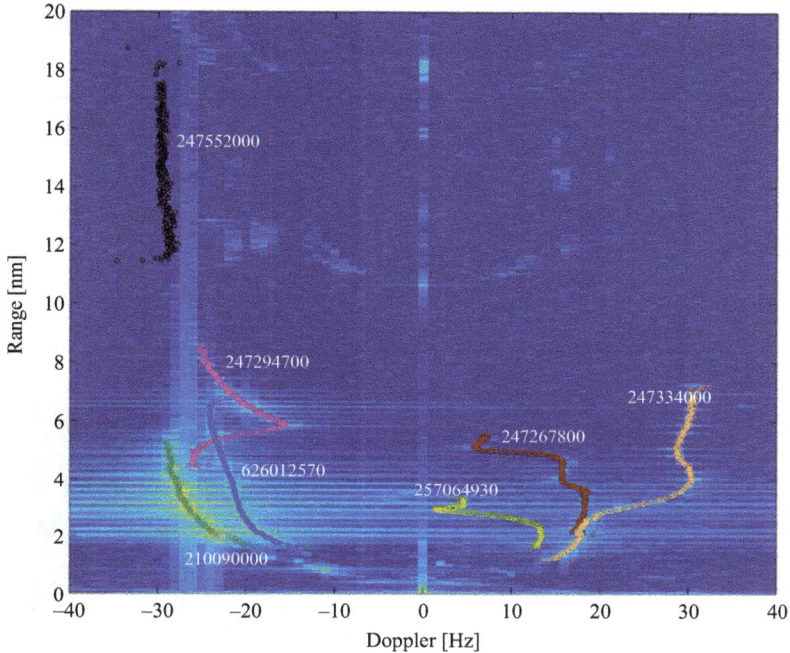

Figure 9.6   UHF band experiment: range-Doppler history map (30 min) with superimposed AIS data

the AIS receiver (coloured lines) are overlapped to the radar tracks (black lines) on a geographic map. The estimated tracks well match the AIS trajectories thus confirming the effective operation of the SMARP demonstrator. The half power beamwidth (HPBW) of the single receiving element of the surveillance array is of about 60° which in Figures 9.7 and 9.8 correspond to the aperture drawn in dotted black line labelled with 230°–275° (less than 60° because toward North the presence of a building limits the antenna visibility). Even though it is expected to achieve reduced radar performance when operating outside the main beam of the surveillance antenna, a number of detections were obtained well beyond this angular sector.

In order to test the Passive ISAR (P-ISAR) capability [29,39] an *ad hoc* experiment has been set up and performed exploiting the UHF band. The tracking results are reported in Figure 9.9 where the trajectories of the vessels acquired by the AIS receiver (coloured lines) are overlapped to the radar tracks (black lines) on a geographic map.

ISAR capability may be activated on a specific target of interest or may be performed simultaneously for all the detected targets. ISAR processing aims at providing well focused e.m. images of the detected targets. ISAR images may provide additional information about the target, such as its size, shape and e.m. properties, which, in turn, may open the door to automatic target classification or automatic target recognition. Two targets have been chosen within the detected ones to show the results of

Table 9.4 P-ISAR UHF band experiment: non-cooperative targets characteristics (Table 9.4)

MMSI	Name	Vessel type	Size	Exploited bandwidth
538090480	Navin Vulture	Cargo	113 m × 17m	UHF
248646000	Loya	Tanker	93 m × 15m	UHF
249830000	Zim Luanda	Cargo	260 m × 32m	UHF
247217500	Grande Colonia	Cargo	176 m × 31m	UHF
247131600	Zeus Palace	Passenger	212 m × 25m	UHF
304317000	Anna Sophie	Cargo	135 m × 23m	UHF
470968000	Al Bahia	Cargo	306 m × 40m	UHF

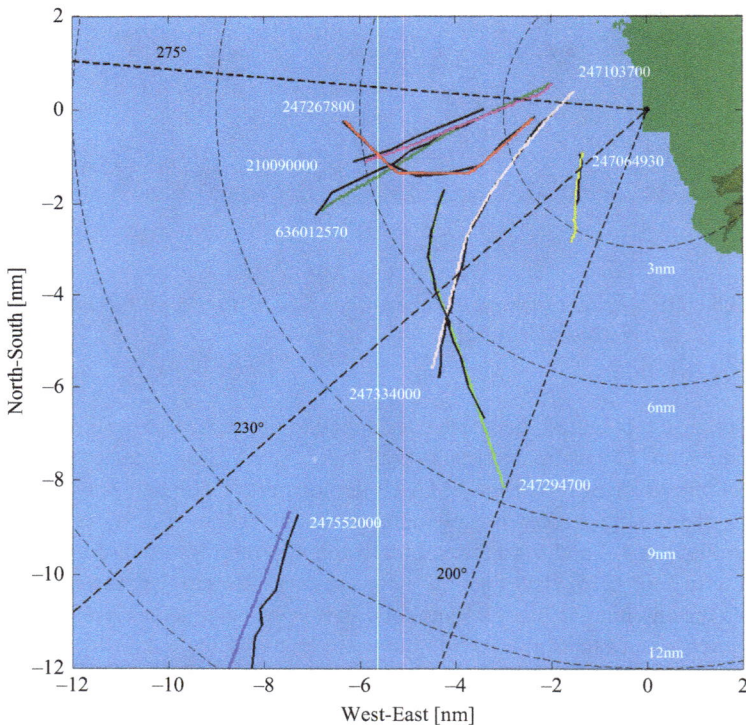

Figure 9.7 UHF band experiment: AIS trajectories (coloured lines) and radar tracks (black lines)

the passive-ISAR algorithm. Specifically, 'Anna Sophie', 'Grande Colonia' vessels have been considered (dimensions shown in Table 9.4).

The two targets underwent complex motion, both because they were manoeuvring and because of the sea state that during the experiments was moderate. Consequently,

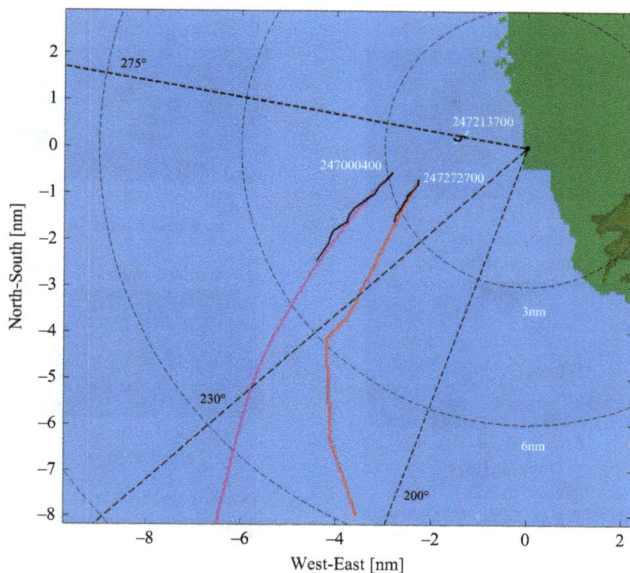

Figure 9.8   *S-band experiment: AIS trajectories (coloured lines) and radar tracks (black lines)*

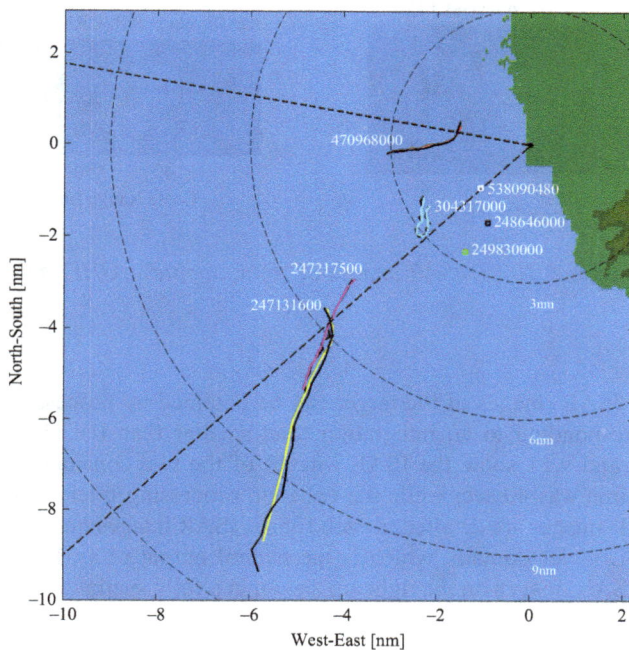

Figure 9.9   *P-ISAR UHF band experiment: AIS trajectories (coloured lines) and radar tracks (black lines)*

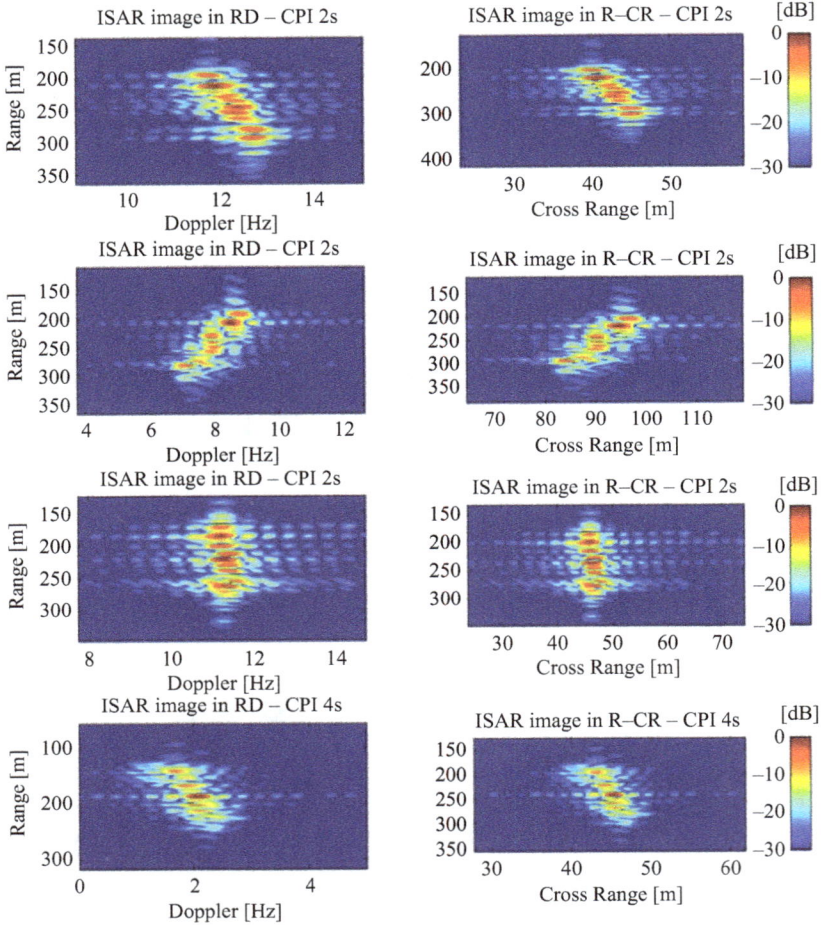

Figure 9.10   P-ISAR images of 'Anna Sophie' – UHF band

well-focused ISAR images of the target can be obtained by processing portions of the data corresponding to an integration time smaller than the observation time. Figures 9.10 and 9.11 show the ISAR images of the two considered vessels. The observation time was 40 sec, while the coherent processing interval (CPIs) used to form the ISAR images are detailed in each figure. ISAR images on the left are shown in the range/Doppler domain, which is the natural output of an ISAR processing. Conversely, ISAR images on the right are shown in a fully spatial coordinate system, namely range/cross-range domain. Since the two-dimensional ISAR image is the result of a projection and convolution of the 3D target reflectivity function onto an unknown 2D plane (known as image projection plane – IPP), the target size could be

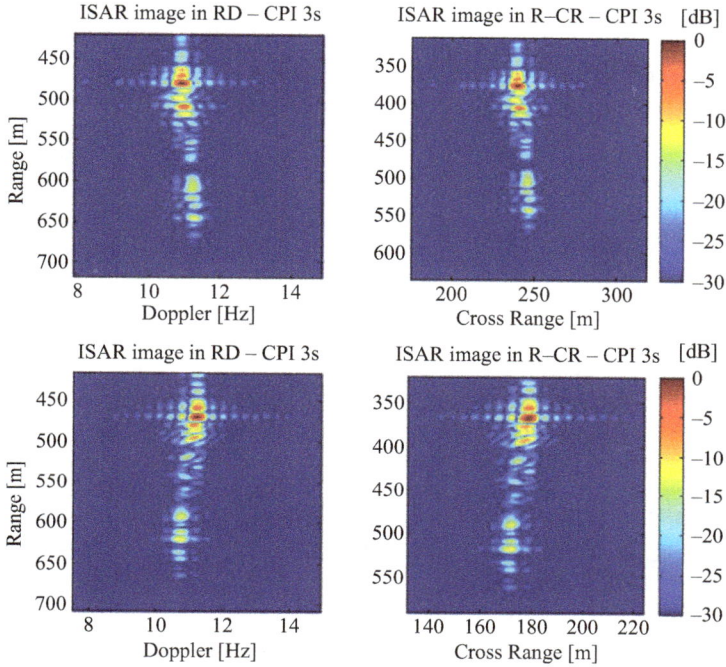

Figure 9.11   P-ISAR images of 'Grande Colonia' – UHF band

Table 9.5   Estimated length of the vessels processed with the P-ISAR technique

Vessel name	Frame	Estimated length	Real length
Anna Sophie	1	119 m	135 m
Anna Sophie	2	115 m	135 m
Anna Sophie	3	129 m	135 m
Anna Sophie	4	109 m	135 m
Grande Colonia	1	178 m	176 m
Grande Colonia	2	168 m	176 m

underestimated and may change over time. This may lead to an incorrect estimation of the target geometrical features. Moreover due to shadowing effects and scintillation, not all scatterers are always visible. In fact, it may occur that a scatterer positioned at the bow or the stern extremes may not respond for a given aspect angle.

Then the use of a sequence of ISAR images instead of a single one may partially overcome this issue and could provide a better estimate of the target length if the maximum estimated size is chosen.

The estimated lengths of the considered vessels are shown in Table 9.5. The overall estimated size of target 'Anna-Sophie' results in 129 m, which is quite accurate. The estimated size of 'Grande Colonia' is instead equal to 178 m. Its true length is 176 m. Since the range resolution is about 6.5 m, then the estimate is consistent with the target length.

9.5 Conclusion

A SDR-based radar system is capable of changing its main specifications by reconfiguring on-the-fly its operational parameters like frequency, bandwidth, digital signal processing blocks and others.

This chapter has been devoted to the advantages of implementing a PBR exploiting the SDR technology. The intrinsic flexibility offered by the SDR paradigm represents a very suitable implementation strategy for this class of radars which operate over a wide range of frequencies with a variety of instantaneous bandwidth. Furthermore, the jointly exploitation of SDR and digital beamforming allow the realization of a passive radar able to achieve the best performance in different operative scenarios and RF environment.

Whereas several advantages can be achieved in a PBR implemented by means of SDR solutions, a huge signal processing load is the main drawback. However, it should be noted that the computational power of nowadays CPUs and FPGAs are making the real-time realization of this implementation possible and effective.

References

[1] Y.-K. Kwag, J.-S. Jung, I.-S. Woo, and M.-S. Park, "Modern software defined radar (SDR) technology and its trends," *Journal of Electromagnetic Engineering and Science*, vol. 14, no. 4, pp. 321–328, 2014.

[2] About GNU Radio. Available from https://www.gnuradio.org/about/ [Accessed 30 April 2021].

[3] T. Debatty, "Software defined RADAR a state of the art," in: *2010 2nd International Workshop on Cognitive Information Processing*, Elba, Italy, 2010, pp. 253–257, doi: 10.1109/CIP.2010.5604241.

[4] B. Godana, A. Barroso, and G. Leus, "Estimating human movement parameters using a software radio-based radar," *International Journal on Advances in Systems and Measurements*, vol. 4, pp. 20–31, 2011.

[5] B. Liu and R. Chen, "Software-defined radar and waveforms for studying micro-Doppler signatures," in: *Proceedings of SPIE: Radar Sensor Technology XVIII*, Baltimore, MD, 2014, p. 907718.

[6] C. W. Rossler, E. Ertin, and R. Moses, "A software defined radar system for joint communication and sensing," in: *Proceedings of IEEE Radar Conference*, Kansas City, MO, 2011, pp. 1050–1055.

[7]   M. Braun, M. Müller, M. Fuhr, and F. J. Jondral, "A USRP-based testbed for OFDM-based radar and communication systems," in: *Proceedings of 22nd Virginia Tech. Symposium on Wireless Communications*, Blacksburg, VA, 2012.

[8]   H. Zhang, L. Li, and K. Wu, "24 GHz software-defined radar system for automotive applications," in: *Proceedings of European Conference on Wireless Technologies (ECWT 2007)*, Munich, 2007, pp. 138–141.

[9]   S. Constanzo, F. Spadafora, O.H. Moreno, F. Scarcella, and G. Di Massa, "Multiband software defined radar for soil discontinuities detection," *Journal of Electrical and Computer Engineering*, vol. 2013, article no. 5, 2013.

[10]   Prabaswara, A., A. Munir, and A.B. Suksmono, "GNU Radio based software-defined FMCW radar for weather surveillance application," in: *International Conference on Telecommunication Systems, Services, and Applications (TSSA)*, October 20–21, 2011, pp. 227–230.

[11]   A. Capria, M. Conti, D. Petri, *et al.*, "Ship detection with DVB-T software defined passive radar," in: *IEEE Gold Remote Sensing Conference*, 2010.

[12]   H. Kuschel and D. O'Hagan, "Passive radar from history to future," in: *International Radar Symposium (IRS) 2010*, 16–18 June, Vilnius Lithuania.

[13]   P. Howland, "Editorial: Passive radar systems," *IEE Proceedings of Radar, Sonar Navigation*, vol. 152, no. 3, pp. 105–106, 2005.

[14]   K. Jamil, M. Alam, M.A. Hadi, and Z.O. Alhekail, "A multi-band multi-beam software-defined passive radar. Part I: system design," in: *IET International Conference on Radar Systems (Radar 2012)*, Glasgow, UK, 2012, pp. 1–4, doi: 10.1049/cp.2012.1580.

[15]   M. Glende, J, Heckenbach, H. Kuschel, *et al.*, "Experimental passive radar systems using digital illuminators (DAB/DVB-T)," in: *Proceedings of the 2014 15th International Radar Symposium (IRS)*, pp. 1–4, 2007.

[16]   D. Cristallini, I. Pisciottano, and H. Kuschel, "Multi-band passive radar imaging using satellite illumination," in: *2018 International Conference on Radar (RADAR)*, Brisbane, QLD, Australia, 2018, pp. 1–6, doi:10.1109/RADAR.2018.8557260.

[17]   H. Kuschel, "Passive radar on fixed and mobile platforms exploiting digital broadcast signals," Workshop at EuRAD 2017, 11–13 October 2017, Nuremberg, Germany.

[18]   M. Malanowski, *Signal Processing for Passive Bistatic Radar*, Artech House, 2019.

[19]   M. Malanowski, K. Kulpa, and J. Misiurewicz, "PaRaDe – PAssive RAdar DEmonstrator family development at Warsaw University of Technology," in: *2008 Microwaves, Radar and Remote Sensing Symposium*, Kiev, Ukraine, 2008, pp. 75–78, doi: 10.1109/MRRS.2008.4669549.

[20]   M. Malanowski, K. Kulpa, P. Samczyński, *et al.*, "Experimental results of the PaRaDe passive radar field trials," in: *2012 13th International Radar Symposium*, 2012, pp. 65–68, doi: 10.1109/IRS.2012.6233290.

[21]   M. Malanowski, K. Kulpa, M. Mordzonek, and P. Samczyński, "PaRaDe – reconfigurable software defined passive radar," in: *Proceedings of NATO Specialist Meeting SET136*, Lisbon, Portugal, June 2009.

[22]   B. Dawidowicz, K. Kulpa, M. Malanowski, J. Misiurewicz, P. Samczyński, and M. Smolarczyk, "PCL systems on moving platforms – challenges and first results," in: *Proceedings of the 3rd FHR Focus Days on PCL*, Wachtberg, Germany, May 2011, pp. 1–41.

[23]   M. Płotka, M. Malanowski, P. Samczyński, K. Kulpa, and K. Abratkiewicz, "Passive bistatic radar based on VHF DVB-T signal," in: *2020 IEEE International Radar Conference (RADAR)*, Washington, DC, USA, 2020, pp. 596–600, doi: 10.1109/RADAR42522.2020.9114859.

[24]   Passive and Active Radar Applications Using NI Equipment [online]. Available from: https://forums.ni.com/t5/Past-NIWeek-Sessions/Passive-and-Active-Radar-Applications-Using-NI-Equipment/ta-p/3518636?profile.language=it [Accessed 30 April 2021].

[25]   D. Petri, "Studio ed analisi della funzione ambiguità in sistemi radar passivi UMTS". Available from: https://etd.adm.unipi.it/t/etd-03292007-230444/, University of Pisa, 2007 [Accessed 30 April 2021].

[26]   D. Petri, A. Capria, M. Martorella, and F. Berizzi, "Ambiguity function study for UMTS passive radar," in: *2009 European Radar Conference (EuRAD)*, Rome, Italy, 2009, pp. 41–44.

[27]   D. Petri, A. Capria, M. Conti, F. Berizzi, M. Martorella, and E. D. Mese, "High range resolution multichannel DVB-T passive radar: aerial target detection," in: *2011 Tyrrhenian International Workshop on Digital Communications – Enhanced Surveillance of Aircraft and Vehicles*, Capri, Italy, 2011, pp. 129–132.

[28]   A. Capria, F. Berizzi, D. Petri, *et al.*, "Multifunction imaging passive radar for harbour protection and navigation safety," *IEEE Aerospace and Electronic Systems Magazine, Special Issue on Passive and Multi-Static Radars for Civil Applications*, vol. 32, no. 2, pp 30–38, 2017.

[29]   D. Olivadese, E. Giusti, D. Petri, M. Martorella, A. Capria, and F. Berizzi, "Passive ISAR with DVB-T signals," *IEEE Transactions on Geoscience and Remote Sensing*, vol. 51, no. 8, pp. 4508–4517, 2013, doi: 10.1109/TGRS.2012.2236339

[30]   D. Olivadese, E. Giusti, D. Petri, *et al.*, "Passive ISAR imaging of ships by using DVB-T signals," in: *IET International Conference on Radar Systems (Radar 2012)*, 2012, pp. 1–4, doi: 10.1049/cp.2012.1617.

[31]   A. Capria, D. Petri, C. Moscardini, *et al*, "Software-defined Multiband Array Passive Radar (SMARP) demonstrator: a test and evaluation perspective," in: *OCEANS 2015 – Genoa*, 2015, pp. 1–6.

[32]   M. Conti, C. Moscardini, and A. Capria, "Dual-polarization DVB-T passive radar: experimental results," in: *2016 IEEE Radar Conference (RadarConf)*, 2016, pp. 1–5, doi: 10.1109/RADAR.2016.7485126.

[33]   OctoClock-G. Available from https://www.ettus.com/wp-content/uploads/2019/01/Octoclock_Spec_Sheet.pdf [Accessed 12 May 2021].

[34]   J. E. Palmer, H. A. Harms, S. J. Searle, and L. Davis, "DVB-T passive radar signal processing," *IEEE Transactions on Signal Processing*, vol. 61, no. 8, pp. 2116–2126, 2013.

[35] C. Moscardini, D. Petri, A. Capria, M. Conti, M. Martorella, and F. Berizzi, "Batches algorithm for passive radar: a theoretical analysis," *IEEE Transactions on Aerospace and Electronic Systems*, vol. 51, no. 2, pp. 1475–1487, 2015.

[36] J.E. Palmer and S.J. Searle, "Evaluation of adaptive filter algorithms for clutter cancellation in passive bistatic radar," in: *2012 IEEE Radar Conference (RADAR)*, 7–11 May 2012, pp. 493–498.

[37] C. Moscardini, M. Conti, F. Berizzi, M. Martorella, and A. Capria, "Spatial adaptive processing for passive bistatic radar," in: *2014 IEEE Radar Conference*, 19–23 May 2014, pp. 1061–1066.

[38] M. Skolnik, *Radar Handbook*, 3rd ed., McGraw-Hill, 2008.

[39] D. Pasculli, A. Baruzzi, C. Moscardini, D. Petri, M. Conti, and M. Martorella, "DVB-T passive radar tracking on real data using Extended Kalman Filter with DOA estimation," in: *2013 14th International Radar Symposium (IRS)*, 2013, pp. 184–189.

Chapter 10

Outlook on future trends

*Diego Cristallini[1], Daniel W. O'Hagan[1], Thomas Weyland[2]
and Rodrigo Blázquez-García[1]*

This chapter provides a brief outlook into the future trends in passive radar. Particularly, it is mentioned how passive radars constitute a key component in system-of-systems architectures. Afterwards, two sections are devoted to preliminary research activities aiming at investigating the potentials of new illuminators of opportunity for passive radar.

10.1 Passive radar in system-of-systems

A trend in radar and sensing more generally is towards all-domain, SEa Air Land Space and Cyber (SEALS-C) System-of-Systems (SoS). Similarly, a trend in mobile communications is towards higher frequencies, wider bandwidths, smaller cells and ever-higher data rates for the delivery of high definition content [1]. Electromagnetic spectrum congestion presents a challenge to both radar and communications. It is therefore imperative to extract valuable information from sensor assets when resources are constrained. In this context, passive bi-/multi-static radar is one example of a highly relevant technology not requiring additional spectrum allocation. Passive SAR, as demonstrated throughout this book, typically operates at a much smaller RF bandwidth than conventional monostatic SAR, but through advanced signal processing, information-rich images can be constructed. Bi-/Multi-static radar, particularly in passive mode, is a know-how of increasing importance as SoS emerge. There are numerous SoS initiatives at varying stages of realisation throughout the world, such as the NATO Alliance Future Surveillance and Control (AFSC) [2], Future Combat Air System (FCAS) and the Boeing Airpower Teaming System (BATS) [3]. These Manned-Un-manned teaming (MUT) SoS concepts could potentially facilitate a passive radar mode of operation for covert surveillance and imaging. For example, a platform (node) within the SoS constellation could serve as a receiver for a wide selection of possible terrestrial or space-borne illuminators of opportunity, such as

[1] Fraunhofer FHR, Germany
[2] RWTH Aachen, Germany

DVB-T2, DVB-S2 and Starlink. Alternatively, or additionally, stand-off illumination could be provided by another platform in the constellation for a more cooperative form of airborne bi-/multi-static surveillance. The concept of SoS has existed from the earliest days of sensing, but the field is energised nowadays by advances in peripheral technology domains, such as flexible programmable hardware, chip-scale atomic clocks, cloud computing and Artificial Intelligence (AI). Another key element of many SoS concepts is the ability to establish a Common Time Reference (CTR) between sensing nodes. Open source protocols like White Rabbit have been demonstrated for the synchronisation of spatially distributed static nodes [4], but achieving wireless synchronisation between dynamically moving platforms in a GNSS denied environment remains a technical challenge – and a work-in-progress!

A SoS carries far greater complexity than stand-alone sensors. The increased complexity is generally justified provided the resultant SoS significantly enhances the Intelligence Surveillance and Reconnaissance (ISR) quotient over-and-above that provided by non-SoS solutions. That is, future SoS is expected to provide a dramatically increased "information value" over what is presently achievable. A simplified example will illuminate one aspect of the added information value case: a monostatic radar illuminates a target and receives returns from a single angular aspect (multipath not considered). Consider a target that has been designed to be Low Observable (LO) – it is effective at reducing backscatter in the direction of the illuminating monostatic radar. Normally LO targets attempt to dramatically reduce their scattering profile in the head-on direction. LO designers expend effort to manage all reflections, and if unavoidable, then the reflections are orientated to less critical aspects such as towards the sky. Nevertheless, studies have shown that reflections may be received by ground-based receivers positioned at non-frontal aspect angles [5]. Therefore, if spatially dislocated receivers were available, then it may be possible to receive scatter from less critical aspect angles. If the information from the dislocated nodes could be combined in a time-aligned (synchronised) manner, and properly fused and interpreted, then additional information would have been obtained! This diversity leads to increased information that leads to detection persistence, which in turn can lead to a persistent track of the target/s. Further diversity, perhaps multispectral, can result in a far greater information level. Passive radar is an inherently diverse technique that will likely be used strategically in a SoS context. Prior to the realisation of a fully cohesive, synchronised, SoS, intensive work is already underway to continually advance the field of passive radar. Eventually sensing techniques will converge for certain applications. The remainder of this chapter considers specific aspects of current and emerging passive radar research such as the use of the dense network of Starlink satellites, the use of 5G, and eventually 6G, signals for passive radar.

10.2 Fifth-generation – new radio (5G-NR) as new illuminator of opportunity for passive radars

The advancement of mobile broadband communication has been substantial over the last decades, with several new generations of mobile standards that have been deployed up to the newest 5G-NR. Building upon its predecessor 4G long-term evolution (LTE), 5G has new features which are the reasons for the increase in data transmission speed. While 4G LTE operates on sub-6 GHz bands, 5G offers the possibility of operating on frequencies up to 53 GHz [6], frequencies which are attractive for passive radar purposes. A 5G-NR base station allocates frequency spectrum on an on-demand basis, which aids in augmenting the transmission speed. Additionally, 5G NR is highly directional in its transmission scheme, through the use of beam forming in the direction of the mobile equipment. This has one possibly strong advantage in the use of passive radar, as the location of beam forming impacts the synchronisation signals. These signals, which are broadcasted periodically, have a dependency on the direction of transmission. Using these signals in the context of passive radar may allow to further extract information on the location of a target, through the estimation of said location-specific data in the target return. This is unprecedented in other passive radar implementations.

A preliminary analysis of the viability of 5G as an illuminator of opportunity for passive bistatic radar has been carried out in [7] and more recently in [8]. The underlying conclusion of [7] is that there are certain waveforms which are adequate for passive bistatic radar. The author identifies these waveforms as being synchronisation signals, which are broadcasted periodically in multiple locations. These signals are called Synchronisation Sequence Blocks, in short SSBs.

In terms of centre frequency, the author in [7] establishes that two frequency ranges exist, which range from 410 MHz to 7.125 GHz (namely frequency range FR1) and from 24.25 GHz to 52.6 GHz (namely frequency range FR2), respectively.

Additionally, several subcarrier spacings are offered for the SSB in the 5G standard. These influence the bandwidth of the signal, thus producing SSBs which may have bandwidths of values 3600 kHz, 7200 kHz, 14.4 MHz, 28.8 MHz, 57.6 MHz, and 115.2 MHz. Calculating the bistatic range resolution using these yields to equivalent monostatic range resolutions of 41.638 m, 20.819 m, 10.41 m, 5.205 m, 2.602 m, and 1.301 m, respectively. These values are clearly attractive for the use in passive radar, especially the higher bandwidth parametrisations.

The same SSB is broadcasted in Pulse Repetition Intervals (PRIs) of 5 ms, 10 ms, 20 ms, 40 ms, 80 ms, or 160 ms. Using $R_{unambiguous} = \frac{c \cdot PRI}{2}$ (being c the speed of light), this yields to an unambiguous target range of 750 km in the worst case, which evidently does not pose any real threat for the radar processing. On the other hand, these large PRIs values lead to fairly low unambiguous Doppler intervals, namely between 200 Hz and 6.25 Hz.

A self-ambiguity analysis is also conducted in [7]. Specifically, Ref. [7] shows that there are range ambiguities in the zero-Doppler cut of an SSB waveform. The first ambiguity peak appears at around 10 km and is due to the SSBs being transmitted multiple times one after another in different directions. If multiple SSBs are received in one location said range ambiguity peaks may appear. The author states that these peaks fall outside of the coverage range of the used 5G configuration. Note that the analyses in [7] refer to a subcarrier spacing of 30 kHz. More details on the 5G-NR standard and the characteristics of the SSB signal for passive radar usage can be found in [7].

10.2.1   *Experimental verification*

A verification experiment was performed with a stationary passive radar based on 5G as illuminator of opportunity. The experiment was conducted during a Master Thesis [9] in cooperation between RWTH Aachen and Fraunhofer FHR. A suitable location for fulfilling all of the requirements was found at the *Bonner Pfad* in Gimmersdorf near Wachtberg, Germany. The exact coordinates at which the receiving antenna was positioned are 50°38′31.4″ North and 7°08′38.0″ East. A map of the location is presented in Figure 10.1. The baseline between transmitter and receiver location is about 2 km.

10.2.1.1   **Experimental setup of receiver**

A demonstrator was deployed based upon a two-channel Ettus USRP receiver [10], where only one single antenna was used and was connected to a Vivaldi antenna. The block diagram of the measurement receiver chain is depicted in Figure 10.2, while the setup is pictured in Figure 10.3.

The antenna was pointed at 143° North, to monitor the area in which the cooperative targets were to be active. The 5G-NR tower (gNB) was located at exactly 12° North from the position of the receiver. That means, that the direct signal could be

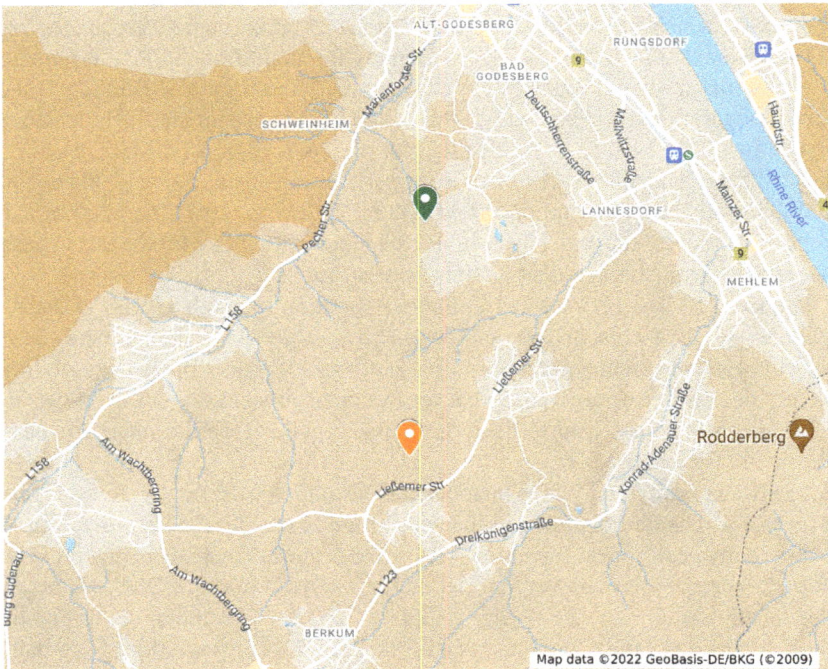

Figure 10.1   *Map of the measurement setup. The green pin marks the 5G base station and the orange pin marks the surveillance antenna.*

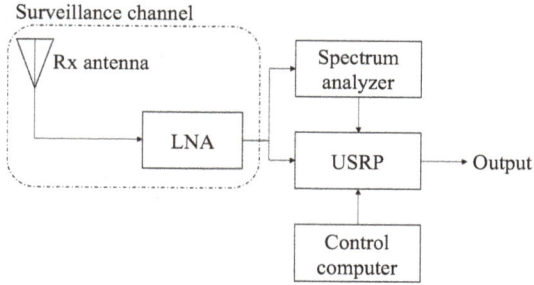

Figure 10.2   Block diagram of the measurement receiver chain

Figure 10.3   Surveillance Vivaldi antenna deployed during the experiment. The Ettus USRP can be seen in the bottom left of the picture.

received attenuated through the skirts of the Vivaldi antenna azimuthal pattern. Nevertheless, the proximity to the 5G base station ensured a clear reception of the direct signal, so that a demod/remod scheme for the SSB signal could be implemented, according to the processing block shown in Figure 10.4.

Two cooperative targets were deployed, a *DJI Inspire 2* drone, and a car, namely a *Jeep Renegade* to be specific. The latter drove up and down the *Bonner Pfad*, while the drone flew perpendicularly to the car. Both a drone and the car are visible in Figure 10.5. The GPS data has been recorded for the cooperative car target. From the data, the bistatic range and the bistatic Doppler frequency of the target can be calculated and used as a baseline.

Figure 10.6 shows a sample spectrogram of the data recorded from the reference channel, where the (seven) SSB signals transmitted in the different SSB beams are clearly recognizable (cf. with Figure 3.3 in [9]). As is apparent, the passive radar

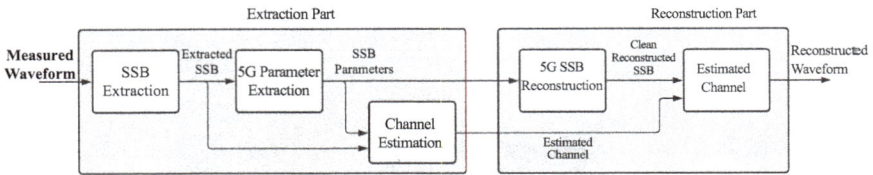

Figure 10.4   Block diagram used for the demod/remod

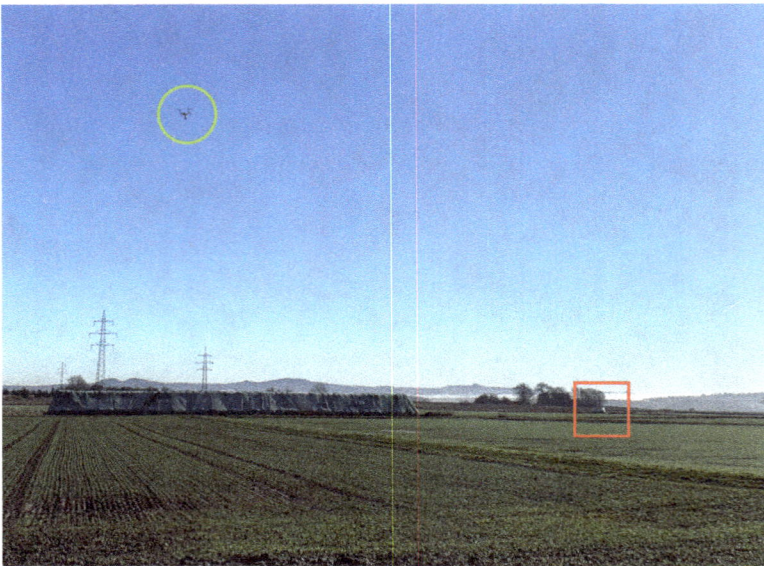

Figure 10.5   Surveillance area, including a drone (green circle) and the car (red box) as cooperative targets

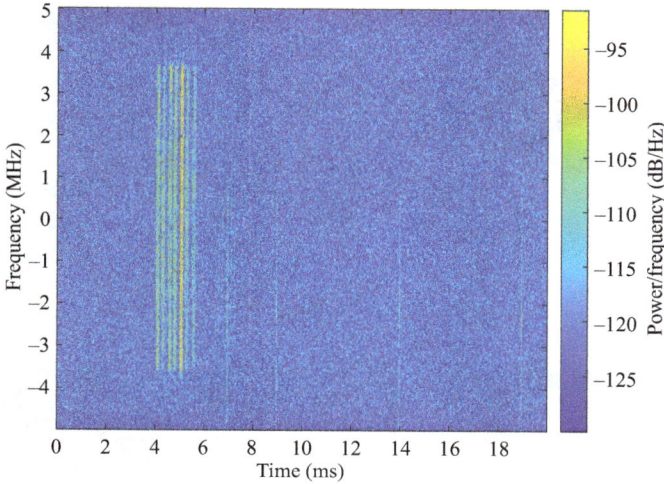

Figure 10.6 Spectrogram of a part of the measured data during the measurement campaign

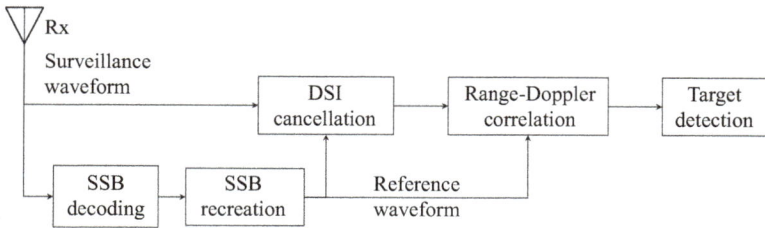

Figure 10.7 Block diagram for the target detection process using a locally recreated reference signal

receiver is located in the fifth SSB beam, which is received with significant higher power with respect to other SSB beams.

10.2.1.2 Target detection

Several minutes of recordings were analysed. Figure 10.8 shows several range-Doppler maps of different time instants. The range-Doppler maps are obtained by cross-correlation between the recreated reference signal and the received surveillance signal as sketched in Figure 10.7. As is apparent from Figure 10.7, this scheme requires a preliminary synchronization, decoding and recreation of the SSB signal. Also a direct signal interference (DSI) cancellation stage has been implemented based on the ECA+ approach described in [11].

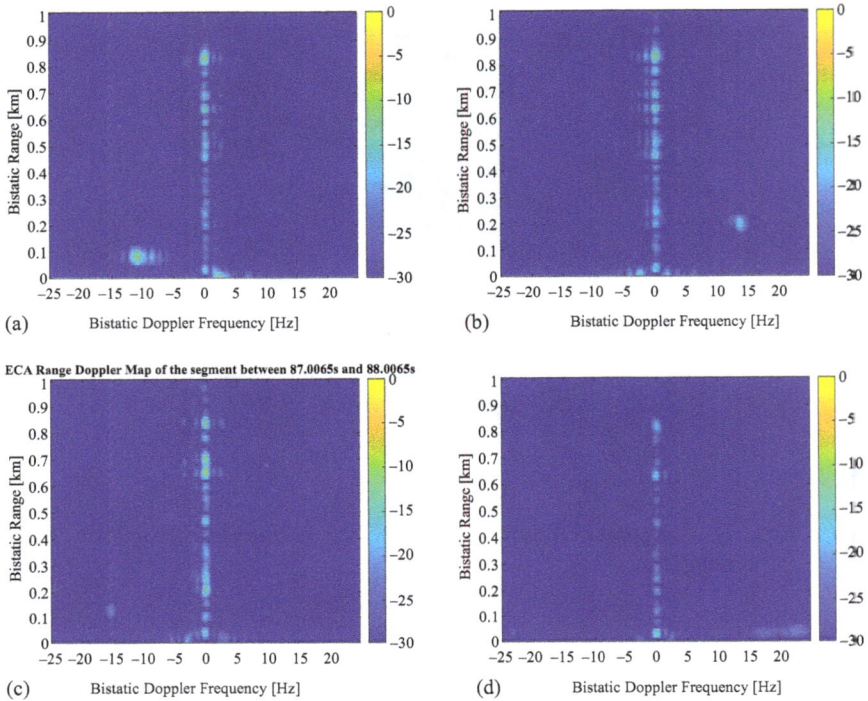

Figure 10.8   *Range Doppler maps of detected targets. Range-Doppler data is in dB and it is normalised in amplitude to the maximum value obtained before ECA+ implementation. (a) Range-Doppler Map at time measurement $t = t_0 + 8$ s. (b) Range-Doppler Map at time measurement $t = t_0 + 16$ s. Range-Doppler Map at time measurement $t = t_0 + 87$ s. (d) Range-Doppler Map at time measurement $t = t_0 + 92$ s.*

Range-Doppler maps in Figure 10.8 show a target moving away from the antenna in the beginning of the recording (<30 s) and moving towards the antenna later in the recording (> 80 s). Figure 10.8 shows the time instants of 8 s, 16 s, 87 s, and 92 s after the initial time of recording t_0, respectively. Here, the cooperative target is clearly visible. This also aligns with the bistatic range data obtained from the GPS location of the car.

It must also be noted that the target disappears in the middle part of the recording. This is assumed to be due to the comparably small radar cross section (RCS) of the car when it is faced parallel to the receiver, i.e., the back or front is facing the receiver.

Unfortunately, throughout the entire recordings of the measured data, the cooperative drone has not been detected. It is assumed that the radar cross section of the drone is too small to be properly detected.

Several initially unknown targets are detected at around 650 m in Figure 10.9. Plotting the bistatic ellipse onto the map from Figure 10.1 answers the question of

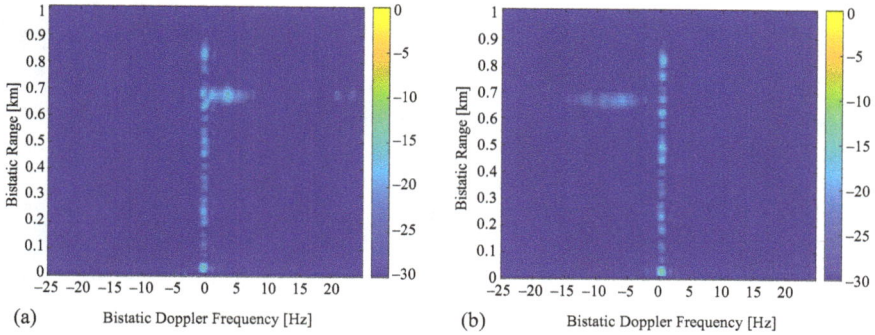

Figure 10.9   *Range-Doppler maps of the measurement data with targets of opportunity at a bistatic range of 650 m. Range-Doppler data is in dB and normalised to the maximum value obtained before ECA implementation. (a) Time of measurement $t = t_0 + 47$ s. (b) Time of measurement $t = t_0 + 61$ s.*

the origin of said targets. As can be seen in Figure 10.10, the targets are actually targets of opportunity in form of passing vehicles on the *Liessener Straße*. The fact that these targets are visible at larger bistatic ranges, when compared to the vanishing car at around 350 m, may be down to two reasons. First, the vehicles passing on the *Liessener Strasse* are driving perpendicularly to the receiver. Since this way, the target vehicles are showing their side face to the receiver, their RCS is in most cases larger. Second, the vehicles have a larger RCS, since they may be trucks or buses. Several of these large vehicles were driving along the street, as was noticed during the time of recording. However, the street was not filmed at the time of recording, meaning that the second theory cannot be verified.

10.3   Broadband low Earth orbit satellite constellations

Thanks to the reduction of satellite-launching costs, another important trend in the communication sector is the deployment of Low Earth Orbit (LEO) satellite constellations, usually referred to as mega-constellations, comprising hundreds of communication satellites to provide Fixed-Satellite Services (FSS). In recent years, several companies such as Starlink, OneWeb, Amazon (under the name of Project Kuiper), Telesat and Keppler Communications have applied to the US Federal Communications Commission (FCC) for deployment and broadcast licenses to offer global broadband internet services using LEO constellations at altitudes below 2,000 km. In fact, by June 2022, Starlink and OneWeb had already launched more than 2,300 and 420 satellites, respectively, out of their initial planned constellations of 4,408 and 720 satellites, with the aim to offer low-latency and high-speed global internet access not only for stationary receivers but also for moving ones (e.g. long-haul flights).

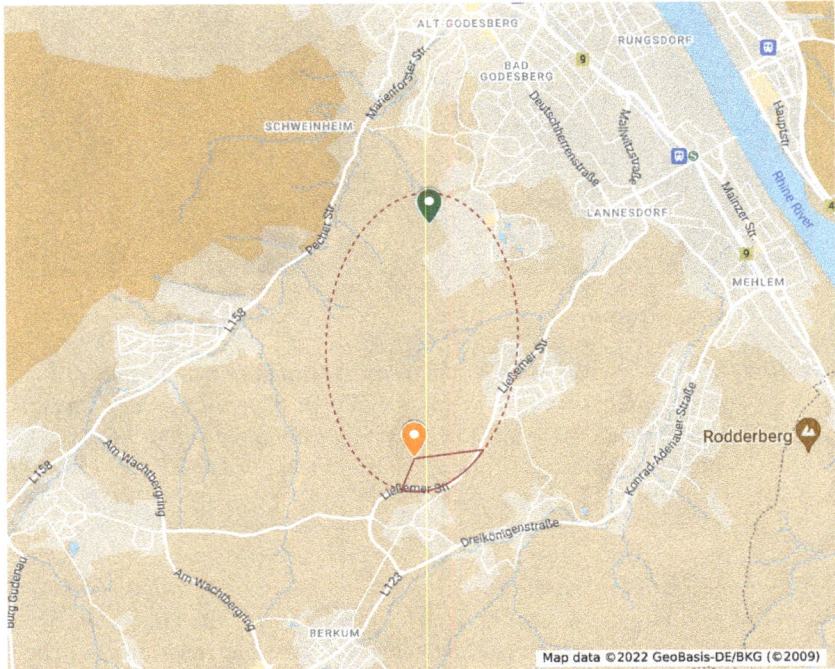

Figure 10.10   Map of the measurement setup with an overlaid bistatic ellipsoid (dashed) of 650 m and the main antenna beam (solid). The green pin marks the 5G base station and the orange pin marks the surveillance antenna.

Besides, this trend towards the use of more flexible and smaller LEO satellites is also shown in the high number of enterprises and research centers that have envisioned and analysed the deployment of LEO constellations not only for communication but also for remote sensing and Positioning, Navigation, and Timing (PNT) services.

Therefore, these emerging broadband communication satellite constellations can be considered as novel and potential candidates to be exploited as space-based transmitters of opportunity for passive radar applications. Despite the challenges that space-based transmitters of opportunity pose, mainly due to the low received power level of their signals, several experimental passive radar demonstrators based on Digital Video Broadcasting Satellites (DVB-S), Global Navigation Satellite Systems (GNSS) or Iridium have been developed by universities and research centers showing promising capabilities for imaging and target detection applications [12]. But compared to these other space-based illuminators, the new broadband communication satellite constellations provide additional relevant capabilities that may open new possibilities for the development of bistatic passive radar systems:

- Global and persistent coverage including oceans, poles and remote areas.
- High bandwidth signals (usually around 250 MHz) for enhanced bistatic range resolution (up to 0.6 m for pseudo-monostatic configuration).
- High radiated power levels with lower propagation losses due to the shorter transmitter–receiver distances from LEO satellites, enhancing the Signal-to-Noise Ratio (SNR) and the detection range.
- Predictable transmitter movement based on publicly available orbital data (i.e. Two-Line Element – TLE – sets) enabling passive Synthetic Aperture Radar (SAR) applications.
- Transmitter robustness and redundancy providing spatial diversity and multi-angle illumination thanks to the various and dense constellations.

However, these LEO communication satellites also pose certain challenges to be used as illuminators of opportunity for passive radar mainly due to: (i) the necessity to track them with a narrow-beam reference antenna; (ii) the fast-varying target bistatic ranges and Doppler frequencies induced by the fast relative motion of the LEO satellites; and (iii) the use of non-standardized signals without detailed open information about their properties (i.e. modulation, coding, internal structure, pilots or protocols). Nevertheless, with the aim to promote further research and developments, a preliminary analysis of the use of broadband LEO communication satellites as illuminators of opportunity is presented here, showing their feasibility and potential performance for passive radar applications. This analysis includes an overview of the emerging broadband LEO communication satellite constellations, a proposed passive radar system architecture, its performance estimation and some results of experimental signal acquisitions and ambiguity functions.

10.3.1 Overview of broadband LEO communication satellite constellations

According to the applications approved by the FCC, Starlink [13] and OneWeb [14] are deploying LEO constellations composed initially of 4,408 and 720 satellites, respectively. The main parameters of these constellations, which include polar and non-polar orbits with global coverage, are detailed in Table 10.1, while the satellite positions of the planned constellations projected over a world map at a point in time are shown in Figure 10.11.

The OneWeb constellation is designed to permanently and globally provide at least one visible satellite, while Starlink constellation will provide a high number of visible satellites, specially at middle latitude locations. As an example, Figure 10.12 shows the percentage of time as a function of the number of visible satellites for a receiver deployed at Fraunhofer FHR (50.617°N, 7.132°E) and the specified minimum elevation angle (25° for Starlink and 55° for OneWeb).

To provide broadband internet access, these satellites maintain bidirectional links with user terminals at X/Ku-band and with gateways at Ka-band using Left-Handed or Right-Handed Circular Polarization (LHCP or RHCP). Figure 10.13 shows their frequency plan, noting that the user-downlink band, which is divided into eight 250 MHz channels that can be considered as novel candidates of illumination of opportunity for

Table 10.1   Starlink and OneWeb constellation parameters

Parameter	Starlink				OneWeb
	Shell 1	Shell 2	Shell 3	Shells 4/5	
Orbital planes	72	72	36	6/4	18
Satellites per plane	22	22	20	58/43	32
Altitude	550 km	540 km	570 km	560 km	1,200 km
Inclination	53°	53.2°	70°	97.6°	87.9°
Total number of satellites	1,584	1,584	720	348/172	720

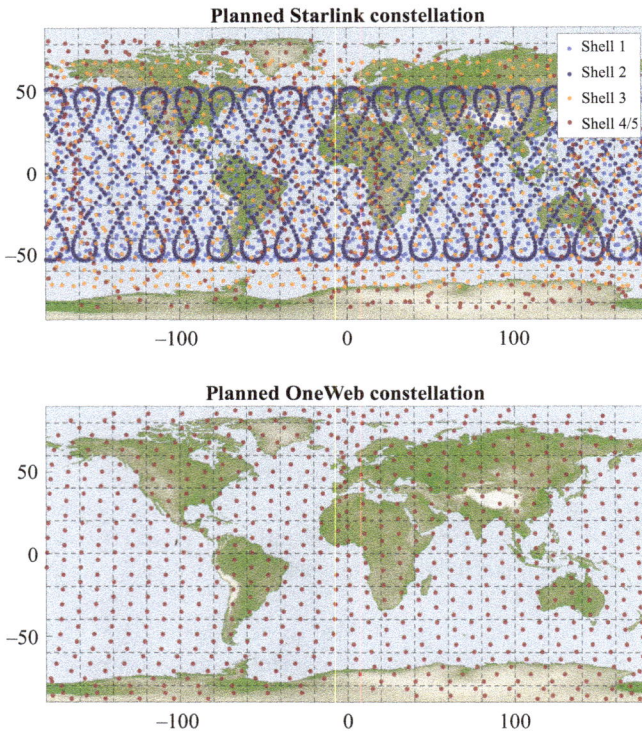

Figure 10.11   Starlink and OneWeb planned LEO satellite constellations

passive radar, shares the same 10.7–12.7 GHz band as geostationary (GEO) DVB-S satellites. Based on their onboard phased-array antennas, Starlink satellites realize a hexagonal cell deployment with simultaneous steerable spot beams whose 3 dB-footprints on the Earth's surface have always a minor axis longer than 20 km. On

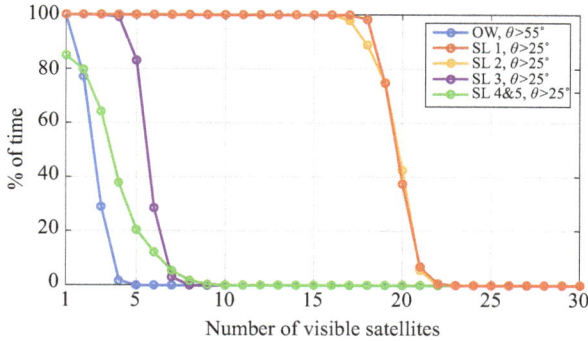

Figure 10.12 *Percentage of time as a function of the number of visible satellites of OneWeb (OW) constellation and the different Starlink (SL) shells for a receiver placed at Fraunhofer FHR (50.617°N, 7.132°E) and the specified minimum elevation angle (θ)*

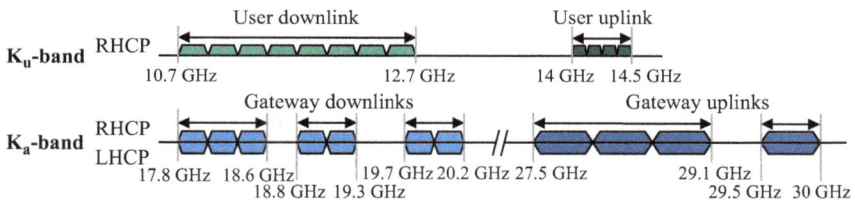

Figure 10.13 *Frequency allocation for broadband LEO communication satellites*

the other hand, OneWeb satellites have 16 user-downlink non-steerable spot beams with a larger footprint on the Earth's surface. In order to avoid interference between spot beams, different frequency channels are assigned to adjacent cells of the same satellite. Besides, the use of LEO satellites requires the implementation of handover protocols to maintain a continuous service.

Depending on the considered constellation, the physical layer of the user downlink is believed to be based on Orthogonal Frequency Division Multiplexing (OFDM) or Single Carrier-Time Division Multiplexed (SC-TDM) signals modulated with Quadrature Phase Shift Keying (QPSK), 8-Phase Shift Keying (8PSK) or 16-Quadrature Amplitude Modulation (16QAM). Regarding the radiated power levels of the user-downlink signals, OneWeb satellites have a maximum Equivalent Isotropically Radiated Power (EIRP) density of −49.4 dBW/Hz (i.e. 34.4 dBW EIRP for a 240 MHz transmission), while Starlink satellites adapt the EIRP density of each spot beam between −56.2 dBW/Hz (nadir) and −52.8 dBW/Hz (maximum offset) (i.e. between 27.6 and 31 dBW EIRP for a 240 MHz transmission) in order to compensate the additional propagation losses when increasing the distance to the Earth's surface in the pointing direction.

10.3.2 Passive radar system architecture

The satellites in these LEO constellations move at a high relative speed with respect to the ground stations. For this reason, a passive radar system architecture based on the use of a narrow-beam tracking antenna for the reference channel is proposed, while the surveillance antenna is pointed towards the area of interest where targets are to be detected, as shown in Figure 10.14. In this way, the direct signal (i.e. user downlink) can be received with enough SNR to be used as a reference signal. The required steerable antennas can be implemented using three-degree-of-freedom (i.e. azimuth, elevation, and skew) mechanical steerable systems or based on low-profile phased or digital planar arrays, which can be mounted onboard moving platforms (e.g. ships or aircrafts). For a near-zenith Starlink satellite pass, which lasts around 2.7 min considering a minimum elevation of 40°, the required azimuth and elevation tracking speed is 5°/s at its maximum.

The inputs of the tracking system are the satellite orbital parameters in TLE format, which are regularly updated and can be used to predict the position of the satellites using orbit propagation methods such as Simplified General Perturbation 4 (SGP4) [15], and the location and attitude data of the platform, which can be retrieved respectively from GNSS and Inertial Measurement Unit (IMU) sensors. If the measurements of these sensors are not accurate enough, a feedback control loop based on the received power may be required to search and maintain the tracking towards the serving satellite. In this regard, the use of digital array antennas for the reference channel can be advantageous since angle of arrival techniques can be implemented and several simultaneous beams can be formed to receive the reference signals from multiple serving satellites, enabling multistatic approaches.

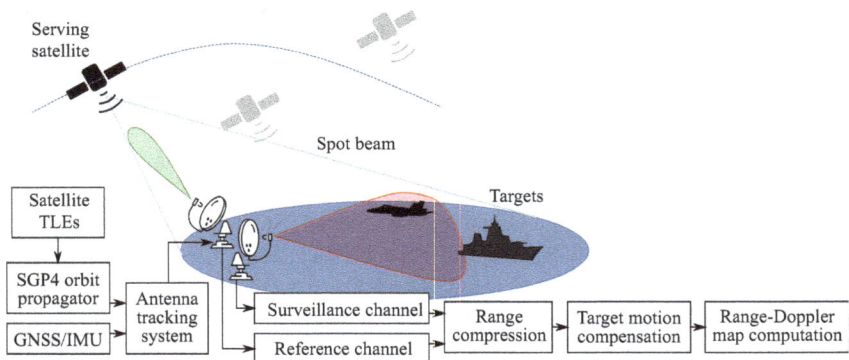

Figure 10.14   Passive radar system architecture based on LEO satellite illuminators of opportunity

Once both coherent reference and surveillance channels are digitized, signal processing for target detection can be applied based on range compression (cross-correlation between surveillance and reference signals), Target Motion Compensation (TMC), and range-Doppler map computation. TMC approaches are required to increase the coherent integration time, compensating for the range and Doppler migration of the reflected signals mainly due to the fast LEO satellite relative motion.

10.3.3 Performance analysis

Considering the characteristics of a Starlink satellite and typical parameters of a passive radar receiver, shown in Table 10.2, a power budget analysis is performed to estimate the maximum detection range of the proposed system. Assuming a free-space propagation model, the SNR of the direct signal in the reference channel is estimated above 12 dB. Besides, Figure 10.15 shows the estimated target SNR considering a perfect coherent integration for different Radar Cross-Section (RCS) as a function of

Table 10.2 System parameters

Parameter	Value
Frequency	11.7 GHz (user downlink)
Bandwidth	250 MHz
EIRP density	-56.22 dBW/Hz (Nadir)
Reference antenna gain	39.5 dB
Surveillance antenna gain	35 dB
Coherent integration time	0.5 s
Rx noise factor	2 dB
Additional losses	4 dB

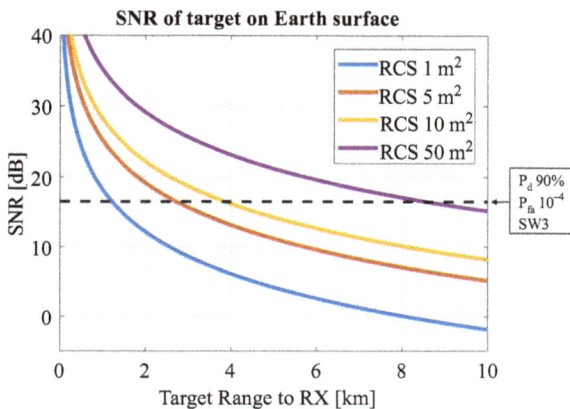

Figure 10.15 Estimated SNR for ground targets of different RCS

their distance to the receiver (RX). Taking into account the required SNR of 16.5 dB for a 90% probability of detection, a 10^{-4} probability of false alarm, a Swerling 3 (SW3) target and a Neyman–Pearson detector, the maximum detection range for a 1 m^2 RCS target is above 1 km. Similar results are obtained when OneWeb satellites are considered.

However, it is also important to note that the power flux density on the Earth's surface of Starlink user downlink signals is approximately -98 dBW/m^2, which is maintained during the satellite pass thanks to the transmitted power adaptation depending on the pointing direction. This power flux density is considerable higher than the one obtained for DVB-S GEO satellites and GNSS signals. As examples, for an Astra 1KR transponder, the estimated power flux density on the Earth's surface is -108 dBW/m^2, while for the Global Positioning System (GPS) L1 signal, it is -140.5 dBW/m^2. Therefore, the proposed system is expected to achieve enhanced detection capabilities when compared to the use of other satellite-based illuminators of opportunity.

On the other hand, the wide bandwidth of user-downlink signals (i.e. around 250 MHz) provides high-range resolution capabilities for passive radar applications. The actual range resolution achieved on the Earth's surface in the orthogonal direction of the iso-range contours depends on the satellite position and changes over time, as shown in Figure 10.16. When the satellite is in a low elevation position, the area between the satellite and the receiver presents a diminished resolution, while the area around the extended baseline (i.e. beyond the receiver) presents a pseudo-monostatic configuration with a resolution close to 0.6 m. In contrast, when the satellite is in a near-zenith position, the achievable range resolution on the Earth's surface around the receiver is mostly uniform with a value around 1.2 m. This spatial dependency can be taken into account when selecting the illuminating satellite among the visible ones in order to optimize the spatial resolution in the area of interest thanks to the spatial diversity provided by the dense satellite constellations.

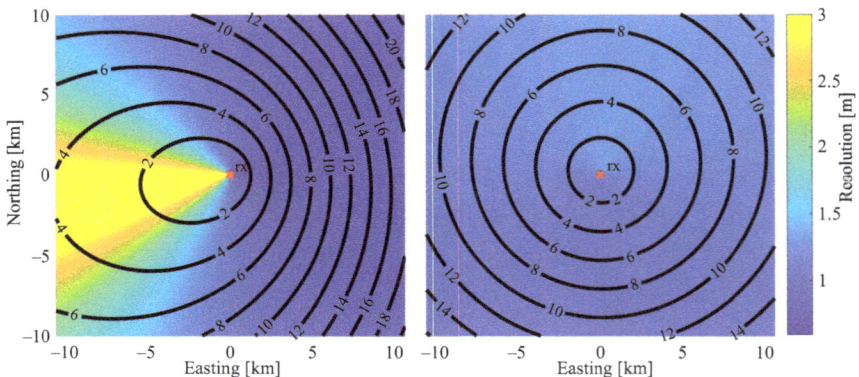

Figure 10.16   Achievable range resolution on the Earth's surface and iso-range contours (black curves, in km) for two positions of a Starlink satellite with: 40° elevation angle (left), and close to 90° elevation, i.e. near-zenith position (right).

However, this fine resolution capability also poses challenges in increasing effectively the coherent integration interval due to the range and Doppler migration effects caused by the movement of the transmitter and, potentially, of the receiver and targets. Therefore, signal processing requires to include target motion compensation techniques in order to cope with these migration effects.

10.3.4 Experimental Starlink signal acquisition

In order to experimentally acquire Starlink user-downlink signals, several measurement campaigns were carried out at Fraunhofer FHR with different receiver configurations.

The first receiver configuration was based on a commercial Low Noise Block (LNB) feed horn (model AK54-XT2) connected to a receiving channel of an Ettus USRP X310 software defined radio [10] with UBX 160 daughterboards in order to digitize the configured band of the intermediate frequency signals. The LNB was pointed towards zenith (i.e. 90° elevation) and a 10 MHz band was recorded with a centre frequency of 11.325 GHz. Although the supply voltage of the LNB was set to 13 V in order to receive the vertical polarization, no polarization axis alignment was performed since Starlink signals are expected to be circularly polarized.

During a 150 s signal recording, 32 Starlink satellites met the elevation angle requirement (i.e. greater than 25°) specified in the constellation application. However, as shown in Figure 10.17, only 4 satellites crossed the 60° −3 dB beam of the LNB

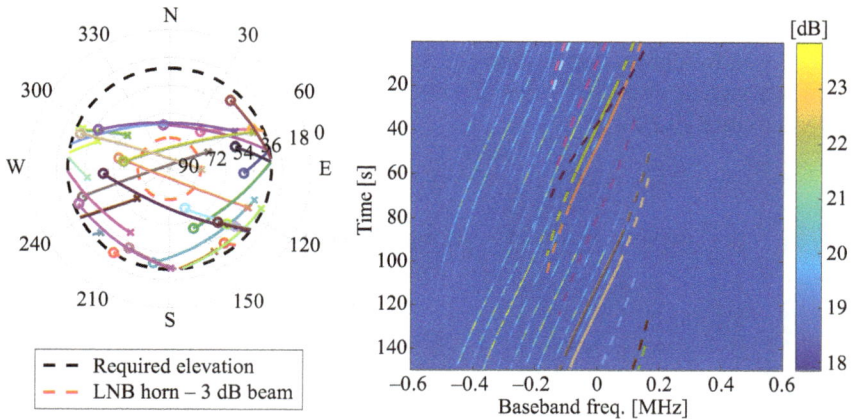

Figure 10.17   *Sky plot of Starlink satellite passes during the measurement (left) and spectrogram of the received signal at a centre frequency of 11.325 GHz (right). On the spectrogram, the time-dependent Doppler frequency, estimated from TLE data, of the satellites that present an elevation above 50° are overlaid (dashed line when the satellite elevation is between 50° and 70° and solid line when the satellite elevation is above 70°).*

feed horn during this signal acquisition. The computed spectrogram of the acquired signal confirms the reception of Starlink beacon signals around the centre frequency of the selected transmission channel, which follow the estimated Doppler frequency pattern based on the predicted trajectories of the satellites. These patterns also show the fast changes (approximately up to 3.4 kHz/s) of the Doppler frequency of the received signals due to the dynamics of the LEO satellites.

As can be seen in the spectrum represented in Figure 10.18, nine beacons or pilot signals are transmitted around the centre frequency of the transmission channel with 44 kHz separation between them. These beacons could be used to point the reference antenna towards the satellite and to estimate and compensate the frequency offset of the received signals. In addition, they show amplitude variations that might be associated to the transmission of low-speed network information such as the identifier of the transmitting satellite or for time and frequency synchronization.

The first receiver configuration with the USRP X310 software-defined radio does not allow the digitization of a complete transmission channel of 250 MHz. For this reason, Fraunhofer FHR has developed an enhanced version of its passive radar system SABBIA based on satellite illuminators with increased bandwidth and enhanced satellite tracking capabilities. With this system, Starlink signals have been received with an appropriate SNR during more than 80 s, tracking the satellites based on their trajectory prediction from TLE data. Figure 10.19 depicts the spectrum and the spectrogram of the received signal for an 80 ms time interval, showing that the transmission channel occupies approximately a 240 MHz band with pilot signals at the centre and boundaries of the channel. During this time interval, the signal presented a pulsed structure with approximately 1.33 ms pulses every 8 ms. It is considered that this might be due to the lack of user data, so the received signal contains mainly pilots, possibly used for channel estimation and synchronization, and signals with network data. These signals give rise to several ambiguities of the autocorrelation function, as shown in Figure 10.20.

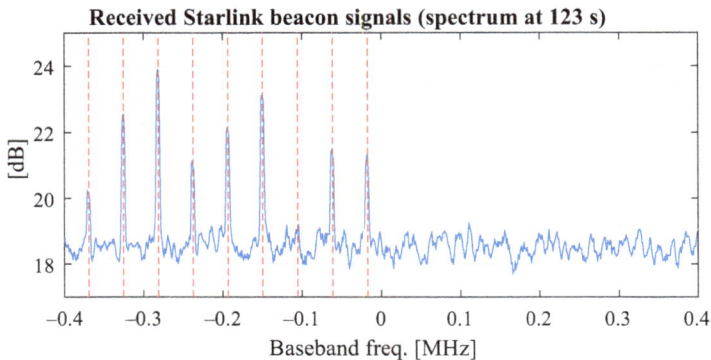

Figure 10.18   Spectrum of the received signal showing the beacon signals around the centre frequency of the transmission channel (11.325 GHz)

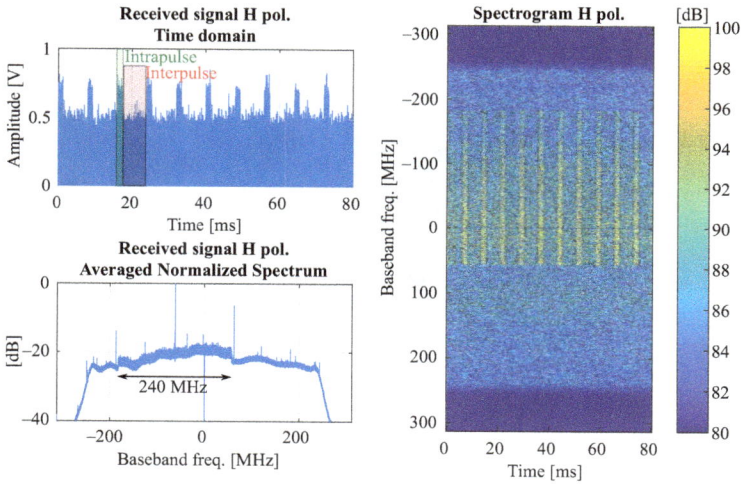

Figure 10.19   Time domain (top left), spectrum (bottom left) and spectrogram
(right) of an 80 ms section of a received Starlink signal showing a
pulsed structure in the transmission channel centered at 11.325 GHz.

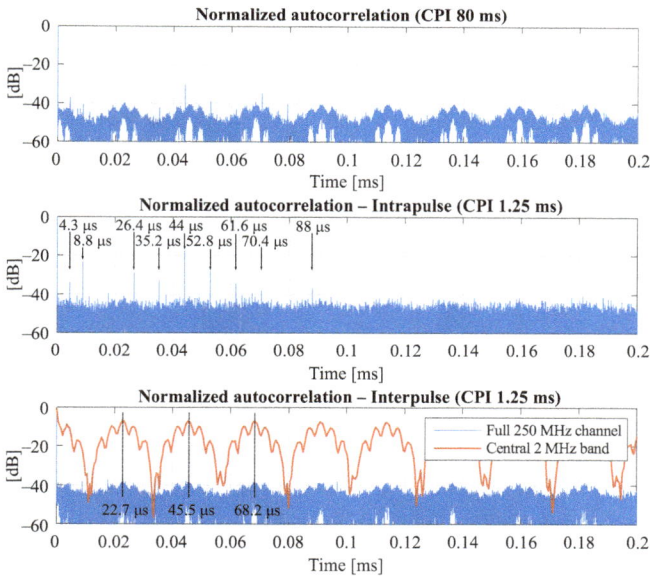

Figure 10.20   Autocorrelation of an 80 ms section (top), a 1.25 ms intrapulse
section (middle) and a 1.25 ms interpulse section (bottom) of the
received Starlink signal with pulsed structure after filtering the 250
MHz transmission channel centered at 11.325 GHz to remove
out-of-band noise. Since in the intrapulse section (bottom), the main
signal components are the pilots transmitted in the center of the
channel and the full-channel autocorrelation is affected by the
autocorrelation of the receiver noise, the autocorrelation when
filtering the center 2 MHz of the channel is also depicted, showing
the periodic sidelobe pattern every 1/44 ms due to the pilots.

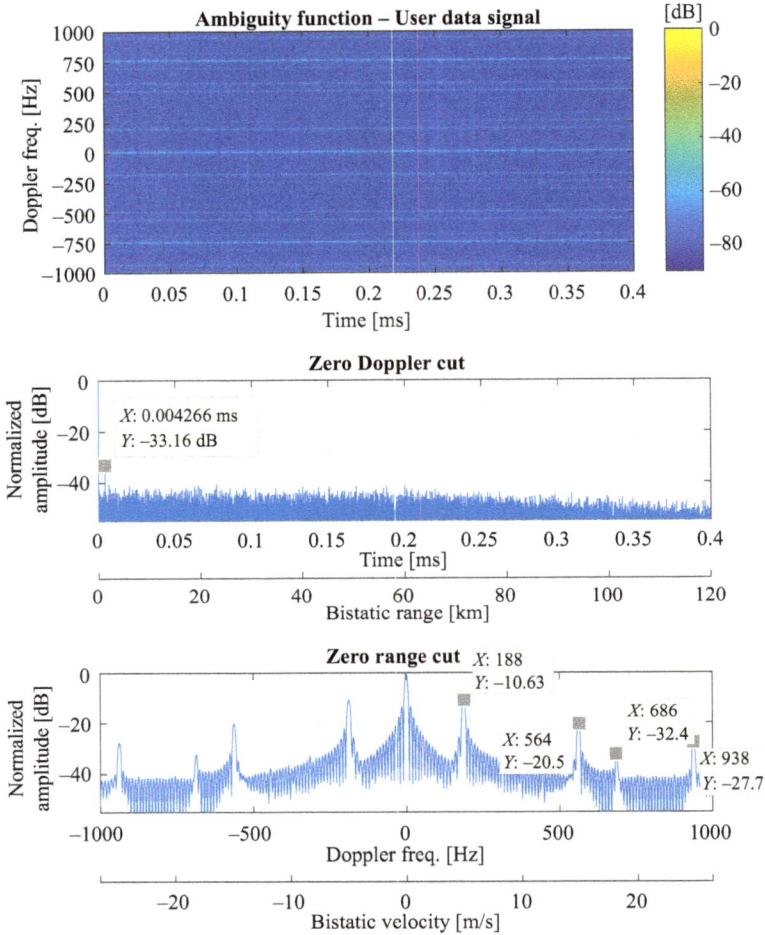

Figure 10.21   *Range-Doppler ambiguity function and zero-Doppler and zero-range cuts of a 0.5 s section of the received Starlink signal with user data after filtering the 250 MHz transmission channel centered at 11.325 GHz to remove out-of-band noise*

However, when the received signal contains user data and does not show a pulsed structure, the computed range-Doppler ambiguity function, depicted in Figure 10.21, shows fewer ambiguities, having good properties for high resolution passive radar applications with a narrow main lobe (3 dB-width of approximately $1/B$ with a bandwidth $B = 240$ MHz) and a low noise floor level. Despite the occurrence of some range and Doppler ambiguities due to the OFDM cyclic prefix and the internal structure of the signal, most of the ambiguities that arise when the signal shows a pulsed structure are not visible when user data is being transmitted. This can be justified because periodic pilot signals during user data transmission may not be dominant and their induced ambiguities may be masked by the noise floor.

This preliminary analysis supports the potential use of broadband communication LEO satellite user-downlink signals as a promising illuminator of opportunity for satellite-based passive radar applications.

References

[1] E. Dahlman, S. Parkvall, and J. Skold, *5G NR: The Next Generation Wireless Access Technology*. Amsterdam, Boston: Academic Press, 2018.

[2] NATO, "Alliance future surveillance and control," July 2020. Available: https://www.nato.int/nato_static_fl2014/assets/pdf/2020/7/pdf/200701-Factsheet_Alliance_Future_Surveil-1.pdf

[3] Boeing, "Boeing airpower teaming system." Available: https://www.boeing.com/defense/airpower-teaming-system/

[4] J. Sandenbergh and M. Inggs, "A summary of the results achieved by the GPS disciplined references of the netrad and nextrad multistatic radars," in *2019 IEEE Radar Conference (RadarConf)*, 2019, pp. 1–6.

[5] H. Kuschel, J. Heckenbach, S. Müller, and R. Appel, "Countering stealth with passive, multi-static, low frequency radars," *IEEE Aerospace and Electronic Systems Magazine*, vol. 25, no. 9, pp. 11–17, 2010.

[6] 3GPP, "NR; Base Station (BS) radio transmission and reception," 3rd Generation Partnership Project (3GPP), TS 38.104, May 2019. Available: https://www.etsi.org/deliver/etsi_ts/138100_138199/138104/15.05.00_60/ts_138104v150500p.pdf

[7] P. R. Singh, "An investigation of passive radar using 5G illuminators of opportunity," Master's thesis, RWTH Aachen University, July 2021.

[8] P. Samczyński, K. Abratkiewicz, M. Płotka, *et al.*, "5G network-based passive radar," *IEEE Transactions on Geoscience and Remote Sensing*, vol. 60, pp. 1–9, 2022.

[9] T. Weyland, "Performance analysis of 5G mobile networks as illuminator of opportunity for passive radar," Master's thesis, RWTH Aachen University, April 2022.

[10] Ettus, *USRP X310 Datasheet*. Available: https://www.ettus.com/wp-content/uploads/2019/01/X300_X310_Spec_Sheet.pdf

[11] S. Searle, D. Gustainis, B. Hennessy, and R. Young, "Cancelling strong Doppler shifted returns in OFDM based passive radar," in *2018 IEEE Radar Conference (RadarConf18)*, 2018, pp. 359–354.

[12] D. Cristallini, M. Caruso, P. Falcone, *et al.*, "Space-based passive radar enabled by the new generation of geostationary broadcast satellites," in *2010 IEEE Aerospace Conference*, 2010, pp. 1–11.

[13]   L. Space Exploration Holdings, "Application for modification of authorization for the SpaceX NGSO satellite system," FCC filing SAT-MOD-20200417-00037, Tech. Rep., April 2020.

[14]   W. S. Limited, "OneWeb non-geostationary satellite system," FCC filing SAT-LOI-20160428-00041, Tech. Rep., April 2016.

[15]   D. A. Vallado, *Fundamentals of Astrodynamics and Applications*, vol. 12. Springer Science & Business Media, 2001.

Index

www.ingramcontent.com/pod-product-compliance
Lightning Source LLC
Chambersburg PA
CBHW050509190326
41458CB00005B/1480